# Studies in Computational Intelligence 444

**Editor-in-Chief**

Prof. Janusz Kacprzyk
Systems Research Institute
Polish Academy of Sciences
ul. Newelska 6
01-447 Warsaw
Poland
E-mail: kacprzyk@ibspan.waw.pl

For further volumes:
http://www.springer.com/series/7092

Michael Glykas (Ed.)

# Business Process Management

## Theory and Applications

 Springer

*Editor*
Michael Glykas
Department of Financial
and Management Engineering
University of the Aegean
Greece
and
Aegean Technopolis
The Technology Park of the
Aegean Region
Greece
and
Faculty of Business and Management
University of Wollongong
Dubai

ISSN 1860-949X                      e-ISSN 1860-9503
ISBN 978-3-642-43576-8              ISBN 978-3-642-28409-0 (eBook)
DOI 10.1007/978-3-642-28409-0
Springer Heidelberg New York Dordrecht London

Printed on acid-free paper

Springer is part of Springer Science+Business Media (www.springer.com)

*To my wife Athina for her support all these years
and to Markos, Aggeliki and Alexandros the three pillars of my life*

# Introduction

## State of the Art in Business Process Management

Michael Glykas[1,2,3]

[1]Financial and Management Engineering Department, University of the Aegean, Greece
[2]Aegean Technopolis, The Technology Park of the Aegean Region, Greece
[3]Faculty of Business and Management, University of Wollongong, Dubai

Business Process Management (BPM) has been in existence for decades. In broad terms BPM uses, complements, integrates and extends theories, methods and tools from other scientific disciplines like: strategic management, information technology, managerial accounting, operations management etc [23, 16, 177, 55, 174, 1, 162, 33, 40, 47, 36, 141, 22, 147, 10, 135, 27, 37, 12, 25, 4, 170, 35, 13, 20, 3, 28, 145, 111, 32, 150, 23, 146, 15, 26, 31, 7, 38, 42, 5, 153, 17, 70, 167, 2, 164, 29, 159, 11, 30, 12, 39, 18, 6, 172, 22, 8, 179, 14, 24, 19, 21, 9, 34].

During this period the main focus themes of researchers and professionals in BPM were: business process modeling, business process analysis, activity based costing, business process simulation, performance measurement, workflow management, the link between information technology, BPM for process automation etc. [181, 154, 46, 118, 52, 163, 44, 114, 124, 53, 41, 101, 69, 116, 63, 149, 138, 176, 127, 64, 120, 84, 98, 112, 57, 140, 91, 131, 75, 95, 152, 54, 137, 88, 166, 86, 134, 78, 158, 89, 105, 122, 76, 60, 110, 80, 160, 48, 107, 96, 103, 51, 143, 169, 126, 58, 156, 129, 171, 82, 67, 178, 107]. More recently the focus moved to subjects like Knowledge Management, Enterprise Resource Planning (ERP) Systems, Service Oriented Architectures (SOAs), Process Intelligence (PI) and even Social Networks [161, 100, 45, 119, 77, 108, 87, 104, 132, 50, 106, 61, 90, 102, 85, 71, 73, 151, 97, 136, 81, 144, 56, 94, 109, 165, 92, 113, 79, 139, 65, 173, 62, 125, 83, 55, 130, 93, 66, 142, 49, 117, 68, 115, 121, 99, 157, 123, 43, 168, 155, 128, 148, 59, 175, 180]. In this collection of papers we present a review of the work and the outcomes achieved in the classic BPM fields as well as a deeper insight on recent advances in BPM. We also present a review of business process modeling and analysis and we elaborate on issues like business process quality and process performance measurement as well as their link to all other organizational aspects like human resources management, strategy, information technology (being SOA, PI or ERP), other managerial systems, job descriptions etc.

Recent advances to BPR tools are also presented with special focus on information technology, workflow, business process modeling and human resources management tools. Other chapters elaborate on the aspect of business process and organizational costing and their relationship to business process analysis, organizational change and reorganization. In the final chapters we present some new approaches that use fuzzy cognitive maps and a recently developed software tool for scenario creation and simulation in strategic management, business process management, performance measurement and social networking.

The audience of this book is quite wide. The first chapters can be read by professionals, academics and students who want to get some basic insight into the BPM field whereas the remaining present more elaborate and state of the art concepts methodologies and tools for an audience of a more advanced level.

More specifically in Chapter 1 Krogsti presents a review of *business process modeling* one of the most important areas of BPM. He first describes the main approaches of process modeling which are then classified according to the main modeling perspective being used. The author has identified 8 modeling perspectives namely the: behavioral, functional, structural, goal-oriented, object-oriented, language action, organizational and geographical. He presents characteristic examples of these process modeling perspectives, and finally discusses which perspectives are most appropriate to be used for the achievement of different modeling goals.

In chapter 2 Lohrman and Reichert present a framework for managing *business process quality*. They provide a definition of business process quality and specify appropriate and measurable sets of quality criteria based on organizational quality in specific application areas like finance and medicine. They finally evaluate their framework in real-world business processes. The authors argue that the outer environment of the business process and its associated organizational targets that define business process efficacy and efficiency constitute the two dimensions of business process quality requirements.

In chapter 3 Sezenias et. al. elaborate on the issue of holistic *organizational performance measurement*. They first classify performance measurement techniques used in the literature according to business symptoms or situations like restructuring, rightsizing etc and they then propose both hard measurable and soft quantitative measures according to different organizational business symptoms and organizational elements like organizational structure, processes, job descriptions, roles etc. The resulting methodology and tool called ADJUST provide a series of business analysis ratios tailored for different analysis perspectives like the structural, process, customer, strategic etc.

In chapter 4 Ozcelick examines the *effectiveness of Business Process Reengineering (BPR) projects* in improving firm performance after analyzing a comprehensive data set on large firms in the United States. The performance measures utilized in the chapter are labor productivity, return on assets, and return on equity. Closer examination proves that firm performance increases after BPR projects are finalized, while it remains unaffected during execution. Functionally focused BPR projects on average contribute more to performance than those focusing on a broader cross-functional scope. Later approaches thus present a higher potential of failure.

Bandar Alhaqbani et. al. in Chapter 5 elaborate on the issue of *privacy and security in workflow and business process management* and in particular on 'soft' security requirements like privacy and trust, which in contrast to hard measurable factors are often neglected. They propose and present a privacy aware conceptual model and workflow engine and validate their proposals on a health care case study. They implement these extensions in the YAWL workflow system and test it in health care scenarios.

Plakoytsi et. al. in chapter 6 elaborate on the need for a *performance measurement engine* that integrates concepts from three different application engines, namely, workflow, business process modeling and analysis and human resources management. The authors present a review of state of the art in workflow and business process management tools and describe their software tool. The resulting tool provides links to and integrates with well known business process management, workflow and human resources management tools.

In chapter 7 Mutschler and Reichert elaborate on the issue of *costs of business process management technology*. As process life cycle support and continuous process improvement have become critical success factors in enterprise computing, a variety of process support paradigms, process specification standards, process management tools, and supporting methods have emerged. The authors argue that significant costs are associated with the introduction of BPM approaches in enterprises. They additionally argue that even though existing economic-driven IT evaluation and software cost estimation approaches have received considerable attention during the last decades, it is difficult to apply them to BPM projects. In particular, they are unable to take into account the dynamic evolution of BPM projects caused by the numerous technological, organizational and project-specific factors influencing them. The latter, in turn, often lead to complex and unexpected cost effects in BPM projects making even rough cost estimations a challenge. What is needed is a comprehensive approach enabling BPM professionals to systematically investigate the costs of BPM projects. The authors take into consideration both known and often unknown cost factors in BPM projects and finally introduce the EcoPOST framework which utilizes evaluation models to describe the interplay of technological, organizational, and project-specific BPM cost factors as well as simulation concepts to unfold the dynamic behavior and costs of BPM projects.

In chapter 8 Papa et. al. model and analyze the *process for acquiring, evaluating and measuring intellectual capital*. They present the categories of intellectual capital and they propose a procedure for managing intellectual capital acquisition. They then propose an implementation procedure and a set of measures for measuring a firm's maturity towards intellectual capital management. They finally apply these measures to a real life case study.

Lorenzo-Romero et. al. in chapter 9 elaborate on the relationship between *BPM and Social Networking Sites* (SNS). SNS are second generation web applications allowing the creation of personal online networks. The social networking domain has become one of the fastest growing online environments connecting hundreds of millions of people worldwide. Businesses are increasingly interested in SNS as sources of customer voice and market information but are also interested in SNS as the domain where promising marketing tactics can be applied; SNS can be also used as business process management (BPM) tools due to powerful synergies between BPM and SNS. In order to analyze the factors influencing the acceptance and use of SNS in The Netherlands a Technology Acceptance Model (TAM) was developed and tested. The findings indicate that there is a significant positive effect of the ease of use of SNS on perceived usefulness. Both variables have a direct effect on the intention to use SNS and an indirect effect on the attitude towards the applications. Moreover, their research has shown that intention to use SNS has a direct and positive effect on the degree of final use of SNS. Results demonstrate empirically that the TAM can explain and predict the acceptance of SNS by users as new communication system connecting them to peers and businesses.

In a similar wavelength Stakias et. al. in chapter 10 present the use of *Fuzzy Cognitive Maps (FCMs) as an analysis technique for social networks and their use in BPM*. They present a literature survey of all approaches and all essential principles in social network analysis as well as their limitations. They then present the basic principles of FCMs and how they can be used for social network analysis with the use of the FCM modeler tool.

Continuing on FCMs Papageorgiou in Chapter 11 has compiled a review of *FCM approaches and applications* identified in the literature. She has classified the problems handled by FCMs during the last decade. Two of the most important application sectors are the business and information technology domains, the cornerstones of business process management. This provides a justification for the application of FCMs in business networks as presented by Stakias et. al in the previous chapter and the application of BPM in SNS in chapter 9.

In chapter 12 Groumpos and Karagiannis critically analyse the nature and state of *Decision Support Systems (DSS) theories, research and applications*. They provide a thorough and extensive historical review of DSS focused on the evolution of a number of sub-groupings of research and practice: personal decision support systems, group support systems, negotiation support systems, intelligent decision support systems, knowledge management-based DSS, executive information systems/business intelligence, and data warehousing. The need for new DSS methodologies and tools is investigated. The possibility of using Fuzzy Logic, Fuzzy Cognitive Maps and Intelligent Control in DSS is reviewed and analyzed. A new generic method for DSS is proposed. Basic components of the new generic method are provided and fully analyzed. Case studies are given showing the usefulness of the proposed method.

In Chapter 13 Glykas presents the application of *FCMs in strategic scenario creation and simulation* with special focus on the Balanced Scorecard. Initially they provide a review of limitations of existing approaches that use strategic maps in Balanced Scorecards. They then propose the use of FCMs as an alternative for use in strategic maps and highlight their advantages in comparison to existing approaches. They provide justification of the method and the FCM Modeler tool, used also by Stakias et. al. in chapter 10, in a case study in banking.

Similarly in Chapter 14 Wong et. al. argue that the current means of extracting domain-imposed software requirements through users or domain experts are often suboptimal and create a new domain of interest. They suggest the use of basic research findings as the more objective source and propose an approach that translates research findings into *a UML model based on which the domain-imposed require-*

*ments can be extracted*. By using business project management as the domain of interest, they outline the steps in the said approach and describe the use of the resulting domain model during requirements specification for a Project Management Information System that caters specially to the needs of business projects.

In chapter 15 Bevilacqua et. al describe a *company reorganization plan developed using business process reengineering (BPR) in a major enterprise operating in the domotics sector*, whose core business is the manufacture of cooker top extractor hoods. An *Object-State Transition* approach was used to generate the new "To-Be" models, focusing our attention on the objects (inputs, outputs, controls and resources) and on how they change during the processes. The action taken on the company's organizational structure generated a series of important changes in its internal hierarchy, the reference roles and the responsibilities of the people involved. The case stydy can be summarized in four main points: (1) complete analysis of the company's situation "As-Is" and preparation of IDEF0 diagrams describing the business processes, (2) proposal of changes to the high-priority processes requiring reorganization and implementation of the new "To Be" diagrams, (3) implementation of the new management and organizational solutions.

In Chapter 16 Arapantzi et al. present a very interesting approach to SME *cluster lifecycle process management*. They propose an innovative framework that concentrates on bottom up cluster creation and the management of the whole cluster lifecycle. They then review available software tools for cluster life-cycle management and support and they finally propose a 3D solution for cluster mangements and support build in Second Life. The solution includes a 3D building with classrooms for real time interactions of avatars representing SME participants.

# References

1. Adam Jr., E.E., et al.: An international study of quality improvement approach and firm Performance. International Journal of Operations & Production Management 17(9), 842–873 (1997)
2. Ahmed, N.U., Montagno, R.V., Firenze, R.J.: Operations strategy and organizational performance: an empirical study. International Journal of Operations & Production Management 16(5), 41–53 (1996)
3. Becker, B., Gerhart, B.: The impact of human resource management on organizational performance: progress and prospects. Academy of Management Journal 39, 779–801 (1996)
4. Bititci, U.S., Turner, T., Begemann, C.: Dynamics of performance measurement systems. International Journal of Operations & Production Management 20(6), 692–704 (2000)
5. Bititci, U.S., Carrie, A.S., McDevitt, L.: Integrated performance measurement systems: a development guide. International Journal of Operations & Production Management 17(5), 522–534 (1997)
6. Bevilacqua, R., Thornhill, D.: Process Modelling. American Programmer 5(5), 2–9 (1992)
7. Bradley, P., Browne, J., Jackson, S., Jagdev, H.: Business Process re-engineering (BPR) - A study of the software tools currently available. Computers In Industry 25, 309–330 (1995)

8. Bourne, M., Mills, J., Wilcox, M., Neely, A., Platts, K.: Designing, implementing and updating performance measurement systems". International Journal of Operations & ProductionManagement 20(7), 754–771 (2000)

9. Carr, D.K., Johansson, H.J.: Best Practices in Reengineering: What Works and What Doesn't in the Reengineering Process. McGraw-Hill, Inc., New York (1995)

10. Connelly, M.S., et al.: Exploring the relationship of leadership skill and knowledge to leader performance. Leadership Quarterly 11(1), 65–86 (2000)

11. Davenport, T.H.: Knowledge management and the broader firm: strategy, advantage, and performance. In: Liebowitz, J. (ed.) Knowledge Management Handbook, ch. 2. CRC Press, Boca Raton (1999)

12. Davenport, T.H., Prusak, L.: Working Knowledge. Harvard Business Press, Cambridge (1998)

13. Davidson, W.H.: Beyond Re-Engineering: The Three Phases of Business Transformation. IBM Systems Journal 32(1) (January 1993); reprinted in IEEE Engineering Management Review 23(2), 17–26 (1996)

14. Deming, W.E.: Out of the Crisis, Massachusetts Institute of Technology Center for Advanced Engineering Study, Cambridge, MA (1986)

15. Flapper, S.D.P., Fortuin, L., Stoop, P.P.M.: Towards consistent performance management systems. International Journal of Operations & Production Management 16(7), 27–37 (1996)

16. Ghalayini, A.M., Noble, J.S.: The changing basis of performance measurement. International Journal of Operations & Production Management 16(8), 63–80 (1996)

17. Glazer, R.: Measuring the knower, toward a theory of knowledge equity. California Management Review 40(3) (1998)

18. Gykas, M., Valiris, G.: Formal Methods in Object Oriented Business Modelling. The Journal of Systems and Software 48, 27–41 (1999)

19. Glykas, M., Valiris, G.: Management Science Semantics for Object Oriented Business Modelling in BPR. Information and Software Technology 40(8), 417–433 (1998)

20. Glykas, M., Valiris, G.: ARMA: A Multidisciplinary Approach to Business Process Redesign. Knowledge and Process Management 6(4), 213–226 (1999)

21. Hammer, M., Champy, J.: Reengineering the Corporation: A Manifesto for Business Revolution. HarperBusiness, New York (Hammer 1993)

22. Hansen, G.A.: Tools in Business Process Reengineering. IEEE Software, 131–133 (1994)

23. Lalli, F.: Why You Should Inves. In: Companies That Inves. In: Their Workers. Money, 11 (March 1996)

24. Longenecker, C.O., Fink, L.S.: Improving management performance in rapidly changing organisations. Journal of Management Development 20(1), 7–18 (2001)

25. Lubit, R.: Tacit knowledge and knowledge management: the keys to sustainable competitive advantage. Organizational Dynamics 29(4), 164–178 (2001)

26. Liebowitz, J., Wright, K.: Does measuring knowledge make"cents"? Expert Systems with Applications 17, 99–103 (1999)

27. Manganelli, R.L., Klein, M.M.: The Reengineering Handbook: A Step-by-Step Guide to Business Transformation. In: AMACON, New York (1994)

28. Ndlela, L.T., du Toit, A.S.A.: Establishing a knowledge management programme for competitive advantage in an enterprise. International Journal of Innovation Management 21(2), 151–165 (2001)

29. Neely, A.: Measuring Business Performance – Why, What and How. Economist Books, London (1998)

30. O'Mara, C., Hyland, P., Chapman, R.: Performance measurement and strategic change. Managing Service Quality 8(3), 178–182 (1998)
31. Ould, M.A.: Business Processes- Modelling and Analysis for Reengineering and Improvement. Wiley (1995)
32. Rolstadas, A.: Enterprise performance measurement. International Journal of Operations & Production Management 18(9/10), 989–999 (1998)
33. Tonchia, S.: Linking performance measurement systems to strategic and organisational choices. International Journal of Business Performance Management 2(1) (2000)
34. Valiris, G., Glykas, M.: Critical Review of existing BPR Methodologies: The Need for a Holistic Approach. Business Process Management Journal 5(1), 65–86 (1999)
35. Valiris, G., Glykas, M.: A Case Study on Reengineering Manufacturing Processes and Structures. Knowledge and Process Management 7(2), 20–28 (2000)
36. Valiris, G., Glykas, M.: Developing Solutions for Redesign: A Case Study in Tobacco Industry. In: Zupančič, J., Wojtkowski, G., Wojtkowski, W., Wrycza, S. (eds.) Evolution and Challenges in System Development, Plenum Press (1998)
37. Xirogiannis, G., Stefanou, J., Glykas, M.: A fuzzy cognitive map approach to support urban design. Journal of Expert Systems with Applications 26(2) (2004)
38. Yen, V.C.: An integrated model for business process measurement. Business Process Management Journal 15(6), 865–875 (2009)
39. Trkman, P.: The critical success factors of business process management. International Journal of Information Management 30(2), 125–134 (2010)
40. Herzog, N.V., Tonchia, S., Polajnar, A.: Linkages between manufacturing strategy, benchmarking, performance measurement and business process reengineering. Computers & Industrial Engineering 57(3), 963–975 (2009)
41. Han, K.H., Kang, J.G., Song, M.: Two-stage process analysis using the process-based performance measurement framework and business process simulation. Expert Systems with Applications 36(3), Part 2, 7080–7086 (2009)
42. Lam, C.Y., Ip, W.H., Lau, C.W.: A business process activity model and performance measurement using a time series ARIMA intervention analysis. Expert Systems with Applications 36(3), Part 2, 6986–6994 (2009)
43. Elbashir, M.Z., Collier, P.A., Davern, M.J.: Measuring the effects of business intelligence systems: The relationship between business process and organizational performance. International Journal of Accounting Information Systems 9(3), 135–153 (2008)
44. Jagdev, H., Bradley, P., Molloy, O.: A QFD based performance measurement tool. Computers in Industry 33(2-3), 357–366 (1997)
45. Tan, W.A., Shen, W., Zhao, J.: A methodology for dynamic enterprise process performance evaluation. Computers in Industry 58(5), 474–485 (2007)
46. Jeston, J., Nelis, J.: Chapter 6 - Process performance In their book Management by Process: A Roadmap to Sustainable Business Process Management. Elsevier (2009) ISBN: 978-0-7506-8761-4
47. Cheng, M.-Y., Tsai, H.-C., Lai, Y.-Y.: Construction management process reengineering performance measurements. Automation in Construction 18(2), 183–193 (2009)
48. Škerlavaj, M., Štemberger, M.I., Škrinjar, R., Dimovski, V.: Organizational learning culture—the missing link between business process change and organizational performance. International Journal of Production Economics 106(2), 346–367 (2007)
49. Neely, A., Mills, J., Platts, K., Gregory, M., Richards, H.: Performance measurement system design: Should process based approaches be adopted? International Journal of Production Economics 46-47, 423–431 (1996)
50. Trienekens, J.J.M., Kusters, R.J., Rendering, B., Stokla, K.: Business-oriented process improvement: practices and experiences at Thales Naval The Netherlands (TNNL). Information and Software Technology 47(2), 67–79 (2005)

51. Nudurupati, S., Arshad, T., Turner, T.: Performance measurement in the construction industry: An action case investigating manufacturing methodologies. Computers in Industry 58(7), 667–676 (2007)
52. Forme, F.-A.G.L., Genoulaz, V.B., Campagne, J.-P.: A framework to analyse collaborative performance. Computers in Industry 58(7), 687–697 (2007)
53. Trkman, P., McCormack, K., de Oliveira, M.P.V., Ladeira, M.B.: The impact of business analytics on supply chain performance. Decision Support Systems (in press, corrected proof, available online April 8, 2010)
54. Gregoriades, A., Sutcliffe, A.: A socio-technical approach to business process simulation. Decision Support Systems 45(4), 1017–1030 (2008)
55. Li, Y.-H., Huang, J.-W., Tsai, M.-T.: Entrepreneurial orientation and firm performance: The role of knowledge creation process. Industrial Marketing Management 38(4), 440–449 (2009)
56. Van Der Merwe, A.P.: Project management and business development: integrating strategy, structure, processes and projects. International Journal of Project Management 20(5), 401–411 (2002)
57. Reijers, H.A., van der Aalst, W.M.P.: The effectiveness of workflow management systems: Predictions and lessons learned. International Journal of Information Management 25(5), 458–472 (2005)
58. Trkman, P., McCormack, K., de Oliveira, M.P.V., Ladeira, M.B.: The impact of business analytics on supply chain performance. Decision Support Systems (in press, corrected proof, available online April 8, 2010)
59. Han, K.H., Park, J.W.: Process-centered knowledge model and enterprise ontology for the development of knowledge management system. Expert Systems with Applications 36(4), 7441–7447 (2009)
60. Jazayeri, M., Scapens, R.W.: The Business Values Scorecard within BAE Systems: The evolution of a performance measurement system. The British Accounting Review 40(1), 48–70 (2008)
61. Gewald, H., Dibbern, J.: Risks and benefits of business process outsourcing: A study of transaction services in the German banking industry. Information & Management 46(4), 249–257 (2009)
62. Carnaghan, C.: Business process modeling approaches in the context of process level audit risk assessment: An analysis and comparison. International Journal of Accounting Information Systems 7(2), 170–204 (2006)
63. Gregoriades, A., Sutcliffe, A.: Workload prediction for improved design and reliability of complex systems. Reliability Engineering & System Safety 93(4), 530–549 (2008)
64. Johnson, M., Kirchain, R.: Quantifying the effects of parts consolidation and development costs on material selection decisions: A process-based costing approach. International Journal of Production Economics 119(1), 174–186 (2009)
65. Tatsiopoulos, I.P., Panayiotou, N.: The integration of activity based costing and enterprise modeling for reengineering purposes. International Journal of Production Economics 66(1), 33–44 (2000)
66. Curtis, B., Kellner, M., Over, J.: Process modeling. Communications of the ACM 35(9), 75–90 (1992)
67. D. Askarany, H. Yazdifar, S. Askary. Supply chain management, activity-based costing and organisational factors. International Journal of Production Economics (in press, corrected proof, available online August 18, 2009 )
68. Tornberg, K., Jämsen, M., Paranko, J.: Activity-based costing and process modeling for cost-conscious product design: A case study in a manufacturing company. International Journal of Production Economics 79(1), 75–82 (2002)

69. Johnson, M.D., Kirchain, R.E.: Quantifying the effects of product family decisions on material selection: A process-based costing approach. International Journal of Production Economics 120(2), 653–668 (2009)
70. Johnson, M., Kirchain, R.: Quantifying the effects of parts consolidation and development costs on material selection decisions: A process-based costing approach. International Journal of Production Economics 119(1), 174–186 (2009)
71. Ng, K.M.: MOPSD: a framework linking business decision-making to product and process design. Computers & Chemical Engineering 29(1), 51–56 (2004)
72. Trkman, P.: The critical success factors of business process management. International Journal of Information Management 30(2), 125–134 (2010)
73. zur Muehlen, M., Indulska, M.: Modeling languages for business processes and business rules: A representational analysis. Information Systems 35(4), 379–390 (2010)
74. Grahovac, D., Devedzic, V.: COMEX: A cost management expert system. Expert Systems with Applications (in press, uncorrected proof, available online May 7, 2010)
75. Rodríguez, A., de Guzmán, I.G.-R., Fernández-Medina, E., Piattini, M.: Semi-formal transformation of secure business processes into analysis class and use case models: An MDA approach. Information and Software Technology (in press, corrected proof, available online April 9, 2010)
76. Jiang, P., Shao, X., Gao, L., Qiu, H., Li, P.: A process-view approach for cross-organizational workflows management. Advanced Engineering Informatics 24(2), 229–240 (2010)
77. Reijers, H.A., Liman Mansar, S.: Best practices in business process redesign: an overview and qualitative evaluation of successful redesign heuristics. Omega 33(4), 283–306 (2005)
78. Huang, Y., Wang, H., Yu, P., Xia, Y.: Property-Transition-Net-Based Workflow Process Modeling and Verification. Electronic Notes in Theoretical Computer Science 159, 155–170 (2006)
79. Lee, S., Kim, T.-Y., Kang, D., Kim, K., Lee, J.Y.: Composition of executable business process models by combining businessrules and process flows. Expert Systems with Applications 33(1), 221–229 (2007)
80. Gašević, D., Guizzardi, G., Taveter, K., Wagner, G.: Vocabularies, ontologies, and rules for enterprise and business process modeling and management. Information Systems 35(4), 375–378 (2010)
81. Dunn, C.: Business Process Modeling Approaches in the Context of Process Level Audit Risk Assessment: An Analysis and Comparison: Discussion comments. International Journal of Accounting Information Systems 7(2), 205–207 (2006)
82. Dreiling, A., Rosemann, M., van der Aalst, W.M.P., Sadiq, W.: From conceptual process models to running systems: A holistic approach for the configuration of enterprise system processes. Decision Support Systems 45(2), 189–207 (2008)
83. Durigon, M.: Business Process Modeling Approaches in the Context of Process Level Audit Risk-Assessment: An Analysis and Comparison. International Journal of Accounting Information Systems 7(2), 208–211 (2006)
84. Liu, D.-R., Shen, M.: Workflow modeling for virtual processes: an order-preservingprocess-view approach. Information Systems 28(6), 505–532 (2003)
85. Shen, J., Grossmann, G., Yang, Y., Stumptner, M., Schrefl, M., Reiter, T.: Analysis of business process integration in Web service context. Future Generation Computer Systems 23(3), 283–294 (2007)
86. Decker, G., Mendling, J.: Process instantiation. Data & Knowledge Engineering 68(9), 777–792 (2009)

87. Tsai, M.-T., Li, Y.-H.: Knowledge creation process in new venture strategy and performance. Journal of Business Research 60(4), 371–381 (2007)
88. Tseng, S.-M.: Knowledge management system performance measure indexqa1sa. Expert Systems with Applications 34(1), 734–745 (2008)
89. Lee, K.C., Lee, S., Kang, I.W.: KMPI: measuring knowledge management performance. Information & Management 42(3), 469–482 (2005)
90. Millie Kwan, M., Balasubramanian, P.: KnowledgeScope: managing knowledge in context. Decision Support Systems 35(4), 467–486 (2003)
91. Chang, M.-Y., Hung, Y.-C., Yen, D.C., Tseng, P.T.Y.: The research on the critical success factors of knowledge management and classification framework project in the Executive Yuan of Taiwan Government. Expert Systems with Applications 36(3), Part 1, 5376–5386 (2009)
92. Kock, N., Verville, J., Danesh-Pajou, A., DeLuca Communication, D.: flow orientation in business process modeling and its effect on redesign success: Results from a field study. Decision Support Systems 46(2), 562–575 (2009)
93. Arias-Aranda, D., Castro, J.L., Navarro, M., Sánchez, J.M., Zurita, J.M.: A fuzzy expert system for business management. Expert Systems with Applications (in press, uncorrected proof, available online May 11, 2010)
94. Zheng, W., Yang, B., McLean, G.N.: Linking organizational culture, structure, strategy, and organizational effectiveness: Mediating role of knowledge management. Journal of Business Research (in press, corrected proof, available online July 12, 2009)
95. Weber, B., Mutschler, B., Reichert, M.: Investigating the effort of using business process management technology: Results from a controlled experiment. Science of Computer Programming 75(5), 292–310 (2010)
96. Niemi, P., Huiskonen, J., Kärkkäinen, H.: Understanding the knowledge accumulation process—Implications for the adoption of inventory management techniques. International Journal of Production Economics 118(1), 160–167 (2009)
97. Yang, Y.J., Sung, T.-W., Wu, C., Chen, H.-Y.: An agent-based workflow system for enterprise based on FIPA-OS framework. Expert Systems with Applications 37(1), 393–400 (2010)
98. Leonardi, G., Panzarasa, S., Quaglini, S., Stefanelli, M., van der Aalst, W.M.P.: Interacting agents through a web-based health serviceflow management system. Journal of Biomedical Informatics 40(5), 486–499 (2007)
99. Nunes, V.T., Santoro, F.M., Marcos, R.S.: A context-based model for Knowledge Management embodied in work processes. Information Sciences 179(15), 2538–2554 (2009)
100. Liao, S.-H., Wu, C.-C.: System perspective of knowledge management, organizational learning, and organizational innovation. Expert Systems with Applications 37(2), 1096–1103 (2010)
101. Lai, M.-C., Lin, Y.-T., Lin, L.-H., Wang, W.-K., Huang, H.-C.: Information behavior and value creation potential of information capital: Mediating role of organizational learning. Expert Systems with Applications 36(1), 542–550 (2009)
102. King, W.R., Rachel Chung, T., Haney, M.H.: Knowledge Management and Organizational Learning. Omega 36(2), 167–172 (2008)
103. Wu, F., Tamer Cavusgil, S.: Organizational learning, commitment, and joint value creation in interfirm relationships. Journal of Business Research 59(1), 81–89 (2006)
104. Hsu, Y.-H., Fang, W.: Intellectual capital and new product development performance: The mediating role of organizational learning capability. Technological Forecasting and Social Change 76(5), 664–677 (2009)

105. Popova, V., Sharpanskykh, A.: Modeling organizational performance indicators. Information Systems 35(4), 505–527 (2010)
106. Hannah, S.T., Lester, P.B.: A multilevel approach to building and leading learning organizations. The Leadership Quarterly 20(1), 34–48 (2009)
107. Yang, C.-W., Fang, S.-C., Lin, J.L.: Organisational knowledge creation strategies: A conceptual framework. International Journal of Information Management 30(3), 231–238 (2010)
108. Liao, S.-H., Chang, W.-J., Wu, C.-C.: An integrated model for learning organization with strategic view: Benchmarking in the knowledge-intensive industry. Expert Systems with Applications 37(5), 3792–3798 (2010)
109. Mouritsen, J., Larsen, H.T.: The 2nd wave of knowledge management: The management control of knowledge resources through intellectual capital information. Management Accounting Research 16(3), 371–394 (2005)
110. Andrade, J., Ares, J., García, R., Pazos, J., Rodríguez, S., Silva, A.: Formal conceptualisation as a basis for a more procedural knowledge management. Decision Support Systems 45(1), 164–179 (2008)
111. Ming-Lang: Implementation and performance evaluation using the fuzzy network balanced scorecard. Computers & Education 55(1), 188–201 (2010)
112. Yüksel, İ., Dağdeviren, M.: Using the fuzzy analytic network process (ANP) for Balanced Scorecard (BSC): A case study for a manufacturing firm. Expert Systems with Applications 37(2), 1270–1278 (2010)
113. Dağdeviren, M., Yüksel, İ.: A fuzzy analytic network process (ANP) model for measurement of the sectoral competititon level (SCL). Expert Systems with Applications 37(2), 1005–1014 (2010)
114. Buytendijk, F., Hatch, T., Micheli, P.: Scenario-based strategy maps. Business Horizons (in press, corrected proof, available online March 4, 2010)
115. Quezada, L.E., Cordova, F.M., Palominos, P., Godoy, K., Ross, J.: Method for identifying strategic objectives in strategy maps. International Journal of Production Economics 122(1), 492–500 (2009)
116. Xirogiannis, G., Glykas, M.: Fuzzy Cognitive Maps in Business Analysis and Performance Driven Change. Journal of IEEE Transactions in Engineering Management 13(17) (2004)
117. Xirogiannis, G., Glykas, M.: Fuzzy Cognitive Maps as a Back-End to Knowledge-Based Systems in Geographically Dispersed Financial Organizations. Knowledge and Process Management Journal 11(1) (2004)
118. Glykas, M., Xirogiannis, G.: A soft knowledge modeling approach for geographically dispersed financial organizations. Soft Computing 9(8), 579–593 (2004)
119. Xirogiannis, G., Glykas, M.: Intelligent Modeling of e-Business Maturity. Expert Systems with Applications 32(2), 687–702 (2006)
120. Xirogiannis, G., Chytas, P., Glykas, M., Valiris, G.: Intelligent impact assessment of HRM to the shareholder value. Expert Systems with Applications 35(4), 2017–2031 (2008)
121. Wu, H.-Y., Tzeng, G.-H., Chen, Y.-H.: A fuzzy MCDM approach for evaluating banking performance based on Balanced Scorecard. Expert Systems with Applications 36(6), 10135–10147 (2009)
122. Thompson, K.R., Mathys, N.J.: The Aligned Balanced Scorecard: An Improved Tool for Building High Performance Organizations. Organizational Dynamics 37(4), 378–393 (2008)
123. Vergidis, K., Turner, C.J., Tiwari, A.: Business process perspectives: Theoretical developments vs. real-world practice. International Journal of Production Economics 114(1), 91–104 (2008)

124. Lam, C.Y., Ip, W.H., Lau, C.W.: A business process activity model and performance measurement using a time series ARIMA intervention analysis. Expert Systems with Applications 36(3), Part 2, 6986–6994 (2009)

125. Turetken, O., Schuff, D.: The impact of context-aware fisheye models on understanding business processes: An empirical study of data flow diagrams. Information & Management 44(1), 40–52 (2007)

126. Footen, J., Faust, J.: Business Process Management: Definitions, Concepts, and Methodologies. The Service-Oriented Media Enterprise, 249–296 (2008)

127. Zelm, M., Vernadat, F.B., Kosanke, K.: The CIMOSA business modelling process. Computers in Industry 27(2), 123–142 (1995)

128. Teng, J.T.C., Grover, V., Fiedler, K.D.: Developing strategic perspectives on business process reengineering: From process reconfiguration to organizational change. Omega 24(3), 271–294 (1996)

129. Noguera, M., Hurtado, M.V., Rodríguez, M.L., Chung, L., Garrido, J.L.: Ontology-driven analysis of UML-based collaborative processes using OWL-DL and CPN. Science of Computer Programming 75(8), 726–760 (2010), OMG. OLC 2.0 Specification, http://www.uml.org

130. Garousi, V.: Experience and challenges with UML-driven performance engineering of a Distributed Real-Time System. Information and Software Technology 52(6), 625–640 (2010)

131. Lima, V., Talhi, C., Mouheb, D., Debbabi, M., Wang, L., Pourzandi, M.: Formal Verification and Validation of UML 2.0 Sequence Diagrams using Source and Destination of Messages. Electronic Notes in Theoretical Computer Science 254, 143–160 (2009)

132. Woodside, M., Petriu, D.C., Petriu, D.B., Xu, J., Israr, T., Georg, G., France, R., Bieman, J.M., Houmb, S.H., Jürjens, J.: Performance analysis of security aspects by weaving scenarios extracted from UMLmodels. Journal of Systems and Software 82(1), 56–74 (2009)

133. Harmon, P., Davenport, T.: Business Process Modeling Notation BPMN CORE NOTATION. Business Process Change, 513–516 (2007)

134. Wong, P.Y.H., Gibbons, J.: Formalisations and applications of BPMN. Science of Computer Programming (September 26, 2009) (available online)

135. Dijkman, R.M., Dumas, M., Ouyang, C.: Semantics and analysis of business process models in BPMN. Information and Software Technology 50(12), 1281–1294 (2008)

136. Wong, P.Y.H., Gibbons, J.: A Relative Timed Semantics for BPMN. Electronic Notes in Theoretical Computer Science 229(2), 59–75 (2009)

137. Bevilacqua, M., Ciarapica, F.E., Giacchetta, G.: Business process reengineering of a supply chain and a traceability system: A case study. Journal of Food Engineering 93(1), 13–22 (2009)

138. Kindler, E.: On the semantics of EPCs: Resolving the vicious circle. Data & Knowledge Engineering 56(1), 23–40 (2006)

139. Mendling, J., Verbeek, H.M.W., van Dongen, B.F., van der Aalst, W.M.P., Neumann, G.: Detection and prediction of errors in EPCs of the SAP reference model. Data & Knowledge Engineering 64(1), 312–329 (2008)

140. Lassen, K.B., van der Aalst, W.M.P.: Complexity metrics for Workflow nets. Information and Software Technology 51(3), 610–626 (2009)

141. Trappey, C.V., Trappey, A.J.C., Huang, C.-J., Ku, C.C.: The design of a JADE-based autonomous workflow management system for collaborative SoC design. Expert Systems with Applications 36(2), Part 2, 2659–2669 (2009)

142. Tsai, C.-H., Huang, K.-C., Wang, F.-J., Chen, C.-H.: A distributed server architecture supporting dynamic resource provisioning for BPM-oriented workflow management systems. Journal of Systems and Software (in press, corrected proof, available online April 13, 2010)

143. Dai, W., Covvey, D., Alencar, P., Cowan, D.: Lightweight query-based analysis of workflow process dependencies. Journal of Systems and Software 82(6), 915–931 (2009)

144. Reijers, H.A., Mans, R.S., van der Toorn, R.A.: Improved model management with aggregated business process models. Data & Knowledge Engineering 68(2), 221–243 (2009)

145. Bertoni, M., Bordegoni, M., Cugini, U., Regazzoni, D., Rizzi, C.: PLM paradigm: How to lead BPR within the Product Development field. Computers in Industry 60(7), 476–484 (2009)

146. Gruhn, V., Laue, R.: What business process modelers can learn from programmers. Science of Computer Programming 65(1), 4–13 (2007)

147. Boiral, O.: Tacit Knowledge and Environmental Management. Long Range Planning 35(3), 291–317 (2002)

148. Werbel, J.D., DeMarie, S.M.: Aligning strategic human resource management and person–environment fit. Human Resource Management Review 15(4), 247–262 (2005)

149. Ould, M.A.: Business processes: Modelling and analysis for re-engineering and improvement, Wiley (1995) ISBN 0471953520

150. Phalp, K.T., Shepperd, M.: Quantitative Analysis of Static Models of Processes. Journal of Systems and Software 52(2-3), 105–112 (2000)

151. Tan, K., Baxter, G., Newell, S., Smye, S., Dear, P., Brownlee, K., Darling, J.: Knowledge elicitation for validation of a neonatal ventilation expert system utilising modified Delphi and focus group techniques. International Journal of Human-Computer Studies 68(6), 354 (2010)

152. Landeta, J.: Crrent validity of the Delphi method in social sciences Original Research Article. Technological Forecasting and Social Change 73(5), 467–482 (2006)

153. Landeta, J., Matey, J., Ruíz, V., Galter, J.: Results of a Delphi survey in drawing up the input–output tables for Catalonia. Technological Forecasting and Social Change 75(1), 32–56 (2008)

154. Chen, M.-K., Wang, S.-C.: The use of a hybrid fuzzy-Delphi-AHP approach to develop global business intelligence for information service firms. Expert Systems with Applications 37(11), 7394–7407 (2010)

155. Akkermans, H.A., Bogerd, P., Yücesan, E., van Wassenhove, L.N.: The impact of ERP on supply chain management: Exploratory findings from a European Delphi study. European Journal of Operational Research 146(2), 284–301 (2003)

156. Tsai, H.-Y., Chang, C.-W., Lin, H.-L.: Fuzzy hierarchy sensitive with Delphi method to evaluate hospital organization performance. Expert Systems with Applications 37(8), 5533–5541 (2010)

157. Nevo, D., Chan, Y.E.: A Delphi study of knowledge management systems: Scope and requirements. Information & Management 44(6), 583–597 (2007)

158. Becker, J., Vilkov, L., Weiß, B., Winkelmann, A.: A model based approach for calculating the process driven business value of RFID investments. International Journal of Production Economics 127(2), 358–371 (2010)

159. Glykas, M.: Work flow and Process Management in Printing and Publishing Firms. International Journal of Information Management 24(6), 523–538 (2004)

160. Xirogiannis, G., Glykas, M.: Intelligent Modeling of e-Business Maturity. Expert Systems with Applications 32(2), 687–702 (2007)

161. Xirogiannis, G., Chytas, P., Glykas, M., Valiris, G.: Intelligent impact assessment of HRM to the shareholder value. Expert Systems with Applications 35(4), 2017–2031 (2008)
162. Xirogiannis, G., Glykas, M., Staikouras, C.: Fuzzy Cognitive Maps in Banking Business Process Performance Measurement. In: Fuzzy Cognitive Maps. STUDFUZZ, vol. 247, pp. 161–200 (2010)
163. Chytas, P., Glykas, M., Valiris, G.: Software Reliability Modelling Using Fuzzy Cognitive Maps. In: Fuzzy Cognitive Maps. STUDFUZZ, vol. 247, pp. 217–230 (2010)
164. Glykas, M.: Fuzzy Cognitive Maps: Advances in Theory, Methodologies, Tools and Applications. STUDFUZZ, vol. 247 (2010)
165. Glykas, M., Xirogiannis, G.: A soft knowledge modeling approach for geographically dispersed financial organizations. Soft Computing - A Fusion of Foundations, Methodologies and Applications 9(8), 579–593 (2005)
166. Xirogiannis, G., Glykas, M.: Fuzzy Causal Maps in Business Modeling and Performance-Driven Process Re-engineering. In: Vouros, G.A., Panayiotopoulos, T. (eds.) SETN 2004. LNCS (LNAI), vol. 3025, pp. 331–341. Springer, Heidelberg (2004)
167. Kerr, D.S., Murthy, U.S.: Beyond brainstorming: The effectiveness of computer-mediated communication for convergence and negotiation tasks. International Journal of Accounting Information Systems 10(4), 245–262 (2009)
168. Muylle, S., Basu, A.: Online support for business processes by electronic intermediaries. Decision Support Systems 45(4), 845–857 (2008)
169. Wilson, C.: Brainstorming. In: User Experience Re-Mastered, pp. 107–134 (2010)
170. Lago, P.P., Beruvides, M.G., Jian, J.-Y., Canto, A.M., Sandoval, A., Taraban, R.: Structuring group decision making in a web-based environment by using the nominal grouptechnique. Computers & Industrial Engineering 52(2), 277–295 (2007)
171. Duggan, E.W., Thachenkary, C.S.: Integrating nominal group technique and joint application development for improved systems requirements determination. Information & Management 41(4), 399–411 (2004)
172. Spencer, D.M.: Facilitating public participation in tourism planning on American Indian reservations: A case study involving theNominal Group Technique. Tourism Management 31(5), 684–690 (2010)
173. Tao, F., Zhao, D., Yefa, H., Zhou, Z.: Correlation-aware resource service composition and optimal-selection in manufacturinggrid. European Journal of Operational Research 201(1), 129–143 (2010)
174. Edwards, H.M., McDonald, S., Young, S.M.: The repertory grid technique: Its place in empirical software engineering research. Information and Software Technology 51(4), 785–798 (2009)
175. Dittrich, R., Francis, B., Hatzinger, R., Katzenbeisser, W.: Modelling dependency in multivariate paired comparisons: A log-linear approach. Mathematical Social Sciences 52(2), 197–209 (2006)
176. Miranda, E., Bourque, P., Abran, A.: Sizing user stories using paired comparisons. Information and Software Technology 51(9), 1327–1337 (2009)
177. Szwed, P., Rene van Dorp, J., Merrick, J.R.W., Mazzuchi, T.A., Singh, A.: A Bayesian paired comparison approach for relative accident probability assessment with covariate information. European Journal of Operational Research 169(1), 157–177 (2006)
178. Mazzuchi, T.A., Linzey, W.G., Bruning, A.: A paired comparison experiment for gathering expert judgment for an aircraft wiring risk assessment. Reliability Engineering & System Safety 93(5), 722–731 (2008)

179. Abdelsalam, H.M.E., Gad, M.M.: Cost of quality in Dubai: An analytical case study of residential construction projects. International Journal of Project Management 27(5), 501–511 (2009)
180. Zhang, H., Xing, F.: Fuzzy-multi-objective particle swarm optimization for time–cost–quality tradeoff in construction. Automation in Construction (in press, corrected proof, available online August 30, 2010)
181. Rantz, M.J., Hicks, L., Petroski, G.F., Madsen, R.W., Alexander, G., Galambos, C., Conn, V., Scott-Cawiezell, J., Zwygart-Stauffacher, M., Greenwald, L.: Cost, Staffing and Quality Impact of Bedside Electronic Medical Record (EMR) in Nursing Homes. Journal of the American Medical Directors Association 11(7), 485–493 (2010)
182. Curkovic, S., Sroufe, R.: Total Quality Environmental Management and Total Cost Assessment: An exploratory study. International Journal of Production Economics 105(2), 560–579 (2007)
183. Ren, T.: Barriers and drivers for process innovation in the petrochemical industry: A case study Original Research Article. Journal of Engineering and Technology Management 26(4), 285–304 (2009)
184. Cheng, M.-Y., Tsai, H.-C., Lai, Y.-Y.: Construction management process reengineering performance measurements. Automation in Construction 18(2), 183–193 (2009)
185. Auzair, S.M., Langfield-Smith, K.: The effect of service process type, business strategy and life cycle stage on bureaucratic MCS in service organizations. Management Accounting Research 16(4), 399–421 (2005)
186. BarNir, A., Gallaugher, J.M., Auger, P.: Business process digitization, strategy, and the impact of firm age and size: the case of the magazine publishing industry. Journal of Business Venturing 18(6), 789–814 (2003)

176. Abdallah, H.M. and M.M. Saleh: Quality in higher learning institutes: government sectors and not-for-profits international journal of Advertising Management (AM), 304–53 (2009).

177. Zhang, B., S.Y. Liu, et al.: Robust optimization approach to emergency location, allocation in earthquake emergency Management structure community proceedings online August (2010).

178. Heng, J.H., Blake, D., Barnes, C.D., Mastren, R.W., Al-Hassan, A., O'Dumolel, O., Cowan, J., Schaffsheynen, L., Zweg, Grundmann, M., Perg, Stoll, et al.: Stalling and quality improvement for Risk factor reduction (R&R) in Nursing Homes. From with the annual conference International (112) 444-462 (2011).

179. Hadock, B., S. Stoot, et al.: The supply and demand and Production, et al.: the comprehensive study. International annual at the conference Production (2012).

180. Loch, C.H. and S.K. et al.: as processes found new in their improvement initiatives. Once Often Research Part I. Journal of Engineering and Technology Management volume (29) 622-641 (2011).

181. Cheng, M.Y., and R. Liu, Y.Y. Construction management process reengineering for Dynamic assessment. Annuals Automation in Construction 18(7), 163–251 (2009).

182. Cook, S.M., Ellen Smith and R.: The state of service processes in business improvement and the study on the service M&L. In service management. Mechanisms et International conference, vol. 9, 201–208.

183. Smith, A. Capplesmith, D.M., Austen, P.: Enhancing service utilities in structure of the free and fair estimate use of the insurance policy for Robert J: Journal of the Quality Manufacturing 10(2), 202 (2008).

# Contents

# Perspectives to Process Modeling

John Krogstie

Norwegian University of Science and Technology (NTNU), Trondheim
krogstie@idi.ntnu.no

**Abstract.** An important area of BPM is the modeling of processes. Processes modeling is done for a number of reasons in relation to BPM, and this chapter will describe main approaches to different types of process modeling. Modeling approaches will be structured according to the main modeling perspective being used. In conceptual modeling in general, one can identify 8 modeling perspectives; behavioral, functional, structural, goal-oriented, object-oriented, language action, organizational and geographical. In this chapter, we will present examples of process modeling according to these different perspectives, and discuss what perspectives are most appropriate to use to achieve the different goals of modeling.

## 1 Introduction

A *process* is a collection of related, structured tasks that produce a specific service or product to address a certain goal for a particular actor or set of actors. Process modeling has been performed relative to IT and organizational development at least since the 70ties. The interest has gone through phases with the introduction of different approaches, including Structured Analysis in the 70ties (Gane and Sarson, 1979), Business Process Reengineering in the late eighties/early nineties (Hammer and Champy, 1993), and Workflow Management in the 90ties (WfMC 2001). Lately, with the proliferation of BPM (Business process management) (Havey, 2005), interest and use of process modeling has increased even further, although focusing primarily on a selected number of modeling approaches.

Models of work processes have long been utilized to learn about, guide and support practice also in other areas. In software process improvement (Bandinelli et al 1995, Derniame 1998), enterprise modeling (Fox and Gruninger, 2000) and quality management, process models describe methods and standard working procedures. Simulation and quantitative analyses are also performed to improve efficiency (Abdel-Hamid and Madnick 1989, Kuntz et al 1998). In process centric software engineering environments (Ambriola et al 1997, Cugola 1998) and workflow systems (WfMC 2001), model execution is automated. This wide range of applications is reflected in current notations, which emphasize different aspects of work.

The traditional way to look on processes is as a function or transformation, according to an IPO (input-process-output) approach. Whereas early process modeling languages had this as a base approach, as process modeling have been integrated with other types of conceptual modeling, also variants of this have appeared, also for modeling of the core processes, as different approaches to modeling is beneficial for achieving different goals. Carlsen (1998) identifies five categories of process modeling languages: transformational, conversational, role-oriented, constraint-based, and system dynamics. In (Lillehagen and Krogstie, 2008) we looked upon these, but also included the OO-perspective due to increased interest in UML and related approaches. Lately, also other approaches have appeared making it important to provide a more comprehensive overview of perspectives to process modeling.

Before looking at the existing process modeling languages, we will discuss potential goals of doing process modeling in the first place. Then we describe different perspectives to modeling, before we provide an overview of modeling languages used for process modeling according to the different perspectives. We will here list both historically important as well as more current approaches, to illustrate the development of process modeling according to the various perspectives. In the conclusion we summarize how modeling according to the different perspectives is beneficial to achieve the various goals.

## 2  Goals of Process Modeling

According to general model theory (Stachowiak 1973) there are three common characteristics of models, namely *Representation, Simplification* and *Pragmatic orientation.*

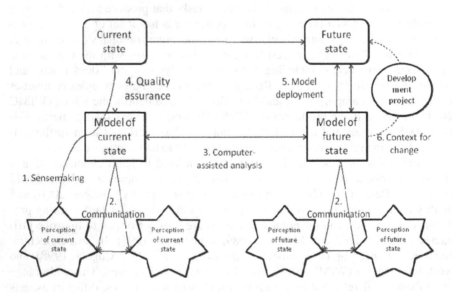

**Fig. 1** Organizational application of modeling

- *Representation*: Models are models of something else
- Simplification: Models possess a reductive trait in that they map a subset of attributes of the phenomenon being modeled rather than the complete set of attributes
- Pragmatic orientation: Models have a substitutive function in that they substitute a certain phenomenon as being conceptualized by a certain subject in a given temporal space with a certain incentive or operation in mind

Process modeling is usually done in some organizational setting. As illustrated in figure 1, one can look upon an organization and its information system abstractly to be in a state (the current state, often represented as a descriptive 'as-is' model) that are to be evolved to some future wanted state (often represented as a prescriptive 'to be' model). Obviously, changes will happen in an organization no matter what is actually planned, thus one might in practice have the use for many different models and scenarios of possible future states (e.g. what happens if we do nothing), but we simplify the number of possible future states in the discussion below.

The state includes the existing processes, organization and computer systems. These states are often modeled, and the state of the organization is perceived (differently) by different persons through these models. This set the scene for different usage areas of conceptual models (adapted from the original description in (Nysetvold and Krogstie 2006)):

1. Human sense-making: The model of the current state can be useful for people to make sense of and learn about the current situation, models being (perceived as) organizational reality.
2. Communication between people in the organization: As discussed by Bråten (1973), models are important for human communication. Thus, in addition to support the sense-making process for the individual, a model can act as a common framework supporting communication between people. This relates both to communication relative to the existing state, and relative to a possible future state. A model can give insight into the problems motivating a development project, and can help the systems developers and users understand the application system to be built. By hopefully helping to bridge the realm of the end-users and the information system, it facilitates a more reliable and constructive exchange of opinions between users and the developers of the system, and between different users. The models both help (but also restrict) the communication by establishing a nomenclature and a definition of the relevant phenomena in the modeling domain.
3. Computer-assisted analysis: To gain knowledge about the organization through simulation or deduction, often by comparing a model of the current state and a model of a future, potentially better state. Moreover, by analyzing the model instead of the domain itself, one might deduce properties that are difficult if not impossible to perceive directly since the model enables one to concentrate on a limited number of aspects at a time.
4. Quality assurance, ensuring e.g. that the organization acts according to a certified process achieved for instance through an ISO-certification process.

5. Model deployment and activation: To integrate the model of the future state in an information system directly and thereby make it actively take part in the actions of the organization. Models can be activated in three ways:
   a) Through people guided by process 'maps', where the system offers no active support.
   b) Automatically, where the system plays an active role, as in most automated workflow systems.
   c) Interactively, where the computer and the users co-operate in interpreting the model. The computer makes decisions about prescribed parts of the model, while the human users resolve ambiguities.
6. To provide a context for a traditional system development project, without being directly activated. This is the traditional usage of conceptual models, where the model represents the wanted future state and act as a prescriptive model including requirements as a basis for design and implementation, acting in the end as documentation of the developed system that can be useful in the future evolution of the system if one manage to keep it up to date.

In process modeling one often differentiate between design-time and run-time models. Whereas all areas above except area 5 can be said to relate to design-time, model, model activation also relates to run-time. Interactive activation is a bit special, since this enables one to at least in principle collapse the differentiation between design time and run-time.

Models produced have different characteristics and need. One dimension is the relevance over time and space. Whereas a model that should act as corporate memory should reflecting the organization and exist as a reference point to be kept up to date over time, sense-making models are often just used within an activity in order to make sense of something in an ad-hoc manner, and will usually not be maintained afterwards. This area has two dimensions:

- Relevance in time: For how long are the models relevant. In one extreme, they are only relevant at a meeting, whereas in the other extreme they are relevant for the life-time of the organization. As an example of a model in-between these, a requirements model for a new IT-systems is typically only relevant in practice in the project developing the system, although ideally it should be relevant for the whole life-cycle of the system.
- Relevance for whom: Is the model only relevant for a small group, for a department in the organization, for the whole organization, or even beyond the organizational boundaries?

Another differentiation is relative to formality of modeling language and approach. The choice of the formality of the modeling practice should be based on where these fit on the line with sense-making and corporate memory as the two extremes. Sense-making initiatives generally require a low level of formality of practice, while when developing a corporate memory, a more formal approach is needed. The choice of methods, tools and languages, as well as the choice of managing practice should reflect the level of need of formality.

Formality can be relative to a number of areas:

- Formality of language: The language used may have a formally defined syntax and semantics. In particular computer-assisted analysis and automatic activation mandate a formal semantics of the language. Most other usage areas is also easier with languages that have a formal syntax, although sense-making with a limited relevance in time for a limited number of actors can often best be supported with languages with a less formal syntax.
- Formality of tool support: The approach is supported more or less well by the used modeling tools. Whereas sense-making and modeling as context for development can be supported by ad-hoc tools (unless there is a need for traceability back to these models), the other goals warrant more formal tool support, especially where time and actor scope extend a little group for a limited time. In particular does this apply for computer-assisted analysis and automatic or interactive activation.
- Formality of modeling process: The modeling can be done rather ad-hoc, or according to a well-defined plan. Note that also modeling for sense-making can be part of a formal modeling process.
- Formality of management
  o Of the modeling language (in case there are language development and adjustments using meta-modeling facilities)
  o Of the modeling tools: Making sure right versions of the tools are available for everyone involved when needed.
  o Of the models: Ensure e.g. versioning, status-tracking of models etc.

Finally, different modeling activities might mandate different expressiveness of the modeling language used.

# 3 Perspectives to Modeling

Modeling languages can be divided into classes according to the core phenomena classes (concepts) that are represented and focused on in the language. We have called this the *perspective* of the language. Another term that can be used, is *structuring principle*. Generally, we can define a structuring principle to be some rule or assumption concerning how data should be structured. We observe that

- A structuring principle can be more or less detailed: on a high level one for instance has the choice between structuring the information hierarchically, or in a general network. Most approaches take a far more detailed attitude towards structuring: deciding what is going to be decomposed, and how. For instance, structured analysis implies that the things primarily to be decomposed are processes, and an additional suggestion might be that the hierarchy of processes should not be deeper than 4 levels, and the maximum number of processes in one diagram 7.
- A structuring principle might be more or less rigid: In some approaches one can override the standard structuring principle if one wants to, in others this is impossible.

A central structuring principle is so-called aggregation principles. Aggregation means to build larger components of a system by assembling smaller ones. Going for a certain aggregation principle thus implies decision concerning

- What kind of components to aggregate.
- How other kinds of components (if any) will be connected to the hierarchical structure.

Some possible aggregation principles are object-orientation, process-orientation, actor-orientation, and goal-orientation. Objects are the things subject to processing, processes are the actions performed, and actors are the ones who perform the actions. Goals are why we do the actions in the first place. Clearly, these four approaches concentrate on different aspects of the perceived reality, but it is easy to be mistaken about the difference. It is not which aspects they capture and represent that are relevant. Instead, the difference is one of focus, representation, dedication, visualization, and sequence, in that an oriented language typically prescribes that (Opdahl and Sindre, 1997):

- Some aspects are promoted as fundamental for modeling, whereas other aspects are covered mainly to provide the context of the promoted ones (focus).
- Some aspects are represented explicitly, others only implicitly (representation).
- Some aspects are covered by dedicated modeling constructs, whereas others are less accurately covered by general ones (dedication).
- Some aspects are visualized in diagrams; others only recorded textually (visualization).
- Some aspects are captured before others during modeling (sequence).

Below we investigate the characteristics of such perspectives in more detail.

## 4   An Overview of Modeling Perspectives

A classic distinction regarding modeling perspectives is between the structural, functional, and behavioral perspective (Olle et al, 1988). Object-orientation analysis appeared as a particular way of combining the structural and behavioral perspective inspired by object-oriented program in the late eighties. (Curtis et. al 1992) included in addition to these three (termed informational, functional, and dynamic) the organizational view, describing who performs the task and where in the organization this is done (Functionally and physically). We notice that a recent overview in the field (Mili et al 2011) uses the perspectives of Curtis et al.

In F3 (Bubenko et al, 1994), it was recognized that a requirement specification should answer the following questions:

- Why is the system built?
- Which are the processes to be supported by the system?
- Which are the actors of the organization performing the processes?

- What data or material are they processing or talking about?
- Which initial objectives and requirements can be stated regarding the system to be developed?

This indicates a need to support what we will term the goal and rule-perspective and the actor/role perspective (as also highlighted by Curtis), in addition to the classical perspectives.

In the NATURE project (Jarke et al, 1993), one distinguished between four worlds: Usage, subject, system, and development. Modeling as we use it here applies to the subject and usage world for which NATURE propose structural (data) models, functional models, and behavior models, and organization models, business models, speech act models, and actor models respectively. The Zachman Framework for enterprise modeling (Zachman 1987) highlight the intersection between the roles in the design process, that is, Owner, Designer and Builder; and the product abstractions, that is, What (material) it is made of, How (process) it works and Where (geometry) the components are, relative to one another. From the very inception of the framework, some other product abstractions were known to exist because it was obvious that in addition to What, How and Where, a complete description would necessarily have to include the remaining primitive interrogatives: Who, When and Why. These three additional interrogatives would be manifest as three additional types of models that, in the case of Enterprises, would depict:

- Who does what work,
- When do things happen (and in what order) and
- Why are various choices made

In addition to perspectives indicated above, the Zachman framework highlight the topological/geographical dimension which have increased in importance over the last decade due to the proliferation of mobile and multi-channel solutions.

Based on the above, to give a broad overview of the different perspectives conceptual modeling approaches accommodate, we have focused on the following:

- Behavioral perspective: Languages in this perspective go back to at least the early sixties, with the introduction of Petri-nets (Petri, 1962). In most languages with a behavioral perspective the main phenomena are states and transitions between states. State transitions are triggered by events (Davis, 1988). A finite state machine (FSM) is a hypothetical machine that can be in only one of a given number of states at any specific time. In response to an input, the machine generates an output, and changes state.
- Functional perspective: The main phenomena class in the functional perspective is the transformation: A transformation is defined as an activity which based on a set of phenomena transforms them to another (possibly empty) set of phenomena. Other terms used are function, process, activity, and task. The best know conceptual modeling language with a functional perspective is data flow diagrams (DFD) (Gane and Sarson, 1979). Another important early example was (IDEF0, 1993).

- Structural perspective: Approaches within the structural perspective concentrate on describing the static structure of a system. The main construct of such languages is the "entity". Other terms used for this role with some differences in semantics are object, concept, thing, and phenomena. Note that objects used in object-oriented approaches are discussed further under the object-perspective below. The structural perspective has traditionally been handled by languages for data modeling. Whereas the first data modeling language was published in 1974 (Hull and King, 1974), the first having major impact was the entity-relationship language of Chen (1976).

- Goal and Rule perspective: Goal-oriented modeling focuses on goals and rules. A rule is something which influences the actions of a set of actors. A rule is either a rule of necessity or a deontic rule (Wieringa, 1989)) A rule of necessity is a rule that must always be satisfied. A deontic rule is a rule which is only socially agreed among a set of persons and organizations. A deontic rule can thus be violated without redefining the terms in the rule. Deontic rules are included in recent standards for rule modeling (OMG 2008). A deontic rule can be classified as being an obligation, a recommendation, permission, a discouragement, or a prohibition (Krogstie and Sindre, 1996). The general structure of a rule is `` if condition then expression" where condition is descriptive, indicating the scope of the rule by designating the conditions in which the rule apply, and the expression is prescriptive. According to Twining & Miers (1982) any rule, can be analyzed and restated as a compound conditional statement of this form. In the early nineties, one started to model high-level rules in so-called rule hierarchies, linking rules of different abstraction levels. The relationships available for this in an early approach of this type, Tempora, were (Seltveit, 1993): Refers-to, Necessitates and motivates, Overrules and suspends.

- Object perspective: The basic phenomena of object oriented modeling languages are similar to those found in most object-oriented programming languages:
  - Object: An object is an "entity" which has a unique and unchangeable identifier and a local state consisting of a collection of attributes with assignable values. The state can only be manipulated by a set of methods defined on the object. The value of the state can only be accessed by sending a message to the object to call on one of its methods. The details of the methods may not be known, except through their interfaces. The happening of an operation being triggered by receiving a message, is called an event.
  - Process: The process of an object, also called the object's life cycle, is the trace of the events during the existence of the object.
  - Class: A set of objects that share the same definitions of attributes and operations compose an object class. A subset of a class, called subclass, may have its special attribute and operation definitions, but still share all definitions of its superclass through inheritance.

- Communication perspective: Much of the work within this perspective is based on language/action theory from philosophical linguistics. In this light, this is often termed the action-perspective (Lind and Goldkuhl, 2003). The basic assumption of language/action theory is that persons cooperate within work processes through their conversations and through mutual commitments taken within them. Speech act theory, which has mainly been developed by Austin and Searle (Austin, 1962, Searle, 1969, Searle, 1979) starts from the assumption that the minimal unit of human communication is not a sentence or other expression, but rather the performance of certain types of language acts. Illocutionary logic (Dignum and Weigand, 1994; Searle and Vanderveken, 1985) is a logical formalization of the theory and can be used to formally describe communication structure.

- Actor and role perspective: The main phenomena of languages within this perspective are actor (alternatively termed agent) and role. The background for modeling of the kind described in this perspective comes both from organizational science, work on programming languages (e.g. actor-languages (Tomlinson and Scheevel, 1989)), and work on intelligent agents in artificial intelligence (e.g. (Genesereth and S. T. Ketchpel, 1994, Shoham, 1994)).

- Geographical perspective: This perspective relates to the topological ordering between the different concepts. The best background for conceptualization of these aspects comes from fields of cartography and CSCW. In CSCW one differentiate between space and place ((Dourish, 2006; Harrison and Dourish, 1996). "Space" describes geometrical arrangements that might structure, constrain, and enable certain forms of movement and interaction; "place" denotes the ways in which settings acquire recognizable and persistent social meaning in the course of interaction.

This is only one way of classifying modeling approaches, and in many cases it will be difficult to classify a specific approach solely according to one perspective. Many modern frameworks and approaches to modeling mix several perspectives in integrated approaches. On the other hand, we have over a long time experienced this as a useful way of ordering the presentation of modeling approaches.

Another way of classifying modeling languages is according to their time-perspective (Sølvberg and Kung, 1993):

- Static perspective: Provide facilities for describing a snapshot of the perceived reality, thus only considering one state. Languages of the structural perspective are usually of this kind.

- Dynamic perspective: Provide facilities for modeling state transitions, considering two states, and how the transition between the states takes place. Languages of the behavioral perspective are often of this type

- Temporal perspective: Allow the specification of time dependant constraints. In general, sequences of states are explicitly considered. Some rule-oriented approaches are of this type.

- Full-time perspective: Emphasize the important role and particular treatment of time in modeling. The number of states explicitly considered at a time is indefinite.

Yet another way of classifying modeling languages is according to their level of formality. Conceptual modeling languages can be classified as semi-formal (having a formal syntax, but no formal semantics) or formal, having a logical and/or executional semantics. The logical semantics used can vary (e.g. first-order logic, description logic, modal logic). Executional/operational semantics indicate that a model in the language can be executed on a computing machine if it is complete. Models can in addition be used together with descriptions in informal (natural) languages and non-linguistic representations, such as audio and video recordings.

Finally, it is important to differentiate the level of modeling; are we modeling types or instances? In traditional conceptual modeling, one are normally only modeling on the type level, whereas in enterprise modeling it is also usual to model on the instance level, often combining concepts on the type and instance level, including process-types and process instances in the same model.

# 5 Process Modeling According to Different Modeling Perspectives

Here we provide an overview of process modeling according to the different modeling perspectives identified in section 3 above. This overview shares some similarities with (Mili et al, 2011). On the other hand, we concentrate here on diagrammatical modeling languages, covering a wider set of perspectives than them.

## 5.1 Behavioral Perspective

As indicated, states (of systems, products, entities, processes) and transformations between states are the central aspect in this perspective, often depicted in finite state machines (FSM). There are two language-types commonly used to model FSM's: State transition diagrams (STD) and state transition matrices (STM). The vocabulary of state transition diagrams is illustrated in Fig. 2 and is described below:

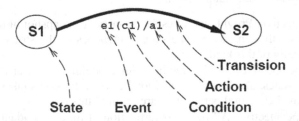

**Fig. 2** Symbols in the state transition modeling language

- State: A system is always in one of the states in the lawful state space for the system. A state is defined by the set of transitions leading to that state, the set of transitions leading out of that state and the set of values assigned to attributes of the system while the system resides in that state.

- Event: An event is a message from the environment or from system itself to the system. The system can react to a set of predefined events.
- Condition: A condition for reacting to an event. Another term used for this is 'guard'.
- Action: The system can perform an action in response to an event in addition to the transition.
- Transition: Receiving an event will cause a transition to a new state if the event is defined for the current state, and if the condition assigned to the event evaluates to true.

A simple example that models the state of a paper during the preparation of a professional conference is depicted in Fig. 3. The double circles indicate end-states. Thus the paper is regarded to be non-existent when under development. When it is received, it is in state 1: Received. Usually a confirmation of the reception of paper is sent, putting the paper in state 2: Confirmed. The paper is sent to a number of reviewers. First it is decide who are to review which paper, providing an even work-load for reviewers, avoiding conflict of interest etc. Then the papers are distributed to the reviewers entering state 3: Distributed. As each review is received the paper is in state 4: Reviewed. Often there would be additional rules relating to the minimum number of reviews that should be received before making a verdict. This is not included in this model. Before a certain time, decisions are made if the paper are accepted, conditionally accepted or rejected, entering state 5, 6, or 7. A conditionally accepted paper needs to be reworked to be finally accepted. All accepted papers have to be sent in following the appropriate format (so-called CRC - Camera Ready Copy). When this is received, the paper is in state 8: Received CRC. When all accepted papers are received in a CRC-form, the proceeding are put together and then published (state 9: Published).

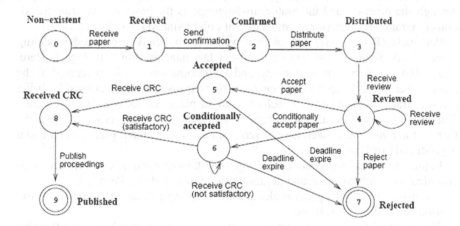

**Fig. 3** Example of a state transition model

It is generally acknowledged that a complex system cannot be comprehensibly described in the fashion depicted in figure 3, because of the unmanageable, exponentially growth of states, all of which have to be arranged in a 'flat' model. Hierarchical abstraction mechanisms where added to traditional STD in Statecharts (Harel, 1987) to provide the language with constructs to support modularity and hierarchy as illustrated in Fig. 4.

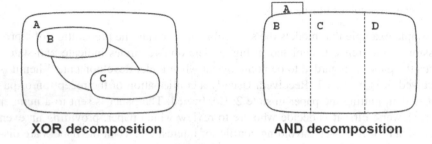

**XOR decomposition**                                     **AND decomposition**

**Fig. 4** Decomposition mechanisms in Statecharts

- XOR decomposition: A state is decomposed into several states. An event entering this state (A) will have to enter one and only one of its sub-states (B or C). In this way generalization is supported.
- AND decomposition: A state is divided into several states. The system resides in all these states (B, C, and D) when entering the decomposed state (A). In this way aggregation is supported.

Statecharts are combined with functional modeling in (Harel et al, 1990). Later extensions of the Statechart-language for object-oriented modeling was developed through the nineties, and the Statechart-language is the basis for the state transitions diagrams in UML (for the modeling of object-states) (Booch et al, 2005).

Petri-nets (Petri, 1962) are another well-known behavior-oriented modeling language. A model in the original Petri-net language is shown in Fig. 5. Here, *places* indicate a system state space, and a combination of *tokens* located at the places determines the specific system state. State transitions are regulated by firing rules: A transition is enabled if each of its input places contains a token. A transition can fire at any time after it is enabled. The transition takes zero time. After the firing of a transition, a token is removed from each of its input places and a token is produced in all output places.

Figure 5 shows how control-flow aspects like precedence, concurrency, synchronization, exclusiveness, and iteration can be modeled in a Petri-net.

The associated model patterns along with the firing rule above establish the execution semantics of a Petri-net.

The classical Petri net cannot be decomposed. This is inevitable by the fact that transitions are instantaneous, which makes it impossible to compose more complex networks (whose execution is bound to take time) into higher level transitions. However, there exists several dialects of the Petri net language (for instance (Marsan, 1985))) where the transitions are allowed to take time, and these approaches provide decomposition in a way not very different from that of a data flow diagram. Timed Petri Nets (Marsan, 1985) also provide probability distributions that can be assigned to the time consumption of each transition and is particularly suited to performance modeling. Other variants include:

- Colored Petri-nets where tokens are named and typed variables.
- Nets with two kinds of input places to a transition: consumption places and reference places. For the former, a token is consumed when a transition fires, whereas the latter is not consumed.
- Nets where transitions have pre- and post-conditions in some logic. For a transition to fire, its precondition must be true, and by the firing its postcondition will become true.

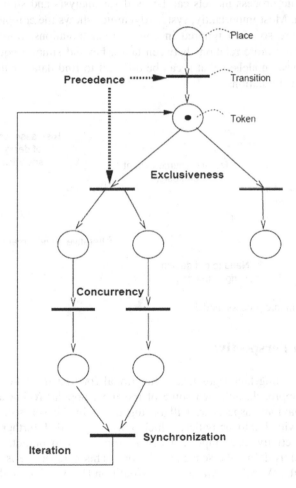

**Fig. 5** Dynamic expressiveness of Petri-nets

Another type of behavioral modeling is based on System Dynamics. Holistic systems thinking (Senge 1990) regards causal relations as mutual, circular and non-linear, hence the straightforward sequences in transformational process models is seen as an idealization that hides important facts. This perspective is also reflected in mathematical models of interaction (Wegner and Goldin 1999). System dynamics have been utilized for analysis of complex relationships in cooperative work arrangements (Abdel-Hamid and Madnick 1989). A simple example is depicted in Fig. 6. It shows one aspect of the interdependencies between design and implementation in a system development project. The more time you spend designing, the less time you have for coding and testing, hence you better get the design right the first time. This creates a positive feedback loop similar to "analysis paralysis" that must be balanced by some means, in our example iterative development.

System dynamic process models can be used for analysis and simulation, but not for activation. Most importantly, system dynamics shows the complex interdependencies that are so often ignored in conventional notations, illustrating the need for articulating more relations between tasks, beyond simple sequencing. A challenge with these models is that it can be difficult to find data for the parameters needed to run simulations.

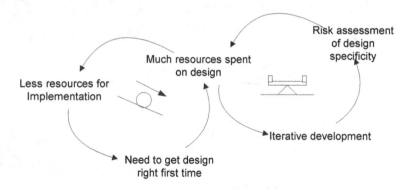

**Fig. 6** A system dynamic process model

## 6 Functional Perspective

Most process modeling languages take a functional (or transformational / input-process-output) approach, although some of the most popular recent approaches also include behavioral aspects as will be discussed in the sub-section below. Processes are divided into activities, which may be divided further into sub-activities. Each activity takes inputs, which it transforms to outputs. Input and output relations thus define the sequence of work. This perspective is chosen for the standards of the Workflow Management Coalition (WfMC 2000), the Internet

Engineering Task Force (IETF) (Bolcer and Kaiser, 1999), and the Object Management Group (OMG 2000) as well as most commercial systems (Fisher 2000). IDEF-0 (1993) and Data Flow Diagram (DFD) (Gane and Sarson 1979) are paradigm examples of this.

DFDs describe a situation using the symbols illustrated in Fig. 7:

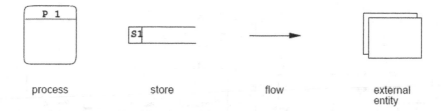

process                   store                    flow                    external
                                                                           entity

**Fig. 7** Symbols in the DFD language

- Process. Illustrates a part of a system that transforms a set of inputs to a set of outputs
- Store. A collection of data or material.
- Flow. A movement of data or material within the system, from one system component (process, store, or external entity) to another;
- External entity. An individual or organizational actor, or a technical actor that is outside the boundaries of the system to be modeled, which interact with the system.

With these symbols, a system can be represented as a network of processes, stores and external entities linked by flows. A process can be decomposed into a new DFD. When the description of the process is considered to have reached a detailed level where no further decomposition is needed, "process logic" can be defined in forms of e.g. structured English, decision tables, and decision trees.

And example from the conference domain is depicted in Fig. 8, where one illustrate the main external actors and tasks relative to the review and evaluation of scientific papers at a conference (cf. Fig 3.)

When a process is decomposed into a set of sub-processes, the sub-processes are co-operating to fulfill the higher-level function. This view on DFDs has resulted in the "context diagram" that regards the whole system as a process which receives and sends all inputs and outputs to and from the system. A context diagram determines the boundary of a system. Every activity of the system is seen as the result of a stimulus by the arrival of a data flow across some boundary. If no external data flow arrives, then the system will remain in a stable state. Therefore, a DFD is basically able to model reactive systems.

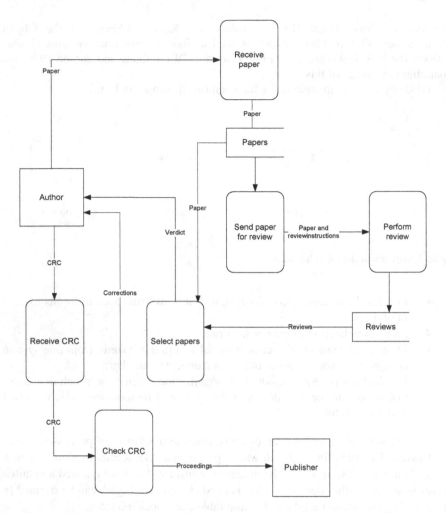

**Fig. 8** DFD of Paper submission and selection

A variant of context-diagrams used in UML is Use-case diagrams (Booch et al, 2005). Developing use case models are often the first step in a project that uses UML. Use cases capture the overall requirements on the system from the perspectives of various *actors*. When use cases represent activities in a process, the limited availability of inter-use-case associations becomes a problem. While specialization (*extends*) and composition (*uses*) can be represented, there is no way of articulating logical precedence relations (work and control flow). This makes use cases mostly suited for initial models and overviews to be used in combination with other languages.

DFD and use cases are semi-formal languages. Some of the short-comings of DFD regarding formality are addressed in the transformation schema presented by Ward (1986). The main symbols of his language are illustrated in Fig. 9.

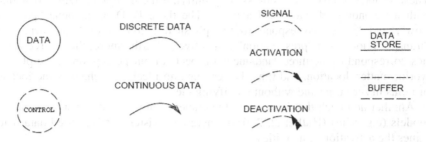

**Fig. 9** Symbols in the transformation schema language

There are four main classes of symbols:

1. Transformations: A solid circle represents a data transformation, which are used approximately as a process in DFD. A dotted circle represents a control transformation which controls the behavior of data transformations by activating or deactivating them, thus being an abstraction on some portion of the systems' control logic.
2. Data flows: A discrete data flow is associated with a set of variable values that is defined at discrete points in time. Continuous data flows are associated with a value or a set of values defined continuously over a time-interval.
3. Event flows: These report a happening or give a command at a discrete point in time. A signal shows the sender's intention to report that something has happened, and the absence of any knowledge on the sender's part of the use to which the signal is put. Activations show the senders intention to cause a receiver to produce some output. A deactivation shows the senders intention to prevent a receiver from producing some output.
4. Stores: A store acts as a repository for data that is subject to a storage delay. A buffer is a special kind of store in which flows produced by one or more transformations are subject to a delay before being consumed by one or more transformations. It is an abstraction of a stack or a queue.

Both process and flow decomposition are supported.

Whereas Ward had a goal of formalizing DFD's and adding more possibilities of representing event-flows (note, one is not able to represent control flows as possible e.g. in behavioral approaches such as Petri-nets), Opdahl and Sindre (1994) worked to adapt data flow diagrams to what they term 'real-world modeling'. Problems they note with DFD in this respect are as follows:

- 'Flows' are semantically overloaded: Sometimes a flow means transportation, other times it merely connects the output of one process to the input of the next.
- Parallelism often has to be modeled by duplicating data on several flows. This is all right for data, but material cannot be duplicated in the same way
- Whereas processes can be decomposed to contain flows and stores in addition to sub-processes, decomposition of flows and stores is not allowed. This makes it hard to deal sensibly with flows at high levels of abstraction (Bubenko, 1988).

These problems have been addressed by unifying the traditional DFD vocabulary with a taxonomy of real-world activity: The three DFD phenomena "process," flow", and "store" correspond to the physical activities of "transformation," "transportation", and "preservation" respectively. Furthermore, these three activities correspond to the three fundamental aspects of our perception of the physical world: matter, location, and time. Hence, e.g., an ideal flow changes the location of items in zero time and without modifying them.

Another approach that is found in the eScience area in data-oriented workflow models (e.g. Scufl (Hull et al, 2006)) where the existence of required data determines the activation of activities.

Given the extensive use of transformational languages, a number of analyses focus on this category (Conradi and Jaccheri 1998, Curtis et al 1992, Green and Rosemann 2000, Lei and Singh 1997). The expressiveness of these languages typically includes decomposition, and data flow, while organizational modeling and roles often are integrated (see also role-oriented modeling below). In approaches which integrate behavioral and functional aspects as discussed further below, we see also a support for control flow. Aspects like timing and quantification, products and communication, or commitments are better supported by other perspectives. User-orientation is a major advantage of functional languages. Partitioning the process into steps, match well the descriptions that people use elsewhere. Graphical input-process-output models are comprehensible given some training, but you can also build models by simply listing the tasks in plain text, or in a hierarchical work breakdown structure. Hence, the models can be quite simple, provided that incomplete ordering of steps is allowed.

## 6.1  Combined Behavioral and Functional Approaches

A number of the recent process modeling notations add control-flow aspects to a functional approach i.e. can be said to somehow combine the functional and behavioral perspectives. Some examples of this are ARIS EPC, UML Activity Diagrams, YAWL, and BPMN.

An Event-driven Process Chain (EPC) (Keller et al, 1992) is a graphical modeling language used for business process modeling. EPC was developed within the framework of Architecture of Integrated Information System (ARIS) (Scheer and Nüttgens, 2000) to model business processes. The language is targeted to describe processes on the level of their business logic, not necessarily on the formal specification level. EPC are supported in tools such as ARIS and Visio.

An event-driven process chain consists of the following elements: Functions; Event; Organization unit; Information, material, or resource object; Logical connectors; Logical relationships (i.e., Branch/Merge, Fork/Join, OR); Control flow; Information flow; Organization unit assignment; Process path.

The strength of EPC lies on its easy-to-understand notation that is capable of portraying business information system while at the same time incorporating other important features such as functions, data, organizational structure, and information resources. However the semantics of an event-driven process chain are not well defined and it is not possible to check the model for consistency and completeness. As demonstrated in (Aalst, 1999), these problems can be tackled by mapping EPC to Petri nets since Petri nets have formal semantics and a number of analysis techniques are provided. In addition, in order to support data and model interchange among heterogeneous BPM tools, an XML-based EPC – EPML (Event-driven Process Chain Markup Language), has been proposed by (Mendeling and Nüttgens, 2006).

The Unified Modeling Language (UML) is a general-purpose visual modeling language that is used to specify, visualize, construct and document the artifacts of a software system (Booch, 2005). It was developed by Booch, Rumbaugh, and Jacobson and was later accepted as standard in object-oriented modeling by the Object Management Group (OMG). The Activity diagram is one of the three diagram types in the UML for modeling behavioral aspect of systems.

The most important concepts in the UML activity diagram are as follows:

- *rounded rectangles* represent activities
- *diamonds* represent decisions
- *bars* represent the start (split) or end (join) of concurrent activities
- a *black circle* represents the start (initial state) of the workflow
- an *encircled black circle* represents the end (final state)

Compared to the other modeling languages in this section, it can be argued that UML activity diagram is more extensible because developers can use the UML stereotype-mechanism to introduce new elements into UML activity diagrams. Moreover, the UML activity diagram is an execution-oriented modeling language. Compared to BPMN below, the UML activity diagram offers less control flow constructs. Last but not least, as illustrated in (Dori, 2002), the lack of first class process primitives is a major limitation of the core of UML.

In 2004, BPMN was presented as the standard business process modeling notation (White 2004). Since then BPMN has been evaluated in different ways by the academic community (Aagesen and Krogstie, 2010) and has become widely supported in industry.

There are a large number of tools supporting BPMN. The tool support in industry has increased with the awareness of the potential benefits of Business Process Management (BPM).

The Business Process Modeling Notation (BPMN version 1.0) was proposed in May 2004 and adopted by OMG for ratification in February 2006. The BPMN 2.0 specification was formally released January 2011. BPMN is based on the revision of other notations and methodologies, especially UML Activity Diagram, UML EDOC Business Process (see under object-perspective), IDEF, ebXML BPSS, Activity-Decision Flow (ADF) Diagram, RosettaNet, LOVeM and EPC.

The primary goal of BPMN was to provide a notation that is readily understandable by all stakeholders, from the business analysts who create the initial draft of the processes, to the technical developers responsible for implementing the technology that will support the performance of those processes, and, finally to the business people who will manage and monitor those processes (White 2004).

Another factor that drove the development of BPMN is that, historically, business process models developed by business people have been separated from the process representations required by systems designed to implement and execute those processes. Thus, it was a need to manually translate the original process models to execution models. Such translations are subject to errors and make it difficult for the process owners to understand the evolution and the performance of the processes they have developed. To address this, a key goal in the development of BPMN was to create a bridge from a user-friendly notation to execution languages. BPMN models are thus designed to be activated through the mapping to BPEL.

BPMN allows the creation of end-to-end business processes and is designed to cover many types of modeling tasks constrained to business processes. The structuring elements of BPMN will allow the viewer to be able to differentiate between sections of a BPMN Diagram using groups, pools or lanes. Basic types of submodels found within a BPMN model can be *private business processes* (internal), *abstract processes* (public) and *collaboration processes* (global).

*Private business processes* are those internal to a specific organization and are the types of processes that have been generally called workflow or BPM processes

*Abstract Processes* represents the interactions between a private business process and another process or participant. Abstract processes are contained within a Pool and can be modeled separately or within a larger BPMN Diagram to show the Message Flow between the abstract process activities and other entities.

*Collaboration process* depicts the interactions between two or more business entities. These interactions are defined as a sequence of activities that represent the message exchange patterns between the entities involved.

The Business Process Diagram (BPD) is the graphical representation of BPMN. Its language constructs are grouped in four basic categories of elements, viz., Flow Objects, Connecting Objects, Swimlanes and Artifacts. The notation is further divided into a core element set and an extended element set. The intention of the core element set is to support the requirements of simple notations and most business processes should be modeled adequately with the core set. The extended set provides additional graphical notations for the modeling of more complex processes.

*Flow Objects* (Fig. 10) contains events, activities and gateways. *Events* are either start events, intermediate events or end events. *Activities* are divided into process, sub-process and tasks and denote the work that is done within a company. *Gateways* are used for determining branching, forking, merging or joining of paths within the process. Markers can be placed within the gateway to indicate behavior of the given construct.

EVENTS        ACTIVITY        GATEWAY

**Fig. 10** BPD elements events (start, intermediate and end), activity and gateway

*Connecting objects* (Fig. 11) are used for connecting the flow objects. *Sequence Flow* defines the execution order of the activities within a process while *Message Flow* indicates a flow of messages between business entities or roles prepared to send and receive them. *Association* is used to associate both text and graphical non-flow objects with flow objects.

**Fig. 11** BPD connection objects: Sequence flow, message flow and association

*Swimlanes* (Fig. 12) are used to denote a participant in a process and acts as a graphical container for a set of activities taken on by that participant. By dividing *Pools* into *Lanes* (thus creating sub-partitioning), activities can be organized and categorized.

**Fig. 12** BPD pool and lanes

*Artifacts* (not illustrated) are data objects, groups and annotations. *Data Objects* are not considered as having any other effect on the process than containing information on resources required or produced by activities. The *Group* construct is a visual aid used for documentation or analysis purposes while the *Text Annotation* is used to add additional information about certain aspects of the model.

Figure 13 shows an example BPMN process summoning participants for a workshop. The workshop organizer sends out the invitations, which are received by the potential participants. The participants evaluate the relevance of the workshop and decide whether they will participate or not. Those who want to participate, sign up for the workshop by informing the organizer.

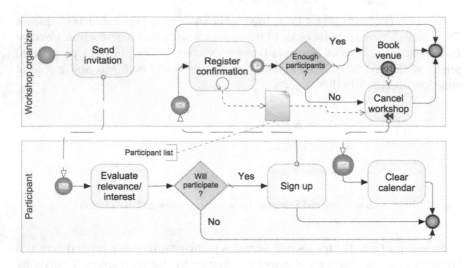

**Fig. 13** BPMN model showing the summons for a workshop

The organizer registers the confirmations from the participants until the deadline for registering, making a list of participants. When the deadline is reached (indicated by the timer event on the looping register confirmation activity), the organizer will see if there are enough participants to conduct the workshop. If there are too few participants, the organizer will inform those participants who signed up that the workshop is canceled, and the registered participants will clear their calendar for the day.

If there is a sufficiently number of participants registered for the workshop, the organizer will try to book a venue. But if there is no venue available, the workshop will have to be canceled by informing registered participants. This is shown using the compensation and undo activity.

## 6.2 Structural Perspective

The structural perspective has traditionally been handled by languages for data modeling, but also includes approaches from semantic networks and the semantic web. In (Chen, 1976), the basic components are:

- Entities. An entity is a phenomenon that can be distinctly identified. Entities can be classified into entity classes;
- Relationships. A relationship is an association among entities. Relationships can be classified into relationship classes which can be looked upon as an aggregation between the related entity-classes cf. section 3.1;
- Attributes and data values. Values are grouped into value classes by their types. An attribute is a function which maps from an entity class or relationship class to a value class; thus the property of an entity or a relationship can be expressed by an attribute-value pair.

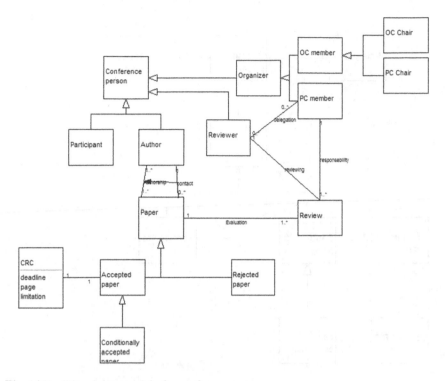

**Fig. 14** Traditional data model of a conference case

Many will argue that structural modeling is fundamentally different from functional (process) modeling, since it focus on the static aspects, whereas process modeling focus on dynamics. It is possible to look at processes and tasks as entities though (like one have done in object-oriented process modeling discussed below, looking at the process instances as the objects) it which case one can model the situation in a similar way as when doing more traditional data-modeling. Figure 14 shows part of a traditional data-model for a conference organization case, relative to involved roles and products in connection to the paper submission, review and selection. Figure 15 depict parts of the conference case where the focus is on tasks performed in connection to this, and one notice many similarity (but also some differences) from when the entities are data/structural constructs and processes, i.e. the inclusion of task outside the scope of the database system, and the possibility of expressing additional rules. Obviously the naming of entities is different.

One find very few attempts for this type of modeling in practice. We look more on such approaches below when discussing OO process models.

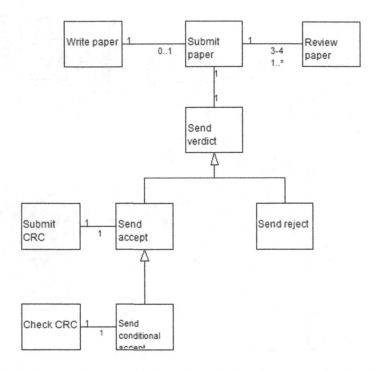

**Fig. 15** Structural process model of conference case

## 6.3   Goal and Rule Perspective

In the workflow area, this perspective is often termed declarative workflow. Constraint based languages (Dourish et al 1996, Glance et al 1996) do not present a course of events, rather they capture the boundaries within which the process must be performed, leaving the actors to control the internal details. Instead of telling people what to do, these systems warn about (deontic) rule violations and enforce constraints. Thus, common problems with over-serialization of the process can be avoided (Glance et al 1996).

A wide variety of declarative modeling approaches has been specified in business process management, from the use of basic Event-condition-action (ECA)-rules (Kappel et al, 1998) to declarative process modeling languages such as DecSerFlow (Aalst and Pesic, 2006), BPCN (Lu et al 2009) and ConDec (Pesic and AAlst, 2006). (Goedertier and Vanthienen, 2009) present an overview of the most common declarative process modeling languages.

Several advantages have been experienced with a declarative, rule-based approach to information systems modeling in general (Krogstie and Sindre, 1996):

- Problem-orientation. The representation of business rules declaratively is independent of what they are used for and how they will be implemented. With an explicit specification of assumptions, rules, and constraints, the analyst has freedom from technical considerations to reason about application problems. This freedom is particularly important for the communication with the stakeholders with a non-technical background.
- Maintenance: A declarative approach makes possible a one place representation of business rules, which is a great advantage when it comes to the maintainability of the specification and system. Moreover, additional business rules can be directly added, without the need to fully redesign the business process. Consequently the design-time flexibility of a declarative business process model can be high.
- Process compliance might be guaranteed when all relevant business rules are represented as mandatory (rules of necessity), which results in traceability and facilitates verification by domain specialists. The impact on the process efficiency of the declarative process model can be influenced by specifying guidelines (deontic rules), which specify an optimal execution path, that can be violated during execution due to unforeseen circumstances.
- Knowledge enhancement: The rules used in an organization, and as such in a supporting computerized information system, are not always explicitly given. In the words of Stamper (1987) ``Every organization, in as far as it is organized, acts as though its members were conforming to a set of rules only a *few of which may be explicit*(our emphasis). This has inspired certain researchers to look upon CIS specification as a process of rule reconstruction, i.e. the goal is not only to represent and support rules that are already known, but also to uncover de facto and implicit rules which are not yet part of a shared organizational reality, in addition to the construction of new, possibly more appropriate ones.

On the other hand, several problems have been observed when using a simple rule-format.

- Every statement must be either true or false, there is nothing in between.
- In most systems it is not possible to distinguish between rules of necessity and deontic rules.
- In many goal and rule modeling languages it is not possible to specify who (which actor) the rules apply to.
- Formal rule languages have the advantage of eliminating ambiguity. However, this does not mean that rule based models are easy to understand. There are two problems with the comprehension of such models, both the comprehension of single rules, and the comprehension of the whole rule-base. Whereas the traditional operational models (e.g. functional models) have decomposition and modularization facilities which make it possible to view a system at various levels of abstraction and to navigate in a hierarchical structure, rule models are usually flat. With many rules such a model soon becomes difficult to grasp, even if each rule should be understandable in itself.

They are also seldom linked to other models of the organization used to understand and develop the information systems.

- A general problem is that a set of rules is either consistent or inconsistent. On the other hand, human organizations may often have more or less contradictory rules, and have to be able to deal with this.

Languages representing rule-based process modeling can potentially provide a higher expressiveness than diagrammatic languages (e.g. the ability to specify temporal requirements) (Lu and Sadiq, 2007), but this might result in process models which are less comprehensible (Fickas, 1989) due to large rule-bases as pointed to above.

Declarative Process Enactment guarantees high run-time flexibility for declarative process specifications that contain only the strictly required mandatory constraints. An individual execution path that satisfies the set of mandatory constraints can be dynamically built for a specific process instance. Process compliance is assured when all mandatory rules are correctly mapped onto mandatory business constraints. During the construction of a suitable execution path little support is provided to the end user (Weske, 2007), which could affect the process effectiveness. In (Krogstie and Sindre, 1996) the idea of differentiating constraints by modality was proposed. Recommendations would guide the user whereas mandatory constraints would ensure compliant behavior. The guidance provided by the deontic rules might depend on explicit domain knowledge or can be learned through process mining (Schonenberg et al, 2008). Lastly, the increased size and complexity of contemporary process models might decrease the potential for process automation since current declarative workflow management systems might have limited efficiency when dealing with large models according to (Aalst et al, 2009).

A graphic depiction is difficult since it would correspond to a visualization of several possible solutions to the set of constraint equations constituting the model. The support for articulation of planned and ongoing tasks is limited. Consequently, constraints are often combined with transformational models (Bernstein 2000, Dourish et al 1996). Alternatively one can have the operational rules also linked to goal hierarchies as in EEML (Krogstie, 2008) and other approaches for goal-oriented modeling (Kavakli and Loucopoulos, 2005).

## 6.4 Object Perspective

UML (Booch 2005) has become the official and de facto standard for object-oriented analysis and design. Consequently, people also apply UML to model business processes. Object orientation offers a number of useful modeling techniques like encapsulation, polymorphism, subtyping and inheritance (Loos and Allweyer 1998, Mühlen and Becker 1999). UML integrates these capabilities with e.g. requirements capture in use case descriptions as described above and behavior modeling in state, activity and sequence diagrams (vs. discussion above). On the other hand, UML is designed for software developers, not for end users. A core challenge thus remains in mapping system-oriented UML constructs to user- and process-oriented concepts (Hommes and Reijswoud 1999). To this problem no general solution exists (Loos and Allweyer 1998, Störle 2001). UML process languages utilize associations, classes, operations, use cases, interaction sequence, or activity diagrams. The lack of a standardized approach reflects the wide range of process modeling approaches in business and software engineering. Many of these are already covered above. One approach which is somewhat similar to how one can use structural modeling is PML (Anderl and Raßler, 2008). A new approach for a process modeling language has been introduced in, which uses object oriented techniques based on looking upon classes in a particular way. Whereas a class is defined by class name, attributes, and methods, in this approach one define this as process name, methods, and resources. The PML process class describes the process in a generic way. It allows one to define all methods with assurances and resources needed for the process. The instantiation of a process is a project. This means, the instance of a process defines the current occurrence of resources, used data models etc. Regarding to connections and dependencies between single process classes, PML features the standard UML-concepts of inheritance and associations. A simple example taken from (Anderl and Raßler, 2008) is shown in Figure 16.

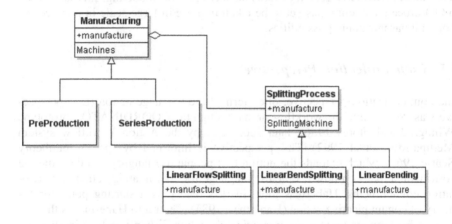

**Fig. 16** Manufacturing process in PML

*Manufacturing* consists of *SplittingProcess*, which has three sub-processes: *LinearFlowSplitting*, *LinearBendSplitting*, and *LinearBending*. To instantiate *Manufacturing* different *SplittingProcess*es are instantiated. But, using the concept of object orientation, not *SplittingProcess* but the sub-processes will be instantiated. The same is true for the resources in *Manufacturing*. Machines should hold the specialized machines, which are linear flow splitting machines, linear bend splitting machines, and linear bending machines. Thus having a generic process description, the product and its manufacturing process is dependent of the used machines, the splitting processes and their order in the manufacturing line.

Another approach is enriching UML with additional concepts. EDOC was an OMG standard modeling framework being an extension of UML 1.4 aimed at supporting component-based enterprise distributed object computing (OMG, 2004), also known as Enterprise Collaboration Architecture. The component collaboration architecture (CCA) views a system as a set of components that interact through a set of ports according to a set of protocols. Components, called ProcessComponents, may be decomposed into a set of collaborating subcomponents. The business process profile is a specialization of CCA aimed at representing business process in terms of a composition of business activities. It enables the expression of:

- Complex dependencies between business tasks (data, control, timing)
- Various time expressions on tasks such as duration and deadlines
- Exception handling
- Associations between business tasks and business roles that perform these tasks
- Selection criteria for selecting entities to fulfill roles at process instantiation.

Some aspects of EDOC was taken further into the work on UML 2, but not those being most particularly geared towards business process modeling. It is safe to say to OO process modeling has yet to be taken into use in large scale in practice, although it has intriguing possibilities.

## 6.5  Communication Perspective

The communication perspective, often termed the language action (LA) perspective was brought into the workflow arena through the COORDINATOR prototype (Winograd and Flores 1986), later succeeded by the Action Workflow system (Medina-Mora et al 1992). This perspective is informed by speech act theory (Searle 1969), which extends the notion that people use language to describe the world with a focus on how people use language for coordinating action and negotiating commitments. Habermas took Searle's theory as a starting point for his theory of communicative action (Habermas, 1984). Central to Habermas is the distinction between strategic and communicative action. When involved in strategic action, the participants try to achieve their own private goals. When they cooperate, they are only motivated empirically (i.e. since they know by experience that it is wise) to do so: they try to maximize their own profit or minimize their own losses. When involved in communicative action, the participants are oriented towards mutual agreement. The motivation for co-operation is thus rational. Illocutionary logic (Dignum and Weigand, 1994; Searle and Vanderveken, 1985) is a logical formalization of the theory and can be used to formally describe the communication structure described by Searle. The main parts of illocutionary logic are the illocutionary act consisting of three parts, illocutionary context, illocutionary force, and propositional context.

The context of an illocutionary act consists of five elements: Speaker (S), hearer (H), time, location, and circumstances.

The illocutionary force determines the reasons and the goal of the communication. The central element of the illocutionary force is the illocutionary point, and the other elements depend on this. Five illocutionary points are distinguished (Searle, 1979):

- Assertives: Commit S to the truth of the expressed proposition (e.g. "It is raining", "A conference will take place in Oslo in November 2011").
- Directives: Attempts by S to get H to do something (e.g. "Close the window"; "Write the article according to these guidelines").
- Commissives: Commit S to some future course of action (e.g. "I will be there", "If you send us your paper before a certain date, it will be reviewed").
- Declaratives: The successful performance guarantees the correspondence between the proposition p and the world (e.g. "The ball is out" (stated by the umpire), "your paper is accepted" (stated by the program committee)).
- Expressives: Express the psychological state about a state of affairs specified in the proposition. (E.g. Congratulations!).

Speech act theory is the basis for modeling of workflow as coordination among people in Action Workflow (Medina-Mora et al, 1992). The basic structure is shown in Fig. 17. Two major roles, customer and supplier, are modeled. Workflow is defined as coordination between actors having these roles, and is represented by a conversation pattern with four phases. In the first phase the customer makes a request for work, or the supplier makes an offer to the customer.

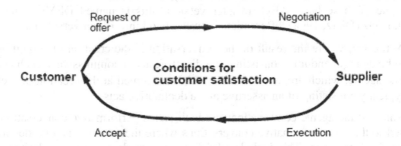

**Fig. 17** Main phases of action workflow

In the second phase, the customer and supplier aim at reaching a mutual agreement about what is to be accomplished. This is reflected in the contract conditions of satisfaction. In the third phase, after the performer has performed what has been agreed upon and completed the work, completion is declared for the customer. In the fourth and final phase the customer assess the work according to the conditions of satisfaction and declares satisfaction or dissatisfaction. The ultimate goal of the loop is customer satisfaction. This implies that the workflow loop have to be closed. It is possible to decompose steps into other loops. The specific activities carried out in order to meet the contract are not modeled.

The main strength of this approach is that it facilitates analysis of the communicative aspects of the process, something we will return to in the next perspective. It highlights that each process is an interaction between a customer and a performer, represented as a cycle with four phases: preparation, negotiation, performance and acceptance. The dual role constellation is a basis for work breakdown, e.g. the performer can delegate parts of the work to other people. Process models may thus spread out. This explicit representation of communication and negotiation, and especially the structuring of the conversation into predefined speech act steps, has also been criticized (Button 1995, De Michelis and Grasso 1994, King 1995, Suchman 1994). The limited support for situated conversations, the danger that explication leads to increased external control of the work, and a simplistic one-to-one mapping between utterances and actions are among the weaknesses. On the other hand, it has been reported that the Action Workflow approach is useful when people act pragmatically and don't always follow the encoded rules of behavior (De Michelis and Grasso 1994), i.e. when the communication models are interactively activated.

Some later approaches to workflow modeling include aspects of both the functional (see Sect. 3.3.3) and language action modeling. In WooRKS (Ader et al, 1994) functional modeling is used for processes and LA for exceptions thus not using these perspectives in combination. TeamWare Flow (Swendson et al, 1994) and Obligations (Bogia, 1995) on the other hand can be said to be hybrid approaches, but using radically different concepts from those found in traditional conceptual modeling.

In addition to the approach to workflow-modeling described above, several other approaches to conceptual modeling are inspired by the theories of Habermas and Searle such as SAMPO (Auramäki, 1992), and ABC/DEMO. We will describe one of these here, ABC (in later versions this is named DEMO (Dietz, 1999)). Dietz (Dietz, 1994) differentiate between two kinds of conversations:

- Actagenic, where the result of the conversation is the creation of something to be done (agendum), consisting of a directive and a commissive speech act.
- Factagenic, which are conversations which are aimed at the creation of facts typically consisting of an assertive and a declarative acts

Actagenic and factagenic conversations are both called performative conversations. Opposed to these are informative conversations where the outcome is a production of already created data. This includes the deduction of data using e.g. derivation rules. A transaction is a sequence of three steps: Carrying out an actagenic conversation, executing an essential action, and carrying out a factagenic conversation.

Speech acts oriented approaches is a subset of the more general area of action-oriented approaches (Lind and Goldkuhl, 2003). All action-based approaches use the idea that each action goes through a number of phases. However, the exact number and definition of these phases differ per approach.

## 6.6 Actor and Role Perspective

Role-centric process modeling languages have been applied for work-flow analysis and implementation. Role Interaction Nets (RIN) (Singh and Rein 1992) and

Role Activity Diagrams (RAD) (Ould 1995) use roles as their main structuring concept. The activities performed by a role are grouped together in the diagram, either in swim-lanes (RIN), or inside boxes (RAD). The use of roles as a structuring concept makes it very clear who is responsible for what. RAD has also been merged with speech acts for interaction between roles (Benson et al 2000). The role-based approach also has limitations, e.g. making it difficult to change the organizational distribution of work. It primarily targets analysis of administrative procedures, where formal roles are important. The use of swimlanes in BPMN and UML Activity Diagrams as described above might also have this effect. Some other approaches worth discussing on this level are REA and e3Value.

The REA language was first described in McCarthy (1982) and then has been developed further in Geerts and McCarthy (1999). REA was originally intended as a basis for accounting information systems and focuses on representing increases and decreases of value in an organization. REA has subsequently been extended to apply to enterprise architectures (Hruby, 2006) and e-commerce frameworks (UMM, 2007).

The core concepts in the REA language are *resource*, *event* and *agent*. The intuition behind this language is that every business transaction can be described as an event where two agents exchange resources. In order to acquire a resource from other agents, an agent has to give up some of its own resource. Basically, there are two types of events: *exchange* and *conversion* (Hruby, 2006). An exchange occurs when an agent receives economic resources from another agent and gives resource back to that agent. A conversion occurs when an agent consumes resources to produce other resources. REA has influenced the electronic commerce standard ebXML, with McCarthy actively involved in the standards committee.

E$^3$Value (Gordijn, 2006) is an actor/role oriented modeling language for inter-organizational modeling. The purpose of this modeling language is to represent how actors of a system create, exchange and consume objects of economic value. The modeling language focuses on the key points of a business model, and to get an understanding of business operations and systems requirements through scenario analysis and evaluation. Through an evaluation, the purpose of e$^3$value is to determine whether a business idea is profitable or not, that is to say by analyzing for each actor involved in the system whether the idea is profitable for them or not.

E3value models give a representation of actors, exchanges, value objects of a business system. Here are the main concepts and the associated graphical representation in e$^3$value (Figure 18).

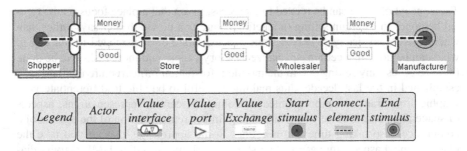

**Fig. 18** Symbols in e$^3$value modeling language

- Actor: Entity that is economically independent in its environment, that is to say supposed to be profitable (for different kind of value, e.g. intellectual, economical...). An actor is identified by its name. *Example: Author, program committee, delegate...*
- Value object: A value object can have a lot of faces: services, product, knowledge... It is exchanged by actors who consider it has an economic value. The value object is defined by its name, which is representative of the kind of object it is. *Example: money, proceedings, paper...*
- Value Port: It belongs to an actor, and allows it to request value objects to the others actors, and so to create an interconnection. Moreover, it is representatives of the external view of $E^3$value, by focusing on external trades and not on internal process. The value port is characterized by its direction ("in" or "out")
- Value interface: It belongs to an actor, and usually groups one "ingoing" value port, and one "outgoing" value port. It introduces the notion of "fair trade" or "reciprocity" in the trade: one offering against one request. *Example: in an Internet provider point of view, the offer ("out") is an Internet connection, the request ("in") is money.*
- Value exchange: It connects two value ports with each other, that is to say it establish a connection between two actors for potential exchange of value object. Because value port is represented by a direction, value exchange is represented by both the *has in* and the *has out* relation.
- Value Transaction: A value transaction links two or more values exchanges to conceptualize the fair-trade exchange between actors. If a value exchange appears in more than one transaction, we call it a multi-party transaction
- Market segment: A market segment group together value interfaces of actors which assign economic value to object equally. It is a simplification for systems where actors have similar value interfaces. An important point is that an actor can be a member of different market segments, because we consider only the value interface. The market segment is identified by the name and a count of the number of members. Example: (Hotels), (travel agencies)

## 6.7 Geographic Perspective

The concept of place can be related to a process, given that a place focus on the typical behavior in a certain setting (a meeting-room or a movie theater say) rather than where this is physically. Whereas some processes are closely related to place (e.g. what can be done in a certain, specialized factory), more and more tasks can be done in more or less any setting due to the mobile information infrastructure that has been established in the last decade, thus making it useful to be able to differentiate geographic from transformation-oriented modeling. In certain representations, aspects of space and place is closely interlinked (e.g. in the representation of the agenda of a conference, also taking time into account). Some approaches that let you take the place-oriented aspects into account exist, e.g. work on extending UML activity diagrams with place-oriented aspects (Gopalakrishnan et al, 2010).

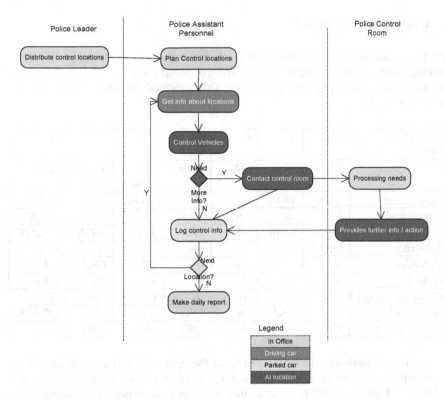

**Fig. 19** Example of modeling of geographic aspects

A specific notations based on UML Activity Diagrams extended using colors to differentiate place-oriented aspects is presented in Fig. 19. The model is based on a simple task in a police traffic control case. The leader of the police allots control locations for each police assistant personnel for controlling the vehicles through those locations. Controlling includes following up things such as driver license, speeding, drunk driving etc. Police Assistant Personnel (PAP) receives info about control locations at the office from the leader. PAP plans and gets info about the control locations while driving the car. After reaching the control location, PAP controls the drivers and vehicles. If he decides he need more info/personnel he contacts through mobile/radio/handheld devices the Police control room. The control room provides necessary info/further actions to PAP at control locations. PAP completes the scheduled task for scheduled hour at particular locations and logs all the info from a parked car. He repeats the task until PAP finishes all control locations for the day. After completion, PAP returns to the office and make a daily report.

Traditional representations of space such as a map have is to a limited degree been oriented towards representation of process knowledge. Some recent approaches do take these aspects more consciously into account, as exemplified by (Nossum and Krogstie, 2009).

# 7 Concluding Remarks

We have summarized the results below of this high-level overview of the field, looking upon approaches according to different perspectives relative to the different usage areas for process modeling, and also indicated the amount of actual use of the approach:

| Area (from Fig.1 ) | 1+2 | 3 | 4 | 5a | 5b | 5c | 6 |
|---|---|---|---|---|---|---|---|
| | S&C | Sim | QA | Man. Act | Work-flow | Inter-active | Req. |
| Behavioral | -/OK | +/OK | -/- | -/- | +/OK | -/- | OK/- |
| Functional | +/+ | -/- | +/OK | +/OK | -/- | OK/- | +/OK |
| Beha+Func | +/OK | +/OK | OK/- | OK/- | +/+ | +/OK | OK/OK |
| Structural | OK/- | -/- | -/- | -/- | -/- | -/- | OK/- |
| Rule/Goal | -/- | OK/- | OK/- | -/- | OK/- | OK/- | OK/OK |
| Object | -/- | OK/- | -/- | -/- | -/- | -/- | OK/- |
| Communication | OK/- | -/- | -/- | -/- | OK/- | OK/- | -/- |
| Actor/Role | +/OK | OK/- | OK/- | OK/- | -/- | -/- | OK/OK |
| Geographical | OK/- | -/- | -/- | OK/- | -/- | OK/- | OK/- |

The legend indicates the 'applicability of the approach' / 'the actual use of the approach' (relative to the usage of modeling for this task). + indicated good applicability or high use, 'OK' is some applicability and use, whereas '-' indicate poor applicability / limited use. Obviously different approaches according to the same perspective can be more or less applicable as partly discussed above, and different languages of a certain perspective would score differently based on the concrete expressiveness and level of formality of the language and modeling approach. Due to space limitations, it is not possible to provide concrete evaluations of all approaches that we mentioned in the previous section, thus we leave it here in this chapter. A more detailed evaluation of some of these aspects is provided in (Mili et al. 2011). From the table, we see that functional and combinations of functional and behavioral approaches are used the most. All other perspectives have potential for use for certain areas, although this often varies relative to concrete needs in the domain for representing particular aspects (such as geographic aspects which in many cases might not be relevant). In particular some of the less traditional approaches appear to have large untapped potential for a richer more appropriate representation of what we term processes and business processes.

# References

Aagesen, G., Krogstie, J.: Analysis and design of business processes using BPMN. In: Handbook on Business Process Management, Springer (2010a)

van de Aalst, W.: Formalization and Verification of Event-driven Process Chains. Information and Software Technology 41, 639–650 (1999) van der Aalst, W.M.P., Pesic, M.: DecSerFlow: Towards a Truly Declarative Service Flow Language. In: Bravetti, M., Núñez, M., Zavattaro, G. (eds.) WS-FM 2006. LNCS, vol. 4184, pp. 1–23. Springer, Heidelberg (2006)

van der Aalst, W.M.P., Pesic, M., Schonenberg, H.: Declarative workows: Balancing between flexibility and support. Computer Science-Research and Development 23(2), 99–113 (2009)

Abdel-Hamid, T.K., Madnick, S.E.: Lessons Learned from Modeling the Dynamics of Software Development. Communications of the ACM 32(12) (1989)

Ader, M., Lu, G., Pons, P., Monguio, J., Lopez, L., De Michelis, G., Grasso, M.A., Vlondakis, G.: Woorks, an object-oriented workflow system for offices. Technical report, ITHACA (1994)

Ambriola, V., Conradi, R., Fuggetta, A.: Assessing Process-Centered Software Engineering Environments. ACM TOSEM 6(3) (1997)

Anderl, R., Raßler, J.: In: Cascini, G. (ed.) Computer-Aided Innovation (CAI). IFIP, vol. 277, pp. 145–156. Springer, Boston (2008)

Auramäki, E.R.H., Lyytinen, K.: Modelling offices through discourse analysis: The SAMPO approach. The Computer Journal 35(4), 342–352 (1992)

Austin, J.L.: How to do things with words. Harvard University Press (1962)

Bandinelli, S., Fuggetta, A., Lavazza, L., Loi, M., Picco, G.P.: Modeling and Improving an Industrial Software Process. IEEE Transactions on Software Engineering 21(5) (1995)

Benson, I., Everhard, S., McKernan, A., Galewsky, B., Partridge, C.: Mathematical Structures for Reasoning about Emergent Organization. In: ACM CSCW Workshop: Beyond Workflow Management, Philadelphia, USA (2000)

Bernstein, A.: How Can Cooperative Work Tools Support Dynamic Group Processes? Bridging the Specificity Frontier. In: ACM CSCW Conference, Philadelphia, USA (2000)

Bogia, D.P.: Supporting Flexible, Extensible Task Descriptions in and Among Tasks. PhD thesis (1995)

Bolcer, G., Kaiser, G.: SWAP: Leveraging the web to manage workflow. IEEE Internet Computing 3(1) (1999)

Booch, G., Rumbaugh, J., Jacobson, I.: The Unified Modeling Language: User Guide Second Edition. Addison-Wesley (2005)

Bråten, S.: Model Monopoly and communications: Systems Theoretical Notes on Democratization. Acta Sociologica, Journal of the Scandinavian Sociological Association 16(2), 98–107 (1973)

Bubenko Jr., J.A., Rolland, C., Loucopoulos, P., DeAntonellis, V.: Facilitating fuzzy to formal requirements modeling. In: Proceedings of the First International Conference on Requirements Engineering (ICRE 1994), Colorado Springs, USA, April 18-22, pp. 154–157. IEEE Computer Society Press (1994)

Bubenko Jr., J.A.: Problems and unclear issues with hierarchical business activity and data flow modelling. Technical Report 134, SYSLAB, Stockholm (June 1988)

Button, G.: What's Wrong with Speach Act Theory. In: CSCW, vol. 3(1) (1995)

Carlsen, S.: Action Port Model: A Mixed Paradigm Conceptual Workflow Modeling Language. In: Third IFCIS Conference on Cooperative Information Systems (CoopIS 1998), New York (1998)

Chen, P.P.: The entity-relationship model: Towards a unified view of data. ACM Transactions on Database Systems 1(1), 9–36 (1976)

Conradi, R., Jaccheri, M.L.: Process Modelling Languages. In: Derniame, J.-C., Kaba, B.A., Wastell, D. (eds.) Promoter-2 1998. LNCS, vol. 1500, pp. 27–52. Springer, Heidelberg (1999)

Cugola, G.: Tolerating deviations in process support systems via flexible enactment of process models. IEEE Transactions on Software Engineering 24(11) (1998)

Curtis, B., Kellner, M.I., Over, J.: Process Modeling. Com. ACM 35(9) (1992)

Davis, A.M.: Software Requirements Analysis & Specification. Prentice-Hall (1990)

Davis, A.M.: A comparison of techniques for the specification of external system behavior. Communications of the ACM 31(9), 1098–1115 (1988)

Derniame, J.-C., Kaba, B.A., Wastell, D. (eds.): Promoter-2 1998. LNCS, vol. 1500. Springer, Heidelberg (1999)

De Michelis, G., Grasso, M.A.: Situating Conversations within the Language/Action Perspective: The Milan Conversation Model. In: ACM CSCW Conference, Chapel Hill, North Carolina, USA (1994)

Dietz, J.L.G.: DEMO: towards a discipline of Organisation Engineering. European Journal of Operations Research (1999)

Dietz, J.L.G.: Integrating management of human and computer resources in task processing organizations: A conceptual view. In: Nunamaker, J.F., Sprague, R.H. (eds.) Proceedings of HICCS 1927, Maui, Hawaii, US, January 4-7. IEEE Computer Society Press (1994)

Dignum, F., Weigand, H.: Communication and deontic logic. In: Wieringa, R., Feenstra, R. (eds.) Working Papers of the International Workshop on Information Systems - Correctness and Reuseability, IS-CORE 1994 (1994)

Dori, D.: Why significant UML change is unlikely. Com. ACM 45, 82–85 (2002)

Dourish, P., Holmes, J., MacLean, A., Marqvardsen, P., Zbyslaw, A.: Freeflow: Mediating between representation and action in workflow systems. In: ACM CSCW Conference, Boston, USA (1996)

Dourish, P.: Re-Space-ing Place: "Place" and "Space" Ten Years On. In: Proc. ACM Conf. Computer-Supported Cooperative Work CSCW 2006, Banff, Canada, pp. 299–308. ACM, New York (2006)

Fickas, S.: Design issues in a rule-based system. Journal of Systems and Software 10(2), 113–123 (1989)

Fischer, L.: Excellence in Practice IV - Innovation and excellence in workflow and knowledge management. In: Workflow Management Coalition. Future Strategies Inc., Florida (2000)

Fox, M.S., Gruninger, M.: Enterprise modeling. AI Magazine (2000)

Gane, C., Sarson, T.: Structured Systems Analysis: Tools and Techniques. Prentice Hall (1979)

Geerts, G.L., McCarthy, W.E.: An Accounting Object Infrastructure for Knowledge-Based Enterprise Models. IEEE Intelligent Systems 14, 89–94 (1999)

Genesereth, M.R., Ketchpel, S.T.: Software agents. Communication of the ACM 37(7), 48–53 (1994)

Glance, N.S., Pagani, D.S., Pareschi, R.: Generalized Process Structure Grammars (GPSG) for Flexible Representation of Work. In: ACM CSCW Conference, Boston, USA (1996)

Goedertier, S., Vanthienen, J.: An overview of declarative process modeling principles and languages. Communications of Systemics and Informatics World Network 6, 51–58 (2009)

Gopalakrishnan, S., Krogstie, J., Sindre, G.: Adapting UML Activity Diagrams for Mobile Work Process Modelling: Experimental Comparison of Two Notation Alternatives. In: van Bommel, P., Hoppenbrouwers, S., Overbeek, S., Proper, E., Barjis, J. (eds.) PoEM 2010. LNBIP, vol. 68, pp. 145–161. Springer, Heidelberg (2010)

Gordijn, J., Yu, Eric, van der Raadt, B.: e-service Design using i* and e3value. IEEE Software (May-June 2006)

Green, P., Rosemann, M.: Integrated Process Modeling: An Ontolocial Evaluation. Information Systems 25(3) (2000)

Habermas, J.: The Theory of Communicative Action. Beacon Press (1984)

Hammer, M., Champy, J.: Reengineering the Corporation: A Manifesto for Business Revolution. Harper Business (1993)

Harel, D.: Statecharts : A visual formalism for complex systems. Science of Computer Programming (8), 231–274 (1987)

Harel, D., Lachover, H., Naamed, A., Pnueli, A., Politi, M., Sherman, R., Shtull-Trauring, A., Trakhtenbrot, M.: STATEMATE: a working environment for thedevelopment of complex reactive systems. IEEE TSE 16(4), 403–414 (1990)

Harrison, S., Dourish, P.: Re-Place-ing Space: The Roles of Space and Place in Collaborative Systems. In: ACM Conf. Computer-Supported Cooperative Work CSCW 1996, Boston, MA, pp. 67–76. ACM, New York (1996)

Havey, M.: Essential Business Process Modelling. O'Reilly (2005)

Hommes, B.-J., van Reijswoud, V.: The quality of business process modelling techniques. In: Conference on Information Systems Concepts (ISCO), Leiden. Kluwer (1999)

Hruby, P.: Model-Driven Design Using Business Patterns. Springer, New York (2006)

Hull, R., King, R.: Semantic database modeling: Survey, applications, and research issues. ACM Computing Surveys 19(3), 201–260 (1987)

Hull, D., Wolstencroft, K., Stevens, R., Goble, C.A., Pocock, M.R., Li, P., Oinn, T.: Taverna: a tool for building and running workflows of services. Nucleic Acids Research 34(Web-Server-Issue), 729–732 (2006)

IDEF-0: Federal Information Processing Standards Publication 183, December 21, Announcing the Standard for Integration Definition For Function Modeling (IDEF-0) (1993)

Jarke, M., Bubenko jr, J.A., Rolland, A.: Sutcliffe, and Y. Vassiliou. Theories underlying requirements engineering: An overview of NATURE at genesis. In: Proceedings of RE 1993, pp. 19–31 (1993)

Kappel, G., Rausch-Schott, S., Retschitzegger, W.: Coordination in workflow management systems a rule-based approach. Coordination Technology for Collaborative Applications, 99–119 (1998)

Kavakli, E., Loucopoulos, P.: Goal Modeling in Requirements Engineering: Analysis and critique of current methods in. In: Krogstie, J., Siau, K., Halpin, T. (eds.) Information Modeling Methods and Methodologies. Idea Group Publishing (2005)

Keller, G., Nüttgens, M., Scheer, A.W.: Semantische Prozeßmodellierung auf der Grundlage Ereignisgesteuerter Prozeßketten (EPK) (1992)

King, J.L.: SimLanguage, Computer Supported Cooperative Work, vol. 3(1) (1995)

Krogstie, J., Dalberg, V., Jensen, S.M.: Process modeling value framework, Enterprise Information Systems. In: Manolopoulos, Y., Filipe, J., Constantopoulos, P., Cordeiro, J. (eds.) Selected Papers from 8th International Conference, ICEIS 2006. LNBIP, vol. 3. Springer (2006)

Krogstie, J., Sindre, G.: Utilizing deontic operators in information systems specifications. Requirement Engineering Journal 1, 210–237 (1996)

Krogstie, J.: Integrated Goal, Data and Process modeling: From TEMPORA to Model-Generated Work-Places. In: Johannesson, P., Søderstrøm, E. (eds.) Information Systems Engineering From Data Analysis to Process Networks, pp. 43–65. IGI Publishing (2008)

Kuntz, J.C., Christiansen, T.R., Cohen, G.P., Jin, Y., Levitt, R.E.: The virtual design team: A computational simulation model of project organizations. Communications of the ACM 41(11) (1998)

Lei, Y., Singh, M.P.: A comparison of workflow metamodels, In: ER Workshop on Behavioral Modeling. LNCS, vol. 1565. Springer, Heidelberg (1997)

Lillehagen, F., Krogstie, J.: Active Knowledge Models of Enterprises. Springer (2008)

Loos, P., Allweyer, T.: Process orientation and object-orientation - An approach for integrating UML with event-driven process chains (EPC), Germany (1998)

Lu, R., Sadiq, S., Governatori, G.: On managing business processes variants. Data & Knowledge Engineering 68(7), 642–664 (2009)

Lu, R., Sadiq, W.: A Survey of Comparative Business Process Modeling Approaches. In: Abramowicz, W. (ed.) BIS 2007. LNCS, vol. 4439, pp. 82–94. Springer, Heidelberg (2007)

Marsan, M.A., et al. (eds.): Proceeding of the International workshop on Timed Petri Nets, Torino, Italy. IEEE Computer Society Press (1985)

McCarthy, W.E.: The REA accounting model: a generalized framework for accounting systems in a shared data environment. The Accounting Review 57, 554–578 (1982)

Medina-Mora, R., Winograd, T., Flores, R., Flores, F.: The Action Workflow approach to workflow management technology. In: ACM CSCW Conference (1992)

Mendling, J., Nüttgens, M.: EPC markup language (EPML): an XML-based interchange format for event-driven process chains (EPC). Information Systems and E-Business Management 4, 245–263 (2006)

Mühlen, M.z., Becker, J.: Workflow management and object-orientation - A matter of perspectives or why perspectives matter. In: OOPSLA Workshop on Object-Oriented Workflow Management, Denver, USA (1999)

Nossum, A., Krogstie, J.: Integrated Quality of Models and Quality of Maps. In: Paper presented at the EMMSAD 2009, Amsterdam, The Netherlands (2009)

Nysetvold, A.G., Krogstie, J.: Assessing Business Process Modeling Languages Using a Generic Quality Framework. In: Siau, K. (ed.) Advanced Topics in Database Research, vol. 5, pp. 79–93. Idea Group, Hershey (2006)

Olle, T.W., Hagelstein, J., MacDonald, I.G., Rolland, C., Sol, H.G., van Assche, F.J.M., Verrijn-Stuart, A.A.: Information Systems Methodologies. Addison-Wesley (1988)

OMG Workflow Management Facility v. 1.2, Object Management Group (2000)

Ould, M.A.: Business Processes - Modeling and Analysis for Re-engineering and Improvement. Wiley, Beverly Hills (1995)

Opdahl, A.L., Sindre, G.: A taxonomy for real-world modeling concepts. Information Systems 19(3), 229–241 (1994)

Opdahl, A.L., Sindre, G.: Facet modeling: An approach to flexible and integrated conceptual modeling. Information Systems 22(5), 291–323 (1997)

Pesic, M., van der Aalst, W.M.P.: A Declarative Approach for Flexible Business Processes Management. In: Eder, J., Dustdar, S. (eds.) BPM Workshops 2006. LNCS, vol. 4103, pp. 169–180. Springer, Heidelberg (2006)

Petri, C.A.: Kommunikation mit automaten. Schriften des Rheinisch-Westfalischen Institut fur Instrumentelle Mathematik an der Universität Bonn (2) (1962) (in German)

Scheer, A.-W., Nüttgens, M.: ARIS Architecture and Reference Models for Business Process Management, pp. 301–304 (2000)

Schonenberg, H., Weber, B., van Dongen, B.F., van der Aalst, W.M.P.: Supporting Flexible Processes through Recommendations Based on History. In: Dumas, M., Reichert, M., Shan, M.-C. (eds.) BPM 2008. LNCS, vol. 5240, pp. 51–66. Springer, Heidelberg (2008)

Searle, J.R.: Speech Acts. Cambridge University Press (1969)

Searle, J.R.: Expression and Meaning. Cambridge University Press (1979)

Searle, J.R., Vanderveken, D.: Foundations of Illocutionary Logic. Cambridge University Press (1985)

Seltveit, A.H.: An Abstraction-Based Rule Approach to Large-Scale Information Systems Development. In: Rolland, C., Cauvet, C., Bodart, F. (eds.) CAiSE 1993. LNCS, vol. 685, Springer, Heidelberg (1993)

Senge, P.: The Fifth Discipline: The Art and Practice of the Learning Organization. Century Business Publishers, London (1990)

Shoham, Y.: Agent Oriented Programming: An Overview of the Framework and Summary of Recent Research. In: Masuch, M., Polos, L. (eds.) Logic at Work 1992. LNCS, vol. 808, pp. 123–129. Springer, Heidelberg (1994)

Singh, B., Rein, G.L.: Role Interaction Nets (RINs); A Process Description Formalism, Technical Report CT-083-92, MCC, Austin, Texas (1992)

Stamper, R.: Semantics. In: Boland, R.J., Hirschheim, R.A. (eds.) Critical issues in Information Systems Research, pp. 43–78. John Wiley & Sons (1987)

Störrle, H.: Describing Process Patterns with UML. In: Ambriola, V. (ed.) EWSPT 2001. LNCS, vol. 2077, p. 173. Springer, Heidelberg (2001)

Suchman, L.: Do categories have politics? In: CSCW, vol. 2(3) (1994)

Swenson, K.D., Maxwell, R.J., Matsymoto, T., Saghari, B., Irwin, I.: A business process environment supporting collaborative planning. Journal of Collaborative Computing 1(1) (1994)

Sølvberg, A., Kung, C.H.: Information Systems Engineering. Springer (1993)

Tomlinson, C., Scheevel, M.: Concurrent Programming. In: Kim, W., Lo-chovsky, F.H. (eds.) Object-oriented Concepts, Databases and Applications. Addison-Wesley (1989)

Twining, W., Miers, D.: How to do things with rules. Weidenfeld and Nicholson (1982)

UMM - UN/CEFACT Modeling Methodology User Guide (2007)

Ward, P.T.: The transformation schema: An extension of the dataflow diagram to represent control and timing. IEEE Transactions on Software Engineering 12(2), 198–210 (1986)

Zatarain-Cabada, R., Goldin, D.Q.: Interaction as a Framework for Modeling. In: Chen, P.P., Akoka, J., Kangassalu, H., Thalheim, B. (eds.) Conceptual Modeling. LNCS, vol. 1565, p. 243. Springer, Heidelberg (1999)

Weske, M.: Business Process Management: Concepts, Languages, Architectures. Springer-Verlag New York Inc (2007)

WfMC Workflow Handbook 2001. Workflow Management Coalition, Future Strategies Inc., Lighthouse Point, Florida, USA (2000)

White. S.A. Introduction to BPMN. IBM Cooperation (2004)

Wieringa, R.: Three roles of conceptual models in information systems design and use. In: Falkenberg, I.E., Lindgren, P. (eds.) Information Systems Concepts: An In-Depth Analysis, pp. 31–51. North-Holland (1989)

Winograd, T., Flores, F.: Understanding Computers and Cognition. Addison-Wesley (1986)

Zachman, J.A.: A framework for information systems architecture. IBM Systems Journal 26(3), 276–291 (1987)

# Understanding Business Process Quality

Matthias Lohrmann and Manfred Reichert

Institute of Databases and Information Systems, University of Ulm, Germany
{matthias.lohrmann,manfred.reichert}@uni-ulm.de

**Abstract.** Organizations have taken benefit from quality management practices in manufacturing and logistics with respect to competitiveness as well as profitability. At the same time, an ever-growing share of the organizational value chain centers around transactional administrative processes addressed by business process management concepts, e.g. in finance and accounting. Integrating these fields is thus very promising from a management perspective. Obtaining a clear understanding of business process quality constitutes the most important prerequisite in this respect. However, related approaches have not yet provided an effective solution to this issue. In this chapter, we consider effectiveness requirements towards business process quality concepts from a management perspective, compare existing approaches from various fields, deduct a definition framework from organizational targets, and take initial steps towards practical adoption. These steps provide fundamental insights into business process quality, and contribute to obtain a clear grasp of what constitutes a good business process.

## 1 Introduction

Since the early 90s, the concept of business process management (BPM) has achieved broad acceptance [3]. Consequently, business processes are increasingly subject to management methods such as planning, monitoring and controlling [35]. These methods generally presume insights into the aspired and actual business performance of the subject matter. For business processes, this means that managers strive to know what constitutes a good process and how to evaluate processes against this standard. Effective concepts to understand and methods to assess *business process quality* are thus a fundamental requirement to further establish BPM as a management practice.

This chapter provides an overview on available approaches to business process quality and argues that these have not achieved full effectiveness yet. In particular, their views on business process quality are not sufficiently aligned

M. Glykas (Ed.): Business Process Management, SCI 444, pp. 41–73.
springerlink.com          © Springer-Verlag Berlin Heidelberg 2013

to organizational needs and targets. We will show that, instead, present approaches mostly define business process quality implicitly by employing certain quality criteria (e.g. related to business process input) without rigorously demonstrating the relation of these criteria to organizational targets. We stipulate that this issue may be addressed by pursuing a deductive approach to derive a concept of business process quality from well-founded premises.

Accordingly, we propose a framework for business process quality as a foundation to guide the development of specific quality attributes, criteria and predicates, for instance with regard to particular application areas. As it rigorously systemizes organizational targets for business processes based on well-founded principles, it can also be used to design or evaluate BPM methodologies, for example in the area of process optimization.

Section 2 of this chapter presents our deductive design methodology including criteria to evaluate the effectiveness of results. In Section 3, we review existing approaches and match them against our effectiveness criteria. Based on basic terms shortly discussed in Section 4, Section 5 elaborates our concept of business process quality, which we apply to a practical example in Section 6. Section 7 concludes the chapter with a discussion of results and directions for future research.

## 2   Deductive Design Methodology

Business processes aim at achieving business objectives of the organization in an economic context [9, 18]. Accordingly, we consider the concept of *business process quality* and associated methods like quality assessment or optimization as means to support this goal. This implies that business process quality is a goal-bound artificial construct. We therefore apply the principles of design science [51, 43, 23] in the methodology set out in this section.

### 2.1   *Effectiveness Criteria*

In design science, the value of design artifacts is to be judged "against criteria of value or utility" [43]. We subsume "value or utility" of an artifact as its *effectiveness*. Consequently, appropriate *effectiveness criteria* constitute an important part of our design methodology. We apply them in the evaluation of existing approaches as well as our results.

Business process quality artifacts are to be employed in the context of BPM activities as defined in [3]: design, enactment, control and analysis of operational processes. Out of these tasks, analysis and control constitute the most relevant fields: the quality of business processes is assessed and analysed (either in the productive stage or even before), and control is exercised by feeding back into design and execution. We therefore derive our effectiveness criteria as comprised in Table 1 from requirements for effective managerial analysis and control [11].

**Table 1** Effectiveness criteria

| Effectiveness criteria | Rationale | Implications |
|---|---|---|
| **EC 1:** Congruence to organizational targets | Explicit feedback loops in management control and performance measurement systems (e.g. [33])<br><br>Content of performance measures impacts managerial behaviour and decisions [19, 47] ("What gets measured, gets done!") | *Comprehensive coverage* of organizational targets for object in question<br><br>*Exclusive coverage* of organizational target aspects for objects in question<br><br>If full congruence cannot be achieved: *transparency on deficiencies* to mitigate defective governance effects |
| **EC 2:** Perceived fairness | Organizations as a social environment: prerequisite for staff motivation and change management<br><br>Performance measures are commonly used for individual target setting and remuneration [56]<br><br>Equivalent provision for financial reporting [26, paragraph 46]: "true and fair view" | *Governance*: quality assessment reflects organizational responsibilities<br><br>*Transparency and retraceability*: accountable managers' understanding the link between status, actions and assessment results, limited complexity |
| **EC 3:** Cost effectiveness | Practical applicability in an economic context<br><br>Equivalent provision for financial reporting [26, paragraph 44] | Avoid large *criteria catalogues* to be evaluated manually<br><br>Avoid manual *inspectment of process instances*<br><br>Formalize to allow for *automated assessment* |

## 2.2 Course of Action

As a preliminary step to detail our motivation for proposing an alternative approach towards business process quality, Section 3 substantiates our claim that available approaches are not fully effective from a management perspective. To this end, we conduct a literature review based on the effectiveness criteria set out in Table 1.

The remaining steps of our methodology are organized around the design processes of *build* and *evaluate* and the design artifact categories of *constructs*, *models*, *methods* and *instantiations* as set out in [43]. Figure 1 delimits design artifacts and processes with regard to business process quality.

In this chapter, we *build* our definition of *business process quality*, which represents a *construct* as it provides common ground to discuss this term. For practical application, it is a means to facilitate the definition of appropriate and measurable quality criteria. Corresponding sets of quality criteria which extend and further specify our definition, for example with respect to application areas such as finance or medicine, constitute *models* because they

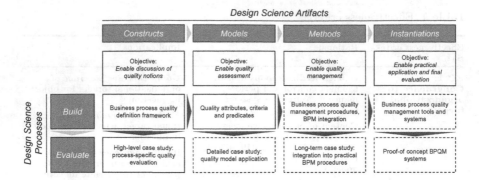

**Fig. 1** Design methodology

essentially relate business process quality to other constructs which can be assessed in practice. We also provide an outlook on possible model content while more formal and rigorous modeling will be subject to future work.

We *evaluate* our results with respect to our effectiveness criteria by way of application to a real-world business process. *Methods* and *instantiations* in the space of business process quality are topics for future work.

## 3   State of the Art

Work related to the quality of business processes can be broadly divided into three categories: general management approaches that are also applicable to business process quality, BPM frameworks, and BPM research addressing individual aspects related to quality.

As stated above, we postulate that existing approaches towards business process quality are not yet fully effective from a management perspective. Therefore, this section first discusses related work and then presents a summary with respect to Effectiveness Criteria EC 1-3 from Table 1.

### 3.1   General Management Approaches

There are many management concepts which are not specific to the field of BPM but might be adapted for our area of research. We shortly discuss two selected approaches because of their wide practical adoption and their special relevance to business process quality.

*Benchmarking* is based on utilizing available experience and knowledge from comparable business processes: *qualitative benchmarking* matches the actual situation against known good practices, which may be documented in frameworks such as CobiT [30]. These practices may relate to organizational structures or directly to business processes or information systems. *Quantitative benchmarking* uses key performance indicators (KPIs) to measure process aspects. This enables comparison to results from peer organizations [5].

---

*Example 1 (Good practices in process design and key performance indicators).*
Consider the process of handling supplier invoices. Here, good practices for qualitative benchmarking include the use of early scanning (also known as "intelligent scanning", see our case example in Section 6) and Electronic Data Interchange (EDI) as IT-based practices, and credit note procedures as an organizational practice. The use of credit note procedures has been described in detail as an example for business process reengineering in [10, 17].

Quantitative key performance indicators include the number of invoices processed per full-time personnel resource and year, the processing cost per invoice, and average cycle time.

---

The *balanced scorecard* approach is used to measure and control organizational performance based on multiple dimensions: the "financial", the "customer", the "innovation and learning", and the "internal business" perspectives [32]. Key performance indicators are specifically developed for the organization and assigned to each dimension to allow for distinct tracking. Compared to traditional financial performance measures, the balanced scorecard recognizes that financials are always backwards-oriented and provide little clarity on an organization's future perspectives. Moreover, organizational goals are often contradictory, for instance when considering cash flow maximization against the need for investments. This issue has been long acknowledged in literature (e.g. [42]), and it is addressed via the multiple dimensions of the balanced scorecard; i.e., the approach does not try to combine everything into one single perspective. Of course, application of the original concept to business processes would require adaptation of even the fundamental scorecard perspectives. However, the basic idea of treating multiple performance dimensions as orthogonal instead of trying to find an absolute single measure of quality may be unavoidable for practical application.

## 3.2  BPM Frameworks

Research on BPM has also led to a wide array of proposals that might be applied to business process quality. A common characteristic of these approaches is that they, as opposed to benchmarking and the balanced scorecard, abstract from the business content of the processes in question. In other words, a person charged with executing the procedures proposed does not necessarily need to be a business subject matter expert.

An attempt to develop a *"Quality of Business Processes (QoBP) framework"* focusing on process models was made by Heravizadeh et al. [22]. Business process quality is defined in terms of 41 quality dimensions which are derived from literature, e.g. in the field of software engineering. The approach does not show the quality dimensions' interrelation to organizational targets or to an overall formal quality definition. This also means that we cannot determine whether the dimensions are complete or how to actually evaluate overall process quality. The quality dimensions are arrayed along the categories of function quality, input / output quality, non-human resource quality, and human resource quality. In our view, this is questionable because it mixes up the quality of a process under consideration with factors not under control of process management. In practical settings, this might lead to issues with the *perceived fairness* effectiveness criterion. The QoBP approach has been presented in more detail in [21]. In this context, quality has been defined as non-functional but distinguishing characteristics of a business process. We do not concur with that view because, from the perspective presented in Section 2, excluding the business objective to be achieved by a process would negate the goal-bound nature of the business process quality construct as a design science artifact.

Heinrich and Paech proposed a *business process quality framework based on software quality* [20]. While work on software quality is not the only source used, the eight main "activity characteristics" with 27 sub-characteristics in [20] have been derived from this field. The "activity characteristics" are amended by four characteristics in the areas of "resource" and "actor". Similar to QoBP, this approach lists various quality characteristics, but it does not integrate them into a comprehensive formal quality definition, leading to similar issues as described above. Moreover, we stipulate that the applicability of software engineering results to design problems in the area of BPM still requires closer analysis.

*Business process reengineering and optimization* constitutes an area which is closely related to optimizing business process quality. [18, 9] provide good examples for the "classic" all-encompassing reengineering view. Reengineering approaches commonly comprise recommended best practices and other informal methods which are mostly based on anecdotal evidence. [53, 49, 38, 31] and, with a focus on well-defined process models, [4] constitute additional examples for optimization based on informal methods. This view is also reflected in the OMG Business Process Maturity Model [54] and other BPM maturity models [52] which suggest criteria to allocate business processes to maturity levels without giving clear evidence on how this structure is devised. While this informal character fits well with practical applicability, we still lack an overarching comprehensive model to ensure causal relations between measures recommended and intended results as well as completeness of coverage of quality aspects.

## 3.3 BPM Approaches Covering Individual Aspects

In the field of BPM, a great number of approaches have been developed to address individual quality aspects of business processes. While they do not aim at an overarching construct of business process quality, they may still provide important methods for practical business process quality management.

There is some related work that deals with the *quality of business process models*: van der Aalst introduced soundness of Workflow Nets [1]. Hallerbach et al. discuss how to ensure soundness for an entire process family [16]. Finally, Reichert et al. enhance these considerations by also considering soundness in the context of dynamic process changes during run-time [48]. Weber et al. developed process model refactoring [59, 60]. Li et al addressed reference model discovery by model merging [36, 37]. Weber et al and Rinderle et al described quality issues in respect to a case-based capturing of knowledge about reusable process adaptations which can be applied in dynamic environments [61, 50]. Ly et al. ensure that both the specification and the enactment of business processes are compliant to global rules and regulations [41]. Becker et al. discussed process model quality focusing on certain stakeholder groups and applications [4]. Gucegioglu and Demirors applied software quality characteristics to business processes [14]. Mendling assessed formal errors in EPC business process models [44] in an automated approach. Cardoso analyzed workflow complexity as one possible measure for process model quality [6], and Vanderfeesten et al. discussed quality metrics in business process modeling [57, 58].

There are also approaches to *formally optimize business process or workflow models*. Examples include [2, 1], where Petri nets are proposed to leverage existing analysis methods, and [24], where various optimization strategies for process designs with given input and output sets per activity are discussed. These approaches are mainly suited to optimize control flow and resource scheduling as they do not address individual activities in terms of necessity, effort or alternatives. They thus constitute important tools but cover only aspects of optimum business process design. We intend further analysis to be part of future work on quality in the process design lifecycle stage.

*Process intelligence, process performance management and business activity monitoring* are closely linked to the quality of process execution. Research in this area is very much driven by practical requirements and tends to take an operational, short-term view as opposed to our rather structural, long-term perspective of business process quality. Exemplary work includes [7, 25] and reflects the close association of this field to industry and tool vendors. Also in the context of process enactment, Grigori et al. have developed a proposal to monitor and manage exceptions in process execution [13].

## 3.4   Evaluation against Effectiveness Requirements

Having reviewed existing approaches to business process quality, we can evaluate them against Effectiveness Criteria EC 1-3 as set out in Table 1. We summarize our conclusions in Table 2.

Note that most approaches do not explicitly state a concise definition of business process quality. Instead, they employ either quality criteria or quality attributes. Quality criteria allow for assessment in the sense that whether the criteria are fulfilled allows to make statements on quality. Quality attributes are properties that may be used to evaluate quality when amended with target or threshold values (they then become quality criteria). Making this distinction may seem overdone at first. However, there are some crucial implications from being able to utilize a formal quality definition and quality criteria as opposed to quality attributes only:

- A short and concise *business process quality definition* as a *construct* facilitates to directly apply the corresponding quality view, for instance by matching against organizational targets on an abstract level. It reduces the risk of misinterpretations and makes the underlying quality notion accessible for straightforward discussion. This reflects the role of *constructs* as defined in [43].
- *Business process quality attributes* allow to discuss what is important to quality. It is possible to discuss each attribute's link to overall organizational targets, but difficult to judge whether organizational targets are properly represented in their entirety. Defining attributes is in themselves not sufficient to assess quality, but they may be amended to constitute quality criteria. "Productivity" or "the capability [...] to enable users to expend appropriate amounts of resources in relation to the effectiveness achieved" constitutes an example for a quality attribute taken from [22].
- *Business process quality criteria* enable us to distinguish between poor and high quality for individual attributes by providing explicit or implicit threshold values. The latter may, for instance, be given by comparison to a peer group. They are therefore required to assess quality. Our "productivity" example for a quality attribute evolves into a quality criterion when "appropriate amounts of resources" are specified.

To clarify our conclusions, we summarize the respective business process quality definitions and the corresponding quality attributes or criteria in our overview on existing approaches. For approaches covering only individual aspects, evaluation against our primary Effectiveness Criterion EC 1 of *congruence to organizational targets* (cf. Table 1) as a whole is obviously not meaningful and therefore omitted.

Our review of present approaches resulted in some recurring issues that substantially affect effectiveness with respect to the criteria we chose to apply:

- There is an overall lack of a clear *definition of business process quality* in the sense of a *construct*. This makes it generally difficult to discuss and evaluate congruence to organizational targets, because completeness and adequacy of attributes or criteria lists remain debatable.

- Generally, BPM approaches tend to employ quality attributes instead of quality criteria. The classic reengineering and optimization approaches are the exception. In themselves, they are thus not sufficient to evaluate the concrete quality of a business process which impacts practical relevance.
- Assuming proper adaptation to the field of BPM, the balanced scorecard approach is the only one where we see high *congruence to organizational targets*: the approach was explicitly developed to accomodate the diverse and possibly conflicting target dimensions encountered in real-world business strategies.
- In all approaches discussed, *perceived fairness* is impacted by a failure to recognize the *organizational environment* of the business process by distinguishing between manageable and non-manageable factors. Non-manageable factors in the organizational environment of a business process comprise, for instance, process input delivered by other ("upstream") business processes. This topic can often be observed in practice when benchmarking results are challenged by management if, for instance, very different organizations are chosen as peers. In this case, an impacted fairness perception due to a lack of consideration for the individual organizational environment leads to impaired acceptance of the entire assessment.

These conclusions provide some guidance to our further progress to design alternative artifacts:

- The current lack of concise definitions of business process quality encourages us to develop such a construct as a first *build* step as set out in our design methodology. The definition should be congruent to organizational targets as this is one of the major deficiencies of present approaches.
- To actually achieve congruence to organizational targets, we employ a deductive approach based on organizational targets for business processes. This methodology differs from existing approaches and will allow to verify congruence to targets at each stage of development.
- In our *build model* step, we place special regard to develop assessible quality criteria instead of mere quality attributes to achieve practical relevance in the analysis and control BPM lifecycle stages.


## 4 Business Processes and Quality: Basic Concepts

As a preliminary step to the development of our concept of business process quality, we have to ensure a common understanding on the basic concepts in the areas of business processes and quality we employ. This is particularly relevant because both terms have been the subject of a great number of attempts to find a definition over time (see, for instance, [39]). This section therefore shortly presents basic terms and definitions we adopt.

**Table 2** Related approaches vs. effectiveness criteria

| Approach | Business process quality definition | Business process quality attributes / criteria | Effectiveness criteria | | |
|---|---|---|---|---|---|
| | | | EC 1: Congruence to organizational targets | EC 2: Perceived fairness | EC 3: Cost effectiveness |
| *General Management Approaches, see Section 3.1* | | | | | |
| **Qualitative benchmarking** | Implicit: degree to which good practices are implemented | Criteria: implementation of good practices known from peer organizations | Low: focus on copying peer strategies without consideration of individual environment | Low: failure to consider organizational constraints (e.g. capital expenditures) | High: easy assessability of good practices implementation |
| **Quantitative benchmarking** | Implicit: degree to which peer key performance indicator values are achieved | Criteria: comparison to key performance indicator values achieved at peer organizations | Low: typically, focus on efficiency measures without consideration of capital expenditures or quality of process input | Low: efficiency measures typically do not reflect non-manageable factors (e.g. capital expenditures or quality of process input) | High: key performance indicators are typically chosen for easy assessability |
| **Balanced scorecard (with adaptations to BPM application)** | Degree to which objectives in target dimensions (typically four) are achieved | Criteria: achievement of objectives defined for measures | High: objectives and measures are derived from organizational targets | Dependent on definition of manageable scorecard dimensions (classic dimensions appropriate for business units) | High: measures are typically chosen for high assessability |
| *BPM Frameworks, see Section 3.2* | | | | | |
| **QoBP framework** | Implicit: degree to which requirements in quality dimensions are fulfilled | Attributes / criteria: fulfilment of requirements in 41 quality dimensions (requirements are not defined) | Low: quality dimensions are not systematically linked to organizational targets, no consideration of target interdependencies | Low: quality requirements do not recognize organizational environment | Low: real-world measurability of attributes not proven, may lead to protracted assessment effort as measures are developed |
| **Business process quality framework based on software quality** | Implicit: degree to which requirements towards quality characteristics are fulfilled | Attributes: twelve main quality characteristics | see QoBP framework | see QoBP framework | see QoBP framework |

**Table 2** (*continued*)

| Approach | Business process quality definition | Business process quality attributes / criteria | Effectiveness criteria | | |
|---|---|---|---|---|---|
| | | | EC 1: Congruence to organizational targets | EC 2: Perceived fairness | EC 3: Cost effectiveness |
| *General Management Approaches, see Section 3.1* | | | | | |
| **Business process reengineering and optimization** | Implicit: all optimization policies have been leveraged | Criteria: implementation of optimization policies / maturity level definitions (similar to qualitative benchmarking, but independent of functional content) | Low: similar to qualitative benchmarking, but peer strategies are replaced with general optimization policies | Low: similar to qualitative benchmarking | High: easy assessability of implementation of recommended practices |
| *BPM Approaches Covering Individual Aspects, see Section 3.3* | | | | | |
| **Quality of business process models** | Implicit: optimization levers for formal model quality are fully utilized | Attributes: measures for model quality (formal definition but coverage of individual aspects only) | n/a | Medium: formal measures allow for objective assessment, but non-manageable factors are not made transparent | Medium: assessment automatable, but formal modeling required first |
| **Business process optimization: formal approaches** | Implicit: formal control flow optimization levers are fully utilized | Attributes: measures for process quality with respect to control flow optimization | n/a | see quality of business process models | Medium: assessment automatable, but formal modeling required first |
| **Process performance management / business activity monitoring** | Implicit: target values for process enactment performance criteria have been achieved | Attributes / criteria: process enactment performance measures without / with target values | n/a | Low: non-manageable factors important e.g. for cycle times are mostly not considered | High: automated assessment tools available to support workflow management systems |

## 4.1 Business Process Concepts

The Workflow Management Coalition (WfMC) defines a *business process* as *"a set of one or more linked procedures or activities which collectively realise a business objective or policy goal, normally within the context of an organisational structure defining functional roles and relationships"* [63]. Summarizing this and other definitions, there is an overall agreement that a business process consists of a set of activities which aims at the realization of a business objective. This definition is very inclusive and covers virtually everything members of an organization undertake to serve organizational purposes. However, quality management in production and (direct) customer service is already well established (see the next section), and quality assessment makes the most sense when its results can be applied in future iterations. We therfore limit the context of our analysis to *repetitive administrative processes*.

Moreover, we can distinguish between a business process model as an abstract notion and business process instances as concrete enactments thereof. The WfMC defines a process instance as *"the representation of a single enactment of a process"* [63]. For the more basic term business process, it remains open whether it refers to a process model or to a set of one or more process instances of one common process model [62]. In most applications, this distinction is made implicitly based on the *business process lifecycle stage* (cf. [3]). For our purposes, we discern between two fundamental lifecycle stages corresponding to the basic interpretations of the term business process.

Table 3 summarizes the fundamental lifecycle stages we use and compares to the business process *management* lifecycle in [3]. Note that our fundamental business process lifecycle for the purpose of quality management excludes the diagnosis stage as comprised in [3] because business process quality assessment is in itself part of this stage. Organizational capabilities in Lifecycle Stage I refer to the organization's ability to actually execute the process model in terms of available resources such as capital goods, personnel etc. The term *actual process model* designates a process model (which may be available as an organizational policy, as an explicit model in a modeling language or just as organizational knowledge) plus its actual implementation in terms of organizational capabilities such as the availability of information systems or machinery.

Contrary to most other BPM applications, business process quality assessment must address both fundamental lifecycle stages. From a management perspective, it makes sense to analyze both the quality of an actual process model *and* the quality of the corresponding process instances. Typically, organizational responsibilities differ for the fundamental lifecycle stages. To reflect this issue, separate results for both analyses are desirable (cf. Effectiveness Criterion EC 2 in Table 1).

**Table 3** Fundamental business process lifecycle stages

| Fundamental lifecycle stage | "Business process"interpretation | Corresponding lifecycle stages in [3] |
| --- | --- | --- |
| **Lifecycle Stage I: Business process design & implementation** | The business process as an abstract process model and its implementation in terms of organizational capabilities (*actual process model*) | Process design, system configuration |
| **Lifecycle Stage II: Business process enactment** | The business process as a set of one or more instances of a common abstract process model | Process enactment |

## 4.2 Quality Concepts

Since the 1950s, quality managment (QM) has become one of the central management concepts adopted by organisations globally. During that time, concepts and notions for quality have evolved from the work of pioneers such as Shewhart, Deming, Crosby, Feigenbaum, Juran and Ishikawa to standardized terminologies and methods that are propagated by trade and governmental bodies (for an overview see [8]). In terms of practical adoption, the definition of quality most widely spread today has been developed by the International Organization for Standardization (ISO) in the ISO 9000 series of standards [28]. As a set of norms in the area of QM for business applications, ISO 9000 has achieved broad acceptance through endorsements by governmental bodies like the European Union and the ISO 9000 certification scheme [15, 45, 29]. For a fundamental definition of quality, we therefore resort to the definition given in the ISO 9000 series of standards: *quality denotes "the degree to which a set of inherent characteristics fulfils requirements".*

The ISO definition, however, does not specify the concrete content of the "requirements". In this respect, various fundamental interpretations or views on quality have been argued. In [40], we gave an overview on these and discussed their fit in the context of BPM based on a classification developed by Garvin [12]. For our discussion, it is sufficient to record that the value-based view on quality, which matches the utility of an object against expenditures incurred, best suits our context, because it can accomodate the whole array of organizational targets. On the other hand, its implementation poses a number of challenges in practice which mostly relate to appraising the actual "value" delivered as well as the actual expenditure incurred when considering issues such as the cost of upstream processes or risk management.

## 5 A Framework for Business Process Quality

As discussed in Section 4.2, quality in itself is an abstract term subject to differing interpretations. However, to be applied in a business context, it should be defined in a way to make it a useful construct for management purposes. Based on our design methodology (cf. Section 2) and the conclusions we made when reviewing existing approaches (cf. Section 3), this section derives a definition of business process quality which aims at achieving this goal.

Based on our analysis of related work, we proposed to deduct a definition of *business process quality* from organizational targets. Accordingly, our reasoning is built along four steps:

1. In terms of design science as described by Simon [51], a business process constitutes an artifact designed to attain goals by acting within its "outer environment". We stipulate that these goals correspond to the organizational targets we refer to in Effectiveness Criterion EC 1 (cf. Table 1). Accordingly, we discuss the outer environment of the business process to focus and structure our field of analysis.
2. We identify and apply organizational targets for the outer environment of the business process. We then discuss how the business process affects the achievement of these targets during its fundamental lifecycle stages.
3. Based on the outer environment of the business process, the associated organizational targets and the respective impact of the business process in the course of its fundamental lifecycle, we state a definition framework for business process quality.
4. We refine the content of the definition framework to obtain a practically applicable model in the sense of the design science paradigm.

Steps 1 and 2 are addressed in Sections 5.1 and 5.2. Step 3 is presented in Section 5.3. We include initial considerations on Step 4 in Section 5.4.

## 5.1   The Outer Environment of the Business Process

When following the methodology set out above, congruence to organizational targets as our most pressing concern is mainly a matter of properly structuring the outer environment of the business process to be able to consider organizational targets comprehensively, but exclusively. Figure 2 summarizes various options.

An initial *common BPM perspective* on the outer environment is based on the concepts of process *input* and process *output* used by many authors (see, for instance, [9, 18]). For our purposes, however, these concepts are not apt to properly structure the outer environment: First, input and output generally overlap as processes if input objects are altered to assume a role as output object as well.

Second, interpretations of process input are prone to omit resources that are not attributable to individual process instances, such as capital goods (cf. [27]), or the availability of staff to execute activities. Usually, there is also no consideration of things affected unintentionally like exposure to litigation risks or pollution. In this case, the outer environment and, consequently,

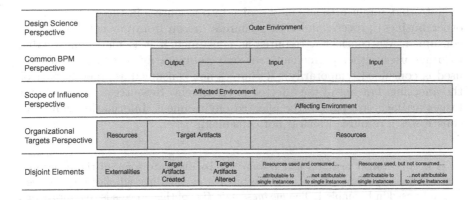

| Design Science Perspective | Outer Environment | | | | | | | |
|---|---|---|---|---|---|---|---|---|
| Common BPM Perspective | | Output | | Input | | | Input | |
| Scope of Influence Perspective | | Affected Environment | | | | | | |
| | | | | Affecting Environment | | | | |
| Organizational Targets Perspective | Resources | Target Artifacts | | | Resources | | | |
| Disjoint Elements | Externalities | Target Artifacts Created | Target Artifacts Altered | Resources used and consumed... | | Resources used, but not consumed... | | |
| | | | | ...attributable to single instances | ...not attributable to single instances | ...attributable to single instances | ...not attributable to single instances | |

**Fig. 2** Outer environment perspectives

organizational targets are not considered comprehensively. Effectiveness Criterion EC 1 is thus impaired.

To obtain a more comprehensive view on the outer environment of a business process, we can assume a *scope of influence perspective*. The business process acts on part of its environment, and a part of its environment acts on the business process. We call these two parts the *affected environment* and the *affecting environment* of the business process. As an example, consider a document which is edited and thus *affected* in the course of the business process, and a piece of information which is *affecting* the business process because it is used to reach a decision. The two parts overlap, but things that belong to neither part are no component of the outer environment of the business process. For quality assessment, only the affected environment is of interest because things that are not affected by the business process are not relevant to judge its quality.[1] However, it is still not possible to state organizational targets for the affected environment without further analysis, because it comprises the intended results of the business process as well as the consumption of economic resources.

We therefore propose an additional *organizational targets perspective* made up of two concepts: A business process interacts with its outer environment by manipulating (i.e., creating and/or altering) *target artifacts* and by using *resources*. The target artifacts involved in a business process are defined by the business objective. The resources involved are defined by the business objective as well as business process design, implementation and enactment. Target artifacts are the part of the outer environment we *strive to alter* while resources are the part we *need to employ or unintentionally affect* to achieve our business objective. Everything beyond these two categories is not relevant to the business process and therefore not part of its outer environment.

---

[1] Note that this proposition contradicts other quality frameworks for business processes which include, for instance, process input characteristics as quality attributes ([22, 20], cf. Section 3).

Note that target artifacts may evolve into resources in the context of another business process, and that resources drawn from are not necessarily
consumed. We consider a resource as consumed if it is made unavailable
to other uses, either permanently or only temporarily (e.g. a plot of land
used is consumed temporarily). Resources not consumed are merely part of
the affecting, but not of the affected environment. Resources consumed and
target artifacts are part of the affected environment. Information generally
constitutes a resource which is not consumed.

---

*Example 2 (Target artifacts and resources).* To illustrate some of the concepts
set out in this section, reconsider the process of handling supplier invoices we
already used in Example 1. The *business objective* of this process is to approve or
reject incoming invoices. They thus constitute the *target artifacts* of the process.
*Resources* involved are affected by business process design, implementation and
enactment.

According to Example 1, available design options comprise early scanning and
EDI. These clearly differ in terms of *resources* such as information systems or
labor required. Accordingly, the resources involved in the business process are
determined by the chosen design option and its implementation. However, both
options pursue the same business objective and work with the same target artifacts.

In the course of the process, the invoices are not created, but merely altered –
in this case, an information item whether the invoice is approved or rejected is
added. This information in turn constitutes a resource for the outgoing payments
process which occurs downstream in the overall process chain.

---

The *disjoint elements* line in Figure 2 depicts a categorization of the
outer environment where each thing in the outer environment belongs to
exactly one category. It is thus comprehensive, free of overlaps and sufficiently expressive to build all other perspectives (for instance, $Output =$
$Target\,Artifacts\,Created \cup Target\,Artifacts\,Altered$).

While the basic content categories as comprised in the disjoint elements
line are universally valid, their concrete content in partially evolves over the
lifecycle of the business process. With respect to the *organizational targets*
perspective in Figure 2, the target artifacts part of the environment remains
stable because the target artifacts of the business process are pre-determined
by the business objective.[2] The resources part, however, is subject to process design & implementation. It therefore evolves with the business process
lifecycle. This occurs in two ways:

---

[2] We do not consider the decision on proper business objectives as part of the
business process lifecycle. Contrary to that, the reengineering advocates of the
90s proposed to rethink the business objectives of an organization as part of
process design and optimization. While we do not share this view, we included
a more detailed discussion on this topic in [40].

- Resources used and affected condense and solidify in the course of the business process lifecycle. Before process design starts, only the general availability of resources to the organization and resources that are *elemental* to the business objective are determined. When process design & implementation are completed, the types of resources used and affected are designated. Once the enactment of the business process has been completed, the environment of the business process is fully determined.
- The share of resources not only used, but consumed by a business process diminishes the more we advance in the business process lifecycle. Note that this closely resembles the concept of marginal cost accounting mostly used in German enterprises [34].

In general, parts of the affected environment during business process design & implementation become parts of the solely affecting environment during business process enactment. We will have to consider this issue in the course of our further investigation.

---

*Example 3 (Resources in the business process lifecycle).* Consider the business process to handle incoming supplier invoices which we already used in our previous examples. As we embark on the design of a corresponding business process, we may consider a number of options to achieve our business objective:

- We might manually send the invoices to the purchasing department and to the department which received goods or services for approval.
- We might implement one or more of the IT-based practices from Example 1.

At this stage, it is still open whether we employ organizational resources to implement an IT-based process or simply stick with more manual effort to distribute and recollect paper documents. However, if our business objective is to check invoices against purchase orders and goods receipts, purchase order information is an elemental resource and will be required regardless of process design. Likewise, if our organizational resources are not sufficient to implement IT solutions, we might have to consider this as a constraint as well.

Once the business process is implemented, however, we know what types of resources will be needed for enactment. The actual quantity per resource type will still depend on the actual number of process instances and their concrete enactment.

Regarding the diminishing share of resources that are actually consumed, consider the implementation of an EDI system. At deploy time, the system is in place regardless whether the business process is executed or not. The business process does not consume the EDI system as a resource. At design time, we get a different picture: whether and how the EDI system has to be implemented depends on the business process's design and will surely impact the consumption of resources.

---

Note that the business objective determines what is to be achieved by the business process in terms of target artifacts, but not how this should be accomplished. Moreover, per definition, direct materials (including information

items) are the only kind of process input to be "embodied" into target artifacts. Accordingly, elemental resources determined by the business objective
always relate to direct materials.

## 5.2   The Impact of the Business Process on Organizational Targets

The quality of a business process as an artifact needs to be assessed in terms
of its impact on its outer environment. Based on our considerations on the
environment of a business process (cf. Section 5.1), we can identify the set
of organizational targets impacted by the business process and thus relevant
to business process quality. We can readily determine *"what the organization
would want to achieve"* with respect to both target artifacts and resources:

- With respect to *target artifacts*, the business objective of the process by
  definition constitutes one or more organizational targets. This aspect is
  typically addressed by conventional quality management approaches' focus
  on the quality of products and services delivered by business processes. It
  corresponds to the notion of *efficacy* as "the ability to produce a desired
  or intended result" [46].
- With respect to *resources*, we may assume that the organization aims to
  act economically (as may be inferred from the term *business* process). Accordingly, resources should be impacted as little as possible. This aspect
  is typically addressed by the focus of process performance management
  approaches on capacity management, cost and time. It corresponds to the
  common management notion of *efficiency*. Note that discussing organizational targets for the common BPM concept of process input would be
  much more difficult.

Assessing business process quality on the basis of relevant organizational
targets amounts to appraising the impact of the business process on the
achievement of the respective targets. To this end, we have to consider that
a business process is enacted within an outer environment which comprises
not only affected, but also affecting elements, i.e. resources used and target
elements to be altered. Thus, the business process cannot "achieve" organizational targets, but merely contribute to their achievement. In other words,
the affecting environment constrains the business process with respect to
achieving organizational targets. To obtain a meaningful assessment of business process quality, we will need to delineate the impact of the affecting
environment from the impact of the business process. Moreover, the affecting
environment and the affected environment evolve with the business process
lifecycle. Thus, the impact of the business process on organizational targets
needs to be discussed specific to differing lifecycle stages as well.

*Example 4 (Impact of business processes vs. affecting environment).* EDI systems
for incoming invoices typically try to match invoices against purchase orders and
goods receipts to determine whether the invoice can be posted and approved for
payment. In this case, purchase order and goods receipt data constitute process
input or resources employed. If one or both elements are missing, the ability of the
business process to check the invoice in time will be impacted. As a result, it may
not be possible to obtain an early payment discount or, worse, the supplier may
decline to make new deliveries. In this case, the achievement of organizational
targets is clearly impeded, but this is not the "fault" of our business process.
Instead, elements of the affecting environment prevent achieving organizational
targets. On the other hand, the EDI process alone cannot ensure timely payments
because effective input of purchasing and goods receipt data is required as well.
To effectively assess the quality of the EDI process, we have to properly delineate
these effects.

As an example for differing requirements to delineate the affecting environment
in the course of the business process lifecycle, consider that EDI operations are
often outsourced to service providers subject to service level agreements. During
design & implementation, this is a deliberate decision under consideration of
the quality of service required. Whether this decision is taken properly should
enter quality assessment. During enactment, however, the availability of the EDI
service becomes part of the affecting environment. When assessing enactment
quality, we need to make sure that our results are not biased by EDI service
failures.

To fulfil Effectiveness Criterion EC 2, we aim to recognize distinct organizational responsibilities for process design and process enactment as
encountered in most organizations. Quality assessment results for business
process design & implementation should therefore not depend on the quality
of business process enactment and vice versa. This implies that the business
process design & implementation lifecycle stage not only determines the types
of resources employed and affected in business process enactment, but also
that business process design & implementation in itself is to be considered
as part of the affecting environment during business process enactment. In
a strict interpretation, this means that business process enactment will in
itself actually not impact the achievement of organizational targets because
the behaviour of the business process is fully determined by its design, its
implementation, the resources used and the target artifacts to be altered.

Of course, this does not match practical requirements because assessing
business process enactment quality is usually understood as assessing the
quality of the human effort involved. Although human effort in principle
constitutes a resource to the business process, we follow this interpretation
for its practical relevance. However, we have to be aware that this decision
implies a certain deviation from a fully stringent approach based on the
business process as an artifact in the sense of Simon.

To summarize and exemplify the evolvement of the outer environment in terms of resources, Figure 3 illustrates the affecting and the affected environment for our fundamental lifecycle stages in terms of common business administration concepts.

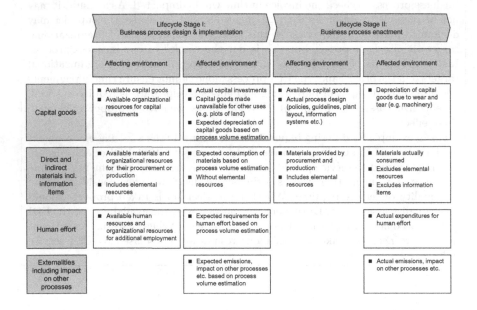

**Fig. 3** Affecting and affected resources in the business process lifecycle

Consider the following explanatory notes:

- As discussed in Section 5.1, target artifacts do not evolve with the business process lifecyle as they are pre-determined by the business objective. They are therefore not included in Figure 3.
- Capital goods refer to property, plant and equipment such as machinery, information systems, etc. In general, this corresponds to resources not attributable to individual process instances. Capital goods are an outcome of the business process design & implementation lifecycle stage.
- Direct materials correspond to resources attributable to individual process instances. For our purpose, this includes information items (as well as special cases like dies, i.e. resouces used, but not consumed). Indirect materials correspond to supplies not attributable to individual process instances.
- Human effort refers to the quantity and quality of labor employed. Note that, as stated above, we do not include human effort in the affecting environment at the enactment stage.

- Externalities refer to unintended impacts caused including emissions and effects on other processes, e.g. when shared resources like machinery are made unavailable. Per definition, externalities are part of the affected environment, but not of the affecting environment.
- Note that the affecting environment for business process enactment also comprises the actual process design, i.e. the results of the process design & implementation stage. This ensures that quality assessment of the enactment stage is not impacted by process design & implementation. We included the actual process design with the capital goods category of resources because it comprises machinery and implemented information systems as well as intellectual property such as policies and guidelines. This inclusion also links both lifecycle stages in terms of their environments: the affected environment of process design & implementation also comprises the affected environment of process enactment, and the affecting environment of process enactment comprises the affecting environment of process design & implementation. The respective impact is "funneled" through the results of the design & implementation stage.

## 5.3 Business Process Quality Based on Organizational Targets

In the previous sections, we made a number of conclusions to guide our definition of business process quality:

1. Business process quality has to be assessed in terms of the impact of the business process on its outer environment. For this purpose, its outer environment can be analyzed in two dimensions: the *affecting* vs. the *affected environment*, and *target artifacts* vs. *resources*.
2. There are differing *organizational targets* with respect to the target artifacts and resources parts of the affected environment. These targets correspond to *business process efficacy* and *business process efficiency*, respectively. As the affected environment will be determined by the business process *and* the affecting environment, the business process cannot achieve these organizational targets, but merely contribute to their achievement.
3. Affecting and affected resources evolve with the *business process lifecycle*. To reflect differing organizational responsibilities, business process quality must be assessable separately for *business process design & implementation* and for *business process enactment*.

Based on these considerations and on the ISO quality definition (cf. Section 4.2), we can derive a definition framework for business process quality:

**Definition 1 (Business process quality framework).**
*Business process efficacy* means the effectiveness of a business
process with respect to achieving its business objective. A business process is efficacious iff its business objective is achieved for a reasonable
set of states of its affecting environment.

*Business process efficiency* means the effectiveness of a business
process with respect to limiting its impact on resources. A business process is efficient iff it limits its impact on resources reasonably considering
the state of its affecting environment.

*Business process design & implementation quality* is the degree to which an actual business process model enables business process
efficacy, achieves business process efficiency during design & implementation, and enables business process efficiency during its enactment.

*Business process enactment quality* is the degree to which a
set of business process instances achieves business process efficacy and
business process efficiency.

According to the outer environment of the business process and the associated organizational targets, business process efficacy and efficiency constitute the two dimensions of business process quality requirements for both
fundamental lifecycle stages. They both take into account the affecting environment, either by demanding achievement of the business objective only
for "a reasonable set of states" of the affecting environment, or by considering the affecting environment in the evaluation of the impact on resources.
A reasonable set of states in this context relates to what can be assumed
regarding the affecting environment under the presumption of due diligence
in upstream processes. This means that the business process, to be effective,
must be able to function in common and expectable business circumstances.
Similarly, reasonably limiting the impact on resources refers to avoiding waste
and diligently managing resources. A more detailed analysis of these topics
(for instance with regard to a special application area) is a core subject of
business process quality *modeling* (see our methodology set out in Section 2).

Note that a business process can be efficacious, but not efficient, whereas
efficiency is only possible if a measure of efficacy is achieved as well: if the
business objective is not achieved, any resources consumed have not been
used reasonably. Table 4 resolves the dimensions of business process quality
in terms of efficacy and efficiency requirements and in relation to fundamental
business process lifecycle stages and their respective affecting environment.

Our definition framework is rather plain and simple. This characteristic
is required to enable straightforward discussion in a business context, for
instance with respect to Garvin's five basic quality notions (cf. Section 4). It

**Table 4** Business process quality requirements

| Fundamental lifecycle stage | Affecting environment constraints | Quality requirements | |
| --- | --- | --- | --- |
| | | Business process efficacy | Business process efficiency |
| **Lifecycle Stage I: Business process design & implementation** | Available organizational resources | Enable achievement of the business objective with respect to the target artifacts | Limit the impact on resources during design & implementation, and enable to limit the impact on resources during enactment |
| **Lifecycle Stage II: Business process enactment** | Actual process design, target entities to be altered, capital goods, direct materials | Achieve the business objective with respect to the target artifacts | Limit the impact on resources |

corresponds to the ISO definition of quality as *"the degree to which a set of inherent characteristics fulfils requirements"* [28]: "inherent characteristics" reflect the design and implementation of the business process during the respective lifecycle stage and the human effort involved during enactment, and the "requirements" are reflected by the quality stipulations we made with respect to business process efficacy and efficiency.

However, due to the high level of abstraction we adopt, it remains difficult to concisely apply our definition to practical examples. While we do not include a fully elaborated formal model of business process quality here, we provide an outlook on quality attributes, criteria and predicates to facilitate a better understanding.

## 5.4  Outlook: Business Process Quality Modeling

In this section, we provide an outlook on possible approaches to business process quality models. As described in Section 1, the main objective of business process quality models is to enable the assessment of business process quality. To this end, quality models basically consist of three components summarized in Figure 4.

Properties of business processes that are apt to determine or measure the impact of the business process on its environment with respect to organizational targets constitute the *quality attributes* of the business process. Quality attributes can assume states we can link to *quality predicates*, i.e. assertions on quality semantically suitable for the respective attribute. These states correspond to *quality criteria*. Accordingly, if a quality attribute assumes a state which fulfills a quality criterion, we may assign the respective quality predicate to the business process. Quality criteria reflect the requirements concept cited in the ISO quality definition.

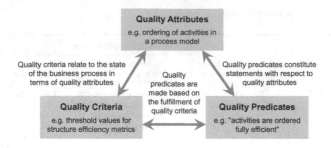

**Fig. 4** Business process quality model components

According to our approach, it is generally desirable to rigorously derive quality attributes by applying Definition 1 and formal definitions of business processes, target artifacts, resources and their interrelation. However, we refer this approach to future work to avoid departing from the scope of this chapter. As an alternative to provide initial practical relevance, we provide an informal quality model along our definition framework. A good mental technique for this is to consider possible deficiencies that might occur. While this is similar to approaches based on listing possible quality attributes without rigorous derivation [22, 20], our extended discussion in Sections 5.1 to 5.3 still provides us with valuable insights and structure. We thus, for instance, avoid including process input properties as quality attributes to the business process.

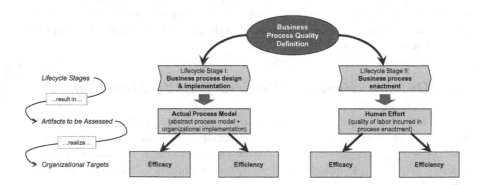

**Fig. 5** Basic quality model deduction

Figure 5 summarizes the basic approach we apply to deduct an initial, non-formalized and simplified quality model: we first consider our lifecycle stages and the resulting artifacts, which are subject to quality assessment. Accordingly, assessing business process quality in Lifecycle Stage I amounts to assessing the quality of the actual process model, and assessing business process quality in Lifecycle Stage II amounts to assessing the quality of human effort during enactment. Both artifacts are then assessed with respect to their impact on the organizational targets of efficacy and efficiency. Additional guidance is provided by the overview on the resources part of the affected environment in Figure 3.

Table 5 lists quality attributes, criteria and predicates we include in our simplified model. Because we do not formally deduct the entire quality model at this stage, we may not yet guarantee its completeness, and we may not give concisely measurable quality criteria. However, the structure along our discussion in the previous sections still allows for a measure of control in this respect, e.g. by considering the system of affected resources in Figure 3.

**Table 5** Simplified quality model

| Ref. | Quality attributes | Quality criteria | Quality predicates |
|------|--------------------|------------------|--------------------|
| *Business process design & implementation efficacy* | | | |
| *A1* | Formal or informal documentation of the business objective | Business objective explicitly modeled or documented as prerequisite to manage efficacy | Transparent and controlled business objective |
| *A2* | Expectations and requirements regarding the actual affecting environment | Expectations regarding the actual affecting environment have been reasonably derived and documented / communicated | Managed affecting environment |
| *A3* | Relation between designated termination states and the business objective | Control flow model conforms to the business objective (e.g., by formal derivation from the business objective) | Efficacious control flow design |
| *A4* | Consideration of procedures to manage deficiencies during business process enactment | Relevant cases covered acc. to affecting environment expectations, procedures comprised in actual process design | Efficacious exception handling |
| *A5* | Relation between capital goods and business process model requirements | Capital goods available according to business process model as far as organizational resources have been available | Efficacious capital expenditures |
| *A6* | Relation between staff capacity and business process model requirements | Staff and procedures available according to business process model as far as organizational resources have been available | Efficacious organizational implementation |
| *Business process design & implementation efficiency* | | | |
| *B1* | Occurrence of non-value-adding activities and execution paths | Control flow explicitly designed to avoid non-value-adding activities and execution paths | Controlled non-value-adding activities and execution paths |

**Table 5** (*continued*)

| Ref. | Quality attributes | Quality criteria | Quality predicates |
|------|--------------------|------------------|--------------------|
| B2 | Occurrence of resource waste in activities | Activities are designed to avoid materials waste (e.g. clippings) and capacity waste (e.g. through idle time for staff or capital goods) | Controlled resource consumption in activities |
| B3 | Modeled sequence of activities: control flow designed to enable early break conditions towards termination states | Avoidance of non-value-adding activities in possible execution paths regarding termination states, early execution of automated checks | Efficient break conditions |
| B4 | Design decisions: employment of capital goods vs. labor to implement automated vs. manual activities | Design decisions have been taken based on explicit business case considerations | Controlled capital goods vs. labor trade-off |
| B5 | Skill requirements: employee skill levels required in manual activities | Design decisions have been taken based on explicit business case considerations, activities and procedures are properly documented and trained | Controlled skill employment |

*Business process enactment efficacy*

| Ref. | Quality attributes | Quality criteria | Quality predicates |
|------|--------------------|------------------|--------------------|
| C1 | Occurrence of deviations from the business process model in manual decisions altering the actual control flow path | Prevalence reasonable with respect to the criticality of the business objective | Efficacious manual decisions in the control flow path |
| C2 | Occurrence of deviations from the business process model in manual manipulations of target artifacts or resources relevant to the control flow in the course of activity execution | Prevalence reasonable with respect to the criticality of the business objective | Efficacious execution of manual activities |
| C3 | Occurrence of time delays in manual execution of activities | Prevalence and severity of time delays reasonable with respect to the criticality of the business objective | Timely execution of manual activities |
| C4 | Occurrence of manual alterations to the actual process model (e.g. overriding of IS customization) in the course of the execution of individual process instances | Prevalence reasonable with respect to the criticality of the business objective | Conformance to the actual process model |

*Business process enactment efficiency*

| Ref. | Quality attributes | Quality criteria | Quality predicates |
|------|--------------------|------------------|--------------------|
| D1 | Occurrence of deviations from the business process model leading to redundant activities caused by manual control flow decisions | Prevalence of redundant activities reasonable with respect to complexity of control flow decisions and additional effort incurred | Efficient execution regarding redundant activities |
| D2 | Occurrence of multiple executions of process instances or activities due to activity execution deficiencies | Prevalence of multiple executions reasonable with respect to complexity of respective tasks and additional effort incurred | Efficient execution regarding multiple executions |

**Table 5** (*continued*)

| Ref. | Quality attributes | Quality criteria | Quality predicates |
|------|--------------------|------------------|--------------------|
| D3 | Occurrence of additional corrective activities due to manually caused deviations or deficiencies | Prevalence of corrective activities reasonable with respect to complexity of respective tasks and additional effort incurred | Efficient execution regarding corrective activities |
| D4 | Occurrence of manual re-allocation of execution responsibility for activities | Prevalence of re-allocated activities reasonable with respect to source (manual vs. automated) and validity of original allocation and additional effort incurred | Efficient execution regarding re-allocated activities |

# 6 Illustrative Case

To illustrate our results, we apply our simplified quality model to a real-world business process in terms of its actual process model and an execution log.

In terms of content, our sample process corresponds to the examples given in the previous sections. Its business objective is to approve or disapprove incoming supplier invoices correctly and timely. In particular, it implements the early scanning design option already mentioned in Example 1. Our execution sample covers a total of 1,130 cases incurred over the period of one week, which have been tracked over the period of 15 weeks (cases not concluded within this timeframe are not considered). Figure 6 presents a BPMN flow chart of the business process model [55]. In addition, we base our evaluation on a central document describing the business process and its technical implementation (the so-called "blueprint").

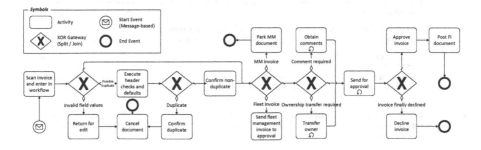

**Fig. 6** Sample process: invoice handling

In Table 6, we apply the quality criteria set out in our simplified quality model in the previous section, and state the respective quality predicates.

**Table 6** Simplified quality model: sample application

| Ref. | Quality analysis | Quality predicates |
|---|---|---|
| *Business process design & implementation efficacy* | | |
| A1 | The business objective has not been formalized or documented in the blueprint, which governs process implementation and enactment | |
| A2 | The expected affecting environment has not been included in the blueprint, but considered informally in actual process design; an evaluation on the expected transactional volume has been conducted | Managed affecting environment |
| A3 | While there is no formal documentation of the business objective, use cases have been described in detail in the blueprint. As use cases have been deducted from available transactional data (cf. A2), we may therefore assume efficacious implementation | Efficacious control flow design |
| A4 | Exception handling routines have not been included in the actual process design | |
| A5 | Actual process execution as per the log sample implies appropriate capital investments according to the process design | Efficacious capital expenditures |
| A6 | Actual process execution as per the log sample implies issues in organizational implementation (cf. C3, D2, D3, D4) due to limited governance of process management over process participants | |
| *Business process design & implementation efficiency* | | |
| B1 | Non-value adding activities occur in the execution path (manual re-allocation of responsibilities), "looping" of check activities is possible | |
| B2 | Capacity waste is avoided through the use of work item lists for all user groups | Controlled resource consumption in activities |
| B3 | All automated checks are designed to occur at the beginning of the control flow sequence | Efficient break conditions |
| B4 | Design option decision (early scanning plus workflow) for the business process is based on an explicit business case consideration | Controlled capital goods vs. labor trade-off |
| B5 | Actual skill employment is based on available resources in the organization instead of documented requirements | |
| *Business process enactment efficacy* | | |
| C1 | Deviations from the business process model do not occur (execution fully controlled by the WfMS) | Efficacious manual decisions in the control flow path |
| C2 | Correct handling of invoice approval is subject to both internal and external audit procedures (risk-based audit approach) | Efficacious execution of manual activities |
| C3 | Total processing time exceeds two weeks in 10% of cases, mainly due to delays in the approval procedure | |
| C4 | Manual alterations to the actual process model do not occur | Conformance to the actual process model |

**Table 6** (*continued*)

| Ref. | Quality analysis | Quality predicates |
|------|------------------|---------------------|
| *Business process enactment efficiency* | | |
| *D1* | Attribute not assessable: redundant activities may occur where approval actions beyond the requirements based on the invoice value are conducted. Due to data protection concerns, we do not analyze this data | n/a |
| *D2* | "Return for edit" occurs in 10% of cases, leading to repeated manual check activities | |
| *D3* | "Return for edit" occurs in 10% of cases, leading to corrective activities in document capturing | |
| *D4* | Manual case ownership transfers occur in 34% of cases | |

In summary, the implications from our case example are twofold: First, we can summarize our assessment with respect to the quality of our sample business process. Second, we are now able to assess our initial design results with respect to the effectiveness criteria set out in Table 1.

With respect to our sample process, quality predicates made imply that the quality of the process largely reflects the chosen design option as a contemporary "best practice". Issues incurred mostly relate to topics where respective approaches have not yet reached practical acceptance (e.g. A1) or to governance issues during the enactment lifecycle phase. This may be due to the fact that, in this case, process management only partially controls process participants as invoice approval is "spread" throughout the organization.

Regarding our effectiveness criteria, we arrive the following conclusions:

- **Effectiveness Criterion EC 1:** *Congruence to organizational targets.* Implications in respect to EC 1 are twofold. On the one hand, we can directly "drill down" from organizational targets to each quality attribute we consider. Accordingly, there are no issues with respect to *exclusive coverage*. On the other hand, we cannot ensure *comprehensive coverage* in our quality model. We stipulate that this restriction is caused by our deviation from a rigid deductive approach when drafting our simplified quality model.

- **Effectiveness Criterion EC 2:** *Perceived fairness.* Our quality model reflects basic organizational governance by adhering to fundamental business process lifecycles. However, as mentioned above with respect to the enactment lifecycle phase, a more fine-grained approach is required for our practical example. Moreover, our "binary" allocation of quality attributes is prone to omit important graduations. While our assessment results still point to issues to be addressed to improve on quality, organizational acceptance might still be impeded by these issues.

- **Effectiveness Criterion EC 3:** *Cost effectiveness.* Our illustrative case
  has shown that the simplified quality model can be applied with very small
  effort, provided that basic information such as, in this case, an implementation blueprint and an expressive execution log sample are available. This
  aspect, however, needs to be tracked as we move into more detailed quality
  models to further accomodate Effectiveness Criteria EC 1 and EC 2.

# 7 Conclusion

Business process quality management constitutes a highly promising area of
research due to the application potential emphasized by the success of quality
management practices in manufacturing and related fields. In this chapter,
we gave an overview on existing approaches. Our considerations have shown
that full effectiveness for management purposes has not been achieved yet.

However, we demonstrated that progress in this field can be made by
applying a rigorous design methodology based on appropriate effectiveness
criteria and due consideration of the outer environment as well as organizational targets. Initial results in the form of our definition framework and an
initial simplified quality framework are promising and readily applicable to
an illustrative case. In line with our design methodology, sample application
also indicates that further and more rigorous research will be required with
respect to quality modeling for business processes.

In the course of future work, we will therefore elaborate a more formal and
detailed quality model, develop appropriate procedures for its application in
the course of business process quality management and methods to achieve
broader integration into BPM tools and methods.

# References

1. van der Aalst, W.M.P.: The application of Petri nets to workflow management.
   J. of Circuits, Systems and Computers 8(1), 21–26 (1998)
2. van der Aalst, W.M.P., van Hee, K.M.: Business process redesign: A Petri-net-
   based approach. Computers in Industry 29(1-2), 15–26 (1996)
3. van der Aalst, W.M.P., ter Hofstede, A.H.M., Weske, M.: Business Process
   Management: A Survey. In: van der Aalst, W.M.P., ter Hofstede, A.H.M.,
   Weske, M. (eds.) BPM 2003. LNCS, vol. 2678, pp. 1–12. Springer, Heidelberg
   (2003)
4. Becker, J., Rosemann, M., von Uthmann, C.: Guidelines of Business Process
   Modeling. In: van der Aalst, W.M.P., Desel, J., Oberweis, A. (eds.) Business
   Process Management. LNCS, vol. 1806, pp. 30–262. Springer, Heidelberg (2000)
5. Camp, R.C.: Benchmarking: the search for industry best practices that lead to
   superior performance. Quality Press (1989)
6. Cardoso, J.: Business Process Quality Metrics: Log-Based Complexity of Workflow Patterns. In: Meersman, R., Tari, Z. (eds.) OTM 2007, Part I. LNCS,
   vol. 4803, pp. 427–434. Springer, Heidelberg (2007)

7. Castellanos, M., Casati, F., Dayal, U., Shan, M.: A comprehensive and automated approach to intelligent business processes execution analysis. Distributed and Parallel Databases 16(3), 239–273 (2004)
8. Dale, B.G.: The Received Wisdom on TQM. In: Managing Quality, 5th edn., pp. 58–73. Wiley-Blackwell (2007)
9. Davenport, T.J.: Process Innovation: Reengineering Work through Information Technology. Harvard Business School Press (1993)
10. Davenport, T.J., Short, J.E.: The new industrial engineering: Information technology and business process redesign. Sloan Mgmt. Review 31(4), 11–27 (1990)
11. Epstein, M.K., Henderson, J.C.: Data envelopment analysis for managerial control and diagnosis. Decision Sciences 20(1), 90–119 (1989)
12. Garvin, D.A.: What does "product quality" really mean? Sloan Mgmt. Review 26(1), 25–43 (1984)
13. Grigori, D., Casati, F., Dayal, U., Shan, M.: Improving business process quality through exception understanding, prediction, and prevention. In: Proc. 27th Int'l Conf. on Very Large Data Bases (VLDB 2001), pp. 159–168. Morgan Kaufmann (2001)
14. Guceglioglu, A.S., Demirors, O.: Using Software Quality Characteristics to Measure Business Process Quality. In: van der Aalst, W.M.P., Benatallah, B., Casati, F., Curbera, F. (eds.) BPM 2005. LNCS, vol. 3649, pp. 374–379. Springer, Heidelberg (2005)
15. Guler, I., Guillén, M.F., Macpherson, J.M.: Global competition, institutions, and the diffusion of organizational practices: The international spread of ISO 9000 quality certificates. Administrative Science Quarterly 47(2), 207–232 (2002)
16. Hallerbach, A., Bauer, T., Reichert, M.: Guaranteeing soundness of configurable process variants in Provop. In: Proc. 11th IEEE Conference on Commerce and Enterprise Computing (CEC 2009). IEEE Computer Society Press (2009)
17. Hammer, M.: Reengineering work: don't automate, obliterate. Harvard Business Review 68(4), 104–112 (1990)
18. Hammer, M., Champy, J.: Reengineering the Corporation. A Manifesto for Business Revolution. HarperBusiness (1993)
19. Healy, P.M.: The effect of bonus schemes on accounting decisions. J. of Accounting and Economics 7(1-3), 85–107 (1985)
20. Heinrich, R., Paech, B.: Defining the quality of business processes. In: Proc. Modellierung 2010. LNI, vol. P-161, pp. 133–148. Bonner Köllen (2010)
21. Heravizadeh, M.: Quality-aware business process management. Ph.D. thesis, Queensland University of Technology, Australia (2009)
22. Heravizadeh, M., Mendling, J., Rosemann, M.: Dimensions of Business Processes Quality (QoBP). In: Ardagna, D., Mecella, M., Yang, J. (eds.) Business Process Management Workshops. LNBIP, vol. 17, pp. 80–91. Springer, Heidelberg (2009)
23. Hevner, A.R., March, S.T., Park, J., Ram, S.: Design science in information systems research. MIS Quarterly 28(1), 75–105 (2004)
24. Hofacker, I., Vetschera, R.: Algorithmical approaches to business process design. Computers and Operations Research 28(13), 1253–1275 (2001)
25. IDS Scheer: Process intelligence white paper: What is process intelligence? (2009), http://www.process-intelligence.com
26. International Accounting Standards Board: Framework for the Preparation and Presentation of Financial Statements (1989), http://eifrs.iasb.org

27. International Accounting Standards Board: International Accounting Standard 16: Property, Plant and Equipment (2003), http://eifrs.iasb.org
28. International Organization for Standardization: ISO 9000:2005: Quality management systems – Fundamentals and vocabulary (2005)
29. International Organization for Standardization: The ISO Survey - 2007 (2008), http://www.iso.org, http://www.iso.org/
30. IT Governance Institute: CobiT 4.1 (2007), http://www.isaca.org
31. Jansen-Vullers, M.H., Kleingeld, P.A.M., Loosschilder, M.W.N.C., Netjes, M., Reijers, H.A.: Trade-Offs in the Performance of Workflows – Quantifying the Impact of Best Practices. In: ter Hofstede, A.H.M., Benatallah, B., Paik, H.-Y. (eds.) BPM Workshops 2007. LNCS, vol. 4928, pp. 108–119. Springer, Heidelberg (2008)
32. Kaplan, R.S., Norton, D.P.: The balanced scorecard: Measures that drive performance. Harvard Business Review 70(1), 71–79 (1992)
33. Kennerley, M., Neely, A.D.: Performance measurement frameworks: a review. In: Business Performance Measurement: Theory and Practice, pp. 145–155. Cambridge University Press (2002)
34. Kilger, W., Pampel, J.R., Vikas, K.: Flexible Plankostenrechnung und Deckungsbeitragsrechnung, 12th edn. Gabler Verlag (2007) (in German)
35. Koontz, H., O'Donnell, C., Weihrich, H.: Essentials of Management, 3rd edn. Tata McGraw-Hill (1982)
36. Li, C., Reichert, M., Wombacher, A.: Discovering Reference Models by Mining Process Variants Using a Heuristic Approach. In: Dayal, U., Eder, J., Koehler, J., Reijers, H.A. (eds.) BPM 2009. LNCS, vol. 5701, pp. 344–362. Springer, Heidelberg (2009)
37. Li, C., Reichert, M., Wombacher, A.: The MinAdept clustering approach for discovering reference process models out of process variants. Int'l J. of Cooperative Information Systems 19(3-4), 159–203 (2010)
38. Liman Mansar, S., Reijers, H.A.: Best practices in business process redesign: use and impact. Business Process Mgmt. J. 13(2), 193–213 (2007)
39. Lindsay, A., Downs, D., Lunn, K.: Business processes – attempts to find a definition. Information and Software Technology 45(15), 1015–1019 (2003)
40. Lohrmann, M., Reichert, M.: Basic considerations on business process quality. Tech. Rep. UIB-2010-03, University of Ulm, Germany (2010)
41. Ly, L.T., Knuplesch, D., Rinderle-Ma, S., Göser, K., Pfeifer, H., Reichert, M., Dadam, P.: SeaFlows Toolset – Compliance Verification Made Easy for Process-Aware Information Systems. In: Soffer, P., Proper, E. (eds.) CAiSE Forum 2010. LNBIP, vol. 72, pp. 76–91. Springer, Heidelberg (2011)
42. March, J.G.: Bounded rationality, ambiguity, and the engineering of choice. Bell J. of Economics 9(2), 587–608 (1978)
43. March, S.T., Smith, G.F.: Design and natural science research on information technology. Decision Support Systems 15(4), 251–266 (1995)
44. Mendling, J.: Detection and prediction of errors in epc business process models. Ph.D. thesis, WU Wien, Austria (2007)
45. Neumayer, J., Perkins, R.: Uneven geographies of organizational practice: explaining the cross-national transfer and diffusion of ISO 9000. Economic Geography 81(3), 237–259 (2005)
46. Oxford University Press: Oxford Dictionaries Online, Keyword: Efficacy, http://oxforddictionaries.com

47. Prendergast, C.: The provision of incentives in firms. J. of Economic Literature 37(1), 7–63 (1999)
48. Reichert, M., Rinderle-Ma, S., Dadam, P.: Flexibility in Process-Aware Information Systems. In: Jensen, K., van der Aalst, W.M.P. (eds.) ToPNoC 2009. LNCS, vol. 5460, pp. 115–135. Springer, Heidelberg (2009)
49. Reijers, H.A., Liman Mansar, S.: Best practices in business process redesign: an overview and qualitative evaluation of successful redesign heuristics. Omega 33(4), 283–306 (2005)
50. Rinderle, S., Weber, B., Reichert, M., Wild, W.: Integrating Process Learning and Process Evolution – A Semantics Based Approach. In: van der Aalst, W.M.P., Benatallah, B., Casati, F., Curbera, F. (eds.) BPM 2005. LNCS, vol. 3649, pp. 252–267. Springer, Heidelberg (2005)
51. Simon, H.A.: The Sciences of the Artificial, 3rd edn. MIT Press (1996)
52. Smith, H., Fingar, P.: Process management maturity models (2004), http://www.bptrends.com
53. Speck, M., Schnetgöke, N.: To-be Modeling and Process Optimization. In: Process Management: A Guide for The Design of Business Processes, pp. 135–163. Springer (2003)
54. The Object Management Group: Business process maturity model (BPMM) (2008), http://www.omg.org
55. The Object Management Group: Business Process Model and Notation: Version 1.2 (2009), http://www.omg.org
56. Towers Watson: Executive compensation flash survey (2010), http://www.towerswatson.com
57. Vanderfeesten, I., Cardoso, J., Mendling, J., Reijers, H.A., van der Aalst, W.M.P.: Quality metrics for business process models. In: Workflow Handbook 2007. Future Strategies, pp. 179–190 (2007)
58. Vanderfeesten, I., Reijers, H.A., van der Aalst, W.M.: Evaluating workflow process designs using cohesion and coupling metrics. Computers in Industry 59(5), 420–437 (2008)
59. Weber, B., Reichert, M.: Refactoring Process Models in Large Process Repositories. In: Bellahsène, Z., Léonard, M. (eds.) CAiSE 2008. LNCS, vol. 5074, pp. 124–139. Springer, Heidelberg (2008)
60. Weber, B., Reichert, M., Mendling, J., Reijers, H.: Refactoring large process model repositories. Computers in Industry (2011) (to appear)
61. Weber, B., Sadiq, S., Reichert, M.: Beyond rigidity - dynamic process lifecycle support: A survey on dynamic changes in process-aware information systems. Computer Science - Research and Development 23(2), 47–65 (2009)
62. Weske, M.: Business Process Management. Springer (2007)
63. Workflow Management Coalition: Terminology & glossary 3.0 (1999), http://www.wfmc.org

# A Holistic Business Performance Measurement Framework

E. Sezenias[1], Alexandros Farmakis[1], Giorgos Karagiannis[1], Evaggelia Diagkou[4], and Michael Glykas[1,2,3]

[1] Financial and Management Engineering Department, University of the Aegean, Greece
[2] Aegean Technopolis, The Technology Park of the Aegean Region, Greece
[3] Faculty of Business and Management, University of Wallangong, Dubai
[4] Department of Interior Design, Technological Education Institution, Athens, Greece

**Abstract.** ADJUST is a performance measurement (PM) tool that integrates human resource management (HRM) in existing approaches of process management and workflow management (WFM) concepts. The ratios that result define and measure key performance indicators in qualitative and quantitative terms and also provide an identification of the relationships between them. These ratios permit enterprises to evaluate performance for continuous improvement in order to respond to increasingly demanding markets. ADJUST provides plans of different reorganisation scenarios for the achievement of objectives, assessment of real time performance and finally reports of deviations from desired planned performance.

Real time monitoring and assessement of performance is provided by interfaces to three kinds of management tool categories:

- *Workflow Management (WFM) Tools*: transactions, business rules, workflow models etc.
- *Process Management Tools (PMT)*: business models, cycle time, primitive cost etc.
- *Human Resource Management (HRM) Tools*: job descriptions, performance measures etc.

Meta-analysis of data analysed in PMT and HRM tools is also provided.

## 1 Introduction and State of the Art in Business Process Modelling

Business process modelling is useful for:

1) *The description of a process* These descriptions can address humans, in which case it is important that they are understandable or machines, in which case there is the need for formality.

M. Glykas (Ed.): Business Process Management, SCI 444, pp. 75–98.
springerlink.com     © Springer-Verlag Berlin Heidelberg 2013

2) *Analysis of a process*: assessement of the *properties* of a process on which rely process re-engineering and improvement. If the process is described in a formal way structural properties can be verified mechanically .Process can be measured and gaps between current and desired performance can be identified.

3) *Enactment of a process*: may occur for simulation purposes, or to provide support for process execution which can range from reacting to the execution of the activities of the process , to driving the execution of the process. Only formal languages make process enactment possible .

Language designers often provide different modelling *views*, which focus on one of the following aspects of the process.

1) The *dynamic (behavioural) view*: it provides sequencing and control information about the process, when certain activities are performed and the way they are performed.

2) *The functional view*: the functional dependencies are presented between the process elements based on the fact that some process elements consume data produced by others. ,

3) *The organizational view* describes *who* performs a given task or function, and *where* in the organization .

4) *The informational view* describes the entities that are produced, consumed or otherwise manipulated by the process such as data, products, etc.

## 1.1  The Fifth Vertical Perspective: Performance Measurement

The need for holistic performance measurement approaches in BPR and the need for linkages between the two subjects have been indicated by many researchers. Current approaches that introduce performance measurement techniques and tools in BPR analyze the business process context, implement the Analytical Hierarchy Process method to BPR and use the contingency theory and  micro and macro process analysis.Nevertheless, they fail to integrate concepts and tools from process management, human resource management and workflow management and don`t analyze business metrics and ratios, which would link strategic performance indicators with organisational and employee objectives and manipulate qualitative measurements

When process reengineering is not linked with other HRM exercises, the employee`s process objectives are conflicting with the objectives of his job descriptions, performance evaluation and career path that is defined by HRM. Employees also utilise workflow management technology through workflow process models,that are not necessarily in accordance with the business models described in BPR and responsibilities or performance measures described in HRM. Concluding, existing performance measurement methodologies and tools depend for successful application to a large extent on the support of external consultants, leading to failure to continue success after the consultants leave the company, are difficult to be used as an integral part of decision making, discourage employee participation and creative thinking and do not show how planned improved performance can be realised.

### 1.1.1 Classification of Tools Related to Performance Management

Users can store and analyze data on product mixes and sales that are keyed according to various criteria in data warehouses. Generic data warehouses must be adapted by the customer to the source system.

With ERP-specific solutions, predefined structures are supplied along with the data warehouse, but they do not provide an interface for process data.

Executive Information Systems are used to navigate through datasets stored in data warehouses or databases. The user must create his or her own analyses and reports. Specialized visualization systems offer partially predefined navigational structures for datasets. Still, all these approaches do not supply any information on process efficiency, such as cycle times and organizational gaps.

A combination of data management and visualization systems are sales information systems and other analysis systems specially conceived for specific functional areas. They use transaction data from application systems and offer standardized display formats for these data. Because they lack suitable collection procedures, however, they do not provide any performance data.

The activity-based cost calculation systems work with process-related performance data. These systems permit the calculation of cost factors based upon measured or estimated throughput times, thus providing interesting information for planning and controlling purposes. These systems do not support data collection, and thus can result in measurement errors in which case , no channels are provided for feedback for the purposes of initiating measures for improvement.

Companies need to ensure that they have an appropriate level of control over the business processes enabled by those applications. They also need to successfully address longer range planning, and more specifically:

- verify that proposed process costs are being adhered to
- verify that scheduled process throughput times are being met
- detect weak points in their processes
- detect points where a potential for time and/or cost-saving improvements exist
- ensure that the business processes implemented are resulting in desired levels of customer satisfaction
- verify that process capacities correspond to the amount of work that needs to be accomplished to meet customer demands
- ascertain whether planning premises were correct.

## 2 The Adjust Framework

ADJUST is a holistic methodology based tool. Its major components are called "Tool Clusters".

Each of the ten clusters is described briefly below:

1  **Requirements** - The customer's needs, requirements, and priorities that must be fulfilled to achieve optimum organizational performance. Analytical tools identify where resources need to be realigned to meet existing and future requirements.

2  **Problem Solving** - The processes for methodically identifying, focusing, prioritizing, and resolving issues related to implementing change within an organization. Analytical tools help the consultant to brainstorm ideas, prioritize issues and opportunities as well as assess the impact of change.

3  **Processes** - The horizontal linkage of activities needed to achieve a desired result. Analytical tools are designed to reengineer, redesign, and optimize business processes.

4  **Outputs** - The internal and external products or services produced by an organization. Analytical tools assess and challenge the internal and external demands for products and services.

5  **Activities** - How effort is spent within an organization at an individual, functional, process, product, or service line level. Analytical tools are designed to estimate the value, efficiency and/or effectiveness of individual and organizational effort.

6  **Structure** - The vertical linkage of organizational functions, processes, and resources within an organization. Analytical tools assess organizational structure issues including the efficiency of reporting relationships, the alignment of functions, and the orientation of resources.

7  **Drivers** - The policy, structural, and operational factors, which impact and influence the operations and cost of an organization. Analytical tools help the consultant identify, determine the impact of, and validate the appropriateness of various drivers.

8  **Inputs** - The human resources and other assets required to produce products or services. Analytical tools help determine the level, mix, and deployment of resources within the organization.

9  **Performance Measures** - The metrics of success or failure, which influence the operations of an organization. Performance measure analyses can be related to financial performance, quality and service levels, operational performance, and/or productivity.

10 **Change Management** - The processes for managing an organization through implementation. The analytical tools within this tool cluster identify and manage factors, which may hinder as well as help to ensure implementation success.

These tool clusters are used in conjunction with one another to achieve a complete view of the organization.

A list of tools used per cluster are described in the table below:

Tools per cluster

| ACTIVITIES | PROCESSES | STRUCTURE |
|---|---|---|
| Mission/Non-Mission (Primary/Secondary) | Process Prioritization | Management Structure Analysis |
| Activity Value Added Analysis | Structured Process Selection | Internal Contact Survey |
| Cost of Quality Analysis | Reengineering Master Planning | Analysis |
| Fragmentation/Concentration Analysis | Visioning | Activity Clustering |
| Fractionalization Analysis | Core Process Analysis & Design | Detailed Organizational Design |
| Equivalent Salary Analysis | Cross Functional Process Analysis & Design | Accountability |
| Work Expectancy Development | Detailed Process Flow Analysis & Design | Framework |
| | Process Profiling | Job Description |
| | Brown Paper Analysis | Definition |
| | Process Critical Path Analysis | |
| | Process Control Risk Assessement | |
| | Methods Analysis | |
| | Cycle Time Reduction Analysis | |
| | Approval Cycle Analysis | |
| | Policy and Procedure Definition | |
| **REQUIREMENTS** | **INPUTS** | **CHANGE MANAGEMENT** |
| | Staffing Trend Analysis | Communication Planning |
| Customer Attribute Baseline Ranking | Staffing Turnover Analysis | Culture Print |
| Customer Attribute | Resource to Workload Matching | Change Propensity Analysis |
| Categorization | Skill Mix Analysis | Force Field Analysis |
| Current vs. Future Requirements | Make vs. Buy Analysis | Attrition Analysis |
| Analysis | | Position Elimination Scoring |
| Client vs. Competitor Analysis | | Relocation/Retraining Analysis |
| External vs. Internal Perceptions | | Implementation Action |
| Analysis | | Planning |
| Acceptable vs. Outstanding | | Management Contracts |
| Analysis | | Implementation Scorekeeping |
| Internal Requirements | | |
| Assessment | | |

| PERFORMANCE MEASURES | OUTPUTS | PROBLEM SOLVING |
|---|---|---|
| Performance Measures | Output Rationalization | Problem Solving Process |
| Framework | Activity Based Product/Service Line | Interviewing |
| Financial Analysis | Costing | "What One Thing" Analysis |
| Benchmarking | Customer Profitability Analysis | Opportunity Logging |
| Best Practices Analysis | Channel Strategy Analysis | Cause and Effect Analysis |
| DRIVER | | Brainstorming |
| Macro Cost Driver Analysis | | Delphi Technique |
| Activity Generator Analysis | | Nominal Group Technique |
| Time Driver Analysis | | List Reduction |
| | | Selection Grid |
| | | Weighted Voting |
| | | Paired Comparisons |

The ADJUST program addresses four business situations:

- **Rightsizing** - Focuses on cost management ,including:
- Declining business volume.
- Inadequate profitability.
- Need to reduce pricing structure.
- Uncompetitive cost structure.
- Excessive layers of management.
- Increasing demand for services.
- Decreasing productivity and efficiency.
  - **Restructuring** - Facilitates refocusing of the organization to meet customer goals including :
- Mismatch between organizational structure and business strategies.
- Inadequate customer or market focus.
- Mismatch of resources to functional workload requirements.
- Decreasing productivity and efficiency.
  - **Reengineering** - Focuses on redesigning and streamlining cross-functional business processes, including:
- Cumbersome business processes and systems.
- Inadequate quality focus.
- Backlogged processes or workload.
- Inadequate service levels.
- Mismatch of resources to process workload requirements.
- Decreasing productivity and efficiency.
  - **Integration** - Focuses on achieving synergies among merged or acquired entities, including:
- Merger or acquisition of another company.
- Need to consolidate divisions or business units.

The table below lists the tools, which are used in each of these business situations.

| TOOL CLUSTER | TOOL | BUSINESS SITUATION | | | |
|---|---|---|---|---|---|
| | | Rightsizing | Reengineering | Restructuring | Integration |
| PROBLEM SOLVING | Problem Solving Process | 2/21 | 2/21 | 2/21 | 2/21 |
| | Interviewing | 2/21 | 2/21 | 2/21 | 2/21 |
| | "What One Thing" Analysis | 2/21 | 2/21 | 2/21 | 2/21 |
| | Opportunity Logging | 2/21 | 2/21 | 2/21 | 2/21 |
| | Cause and Effect Analysis | d/c | 2/21 | d/c | d/c |
| | Brainstorming | 2/21 | 2/21 | 2/21 | 2/21 |
| | Delphi Technique | d/c | d/c | d/c | d/c |
| | Nominal Group Technique | d/c | 2/21 | 2/21 | d/c |
| | List Reduction | 2/21 | 2/21 | 2/21 | 2/21 |
| | Selection Grid | d/c | d/c | d/c | d/c |
| | Weighted Voting | d/c | d/c | d/c | d/c |
| | Paired Comparisons | d/c | d/c | d/c | d/c |
| | Impact Analysis | 2/21 | 2/21 | 2/21 | 2/21 |
| | Mission/Non-Mission (Primary/Secondary) | 2/21 | 2/21 | 2/21 | 2/21 |
| | Activity Value-Added Analysis | 2/21 | 2/21 | 2/21 | 2/21 |
| | Cost of Quality Analysis | d/c | 2/21 | d/c | d/c |
| | Fragmentation/ Concentration Analysis | 2/21 | 2/21 | 2/21 | 2/21 |
| | Fractionalization Analysis | 2/21 | d/c | 2/21 | d/c |
| | Equivalent Salary Analysis | 2/21 | d/c | 2/21 | d/c |
| | Work Expectancy Development | 2/21 | | d/c | d/c |
| | Process Prioritization | a/b | 2/21 | d/c | d/c |
| | Structured Process Selection | d/c | 2/21 | a/b | a/b |
| | Reengineering Master Planning | d/c | 2/21 | a/b | a/b |
| | Visioning | d/c | 2/21 | d/c | d/c |
| | Core Process Analysis & Design | d/c | 2/21 | d/c | 2/21 |
| | Cross-Functional Process Analysis & | d/c | 2/21 | d/c | 2/21 |
| | Detailed Process Flow Analysis & | a/b | 2/21 | d/c | d/c |
| | Process Profiling | a/b | 2/21 | d/c | d/c |
| | Brown Paper Analysis | a/b | 2/21 | a/b | a/b |
| | Process Critical Path Analysis | d/c | 2/21 | a/b | a/b |
| | Process Control Risk Assessment | a/b | 2/21 | a/b | a/b |
| | Methods Analysis | d/c | 2/21 | d/c | d/c |
| | Cycle Time Reduction Analysis | d/c | 2/21 | d/c | d/c |
| | Approval Cycle Analysis | 2/21 | 2/21 | d/c | d/c |
| | Policy and Procedure Definition | a/c | 2/21 | 2/21 | d/c |
| | Management Structure Analysis | d/c | d/c | 2/21 | 2/21 |

| TOOL CLUSTER | TOOL | BUSINESS SITUATION | | | |
|---|---|---|---|---|---|
| REQUIREMENTS | Internal Contact Survey Analysis | q/c | q/c | 2/21 | q/c |
| | Top-Level Organizational Design | q/c | q/c | 2/21 | 2/21 |
| | Activity Clustering | 2/21 | q/c | 2/21 | q/c |
| | Detailed Organizational Design | q/c | q/c | 2/21 | q/c |
| | Accountability Framework | q/c | 2/21 | 2/21 | 2/21 |
| | Job Description Definition | q/c | 2/21 | 2/21 | 2/21 |
| | Customer Attribute Baseline Ranking | q/c | 2/21 | q/c | q/c |
| | Customer Attribute Categorization | q/c | 2/21 | q/c | q/c |
| | Current vs. Future Requirements | q/c | 2/21 | q/c | q/c |
| | Client vs. Competitor Analysis | q/c | 2/21 | q/c | q/c |
| | External vs. Internal Perceptions | q/c | 2/21 | q/c | q/c |
| | | | | | |

The analysis of the business models including developing solutions for redesign, is the main focus of process related performance measures in ADJUST.

Business process analysis is the appropriate approach to research and practice in the management of organisations. The Adjust approach is presented in the following diagram:

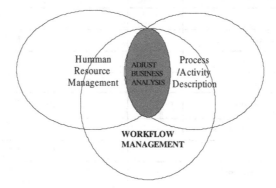

The aim is to integrate human resource management and workflow management with business process reengineering concepts in order to perform business analysis that applies both to the holistic (process) and individualistic (employee) view of the organisation. This is described in the following diagram:

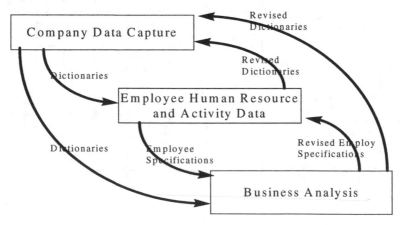

## 3 The Adjust Tool

The holistic perspective of the organisation is described in the company data capture module, where company wide management data in the form of dictionaries are held. The individualistic perspective of the organisation is described in the employee description module, where employee specific data based on the dictionaries provided by the previous module are specified. The final business analysis module provides analytical tools and techniques for achieving the desired performance both in terms of employee and dictionary specifications.

A schematic representation of ADJUST is presented in the following diagram:

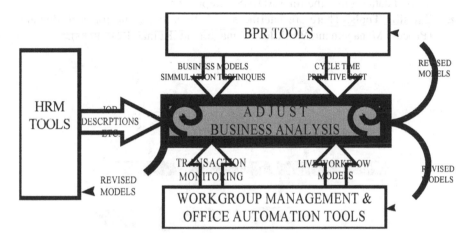

The ADJUST software platform provides the user with a number of tools that store and analyse information stored into repositories belonging to different satellite tools.

The ADJUST platform contains the following tools:

- **WPDL Exporters:** They are to convert the information that is kept in a satellite tool repository form its internal representation to Workflow Process Definition Language (WPDL). Two WPDL exporters have been build in this project: ARIS WPDL Exporter, and ELENIX WPDL Exporter.
- **WPDL Importer:** This tool is checking WPDL files for syntactical errors and then converts the information which will be imported into a temporary BPR or WFMS ADJUST repository.
- **ADJUST Engine:** With this tool a user can perform analysis based on the information stored in the final ADJUST Repository. It contains a number of functionalities such as: reporting, analysis , database management e.t.c.
- **ADJUST Repository API:** It is a set of functions stored in a Dynamic Link Library (DLL) that can be used by any other tool in order to import, export, or update information from and into ADJUST Repository.

- **HRM Interface:** This is an interface to an HRM tool selected (Description/Performance now). The interface tests API's functionality and imports information from Description/Performance Now to ADJUST Repository.
- *Adjust Consistency Checker*: This tool is comparing the information that exists in the temporary ADJUST Repository for inconsistencies. The user can receive a report with inconsistencies and he or she has to go back to the source of the information to make any corrections.
- *Adjust Unification*: This tool gets the consistent information that is stored into temporary ADJUST repositories, performs unification and imports the unified data set into the final ADJUST Repository.
- **Satellite Tools:** There are interfaces with three tools one for each domain (Process Management, Workflow Management, Human Resources).

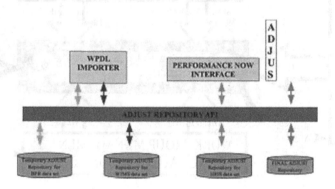

## 4  Case Study in a Telecommunications Company

The ADJUST methodology and framework was applied in a big European Telecoms Operator. Initiatives used to gather the information needed by the methodology guide included:

1  Interviews of key personnel, coordinators and employees
2  Internal official documents stating the missions and objectives of the company's divisions and departments.
3  Identification, through interviews to the responsible people for the departments, of the critical success factors and job descriptions and their relative importance to the business objectives.
4  Reuse, trough the workflow tool, of relevant process data.
5  Validation of the model with all the actors involved.

## 5 HRM Performance Measures and Links

Unnecessary organisational bureaucracy and levels of management of the classical management model, are changing with the availability of information and communication technology.Workers in a company can now be geographically far apart and send e-mails in real time. In making performance measurement systems it is often overlooked that factors concerning the process environment like culture and the people that support the processes can have a significant influence on their success and acceptance .It is argued that Human performance constitutes an important parameter to business process performance and should be modelled based on Human Performance Shaping Factors (PSF) and assessed using Bayesian Belief Networks (BBN).In ADJUST HRM data are included in job descriptions which link with a competency framework that specifies career planning and individual performance measurement.The measures are related to Soft Factors and Knowledge.

### 5.1 Soft Factors

The soft "people"factors framework provides a set of measures of factors which influence employee and team performance. The performance model is based on the interactions between people and organisational culture.

The objectives in a job description are:

✓Clarification of the links between hard factors such as costs, and issues such as culture.
✓Facility for tracking soft factors, performance and variations over time.

Measures are related to:

• Benefits of the investment and improved performance.
• Performance of the employee-team, such as the quality of decisions or actions
• Investment in the employee-team and its environment for improved performance.
• Culture of the organisation in terms of the informal micro-culture and the formal macro-culture
• People factors such as staff turnover and motivators
• Time over which the investment takes place, and performance is tracked.

The soft factors in ADJUST included in the job descriptions are the following:

1.Customer Orientation, 2.Relationships with other employees, 3 Planning and Organisation of Work 4.Problem Solving, 5.Decision Making, 6.Results Orientation, 7.Motivational Skills, 8. Innovation-Flexibility, 9.Communication Skills, 10.Team Work, 11.Leadership,12 Commitment

The values for the subjective measures presented above take values in the scale from 1-10. The data for all these measures are provided from an interface or could be inserted manually by the user of the ADJUST system, providing a graph with the results. Graphs also provide comparisons of Planned (job description level) vs Actual employee level values.

Graphs can also be produced at the company Level, per Department, or for Comparison between different job titles, different employees and different business processes.

## 5.2 Knowledge Management and Business Process Management

### 5.2.1 State of the Art

Operational knowledge sharing ads value to the organisation. Many researchers have linked organisational knowledge creation to business strategy and organisational performance .

Most of these researchers concentrate on the link between business processes and knowledge creation, capture and retrieval while others thers like Chung et. al., Lee et. al.,and Tseng et. al. have created frameworks that assess and measure knowledge readiness and capture in organisations. Kock et. al. conducted a study of a large number of completed BPR projects and have concluded that communication flow in the business process amongst the process participants and external stakeholders is the most critical factor for success. They also concluded that communication flow is very closely coupled with process model quality and its link with the reality and real time execution.

Niemi et. Al have linked the knowledge management with skills, roles and responsibilities, performance measurement the rewards and incentives systems.

In ADJUST the importance of knowledge management is recognised and knowledge is handled both at the holistic (strategic and process level) as well as at the individualistic level of job descriptions with workflows and communication flows being described not only at process level but also amongst job descriptions (structural concepts).

### 5.2.2 Explicit Knowledge and Tacit Knowledge

Knowledge in general is dynamic and represents a capacity to act and adapt within a changing environment. There are two knowledge types that are fundamental to an understanding of operational knowledge, i.e. explicit (or articulated) and implicit (or tacit) knowledge . Tacit knowledge is difficult to articulate, and is mostly internalised. It is learnt through concrete examples, experience and practice. In contrast, explicit knowledge can be formally represented and communicated through languages. All knowledge relies on being tacitly understood and applied. But tacit knowledge can be made explicit by externalisation, just as explicit knowledge can become tacit by internalisation. Tacit knowledge can also be communicated from one holder to another without becoming explicit.

### 5.2.3  Agents and Objects as Knowledge Carriers

There are two distinct types of knowledge carriers, namely agents (individuals and groups of individuals) and objects (inanimate things such as manuals, drawings, logbooks and computer records). Codified knowledge, namely encoded data and information is the knowledge that is recorded in an explicit format, in text, pictures or speech which requires human reasoning to be made utilised. Uncodified knowledge, on the other hand, represents tacit knowledge that resides in human agents. Part of it can be conceptualised and formalised. The knowledge of agents includes the rules for doing certain tasks and the formulae acquired through experience and learning. This knowledge then becomes increasingly explicit and can be transmitted through formal languages or speeches, and so be externalised and codified.

Knowledge that resides in objects is codified and explicit. Such objects are critical in process operations.and the knowledge content of such carriers must be clear and unambiguous, so that it can serve as a platform for common interpretation. Knowledge objects such as a night order books, area log books, written safety procedures, must be made available for knowledge sharing in the operational environment.To ensure effective sharing of knowledge, it is necessary to establish prioritisation, categorisation and consolidation of information.

### 5.2.4  Declarative Knowledge and Procedural Knowledge in Organisationa Learning

ACT theory distinguishes two types of knowledge involved in performing a specific task: declarative knowledge which is essentially explicit and covers the facts and rules relevant to the task and procedural knowledge which is mostly tacit and is concerned with knowing how to perform the task effectively.The process of acquiring expertise or specialised knowledge occurs through knowledge conversion from declarative form into procedural form. This conversion takes time and occurs through a variety of learning mechanisms such as composition and proceduralisation.

Many researchers have focused on the link between knowledge management and organisational learning According to Liao et. al. knowledge management is closely related to organisational innovation and learning. Human activity systems embody inherent incoherence such as ambiguities, uncertainties and contradictions. These have to be harmonised for knowledge to be shared and organisational learning facilitated. Therefore it is necessary to pay attention to contextual influences such as social, cultural and organisational factors, which affect knowledge sharing and organisational leaning. Lai et. al. [101] have concentrated on the link of information capital and organisational learning and how this may affect information behaviour and create information value.

In ADJUST the several types of knowledge are measured. Some indicative include cultural, organisational, process, team based, process, product, technology and innovation etc. These knowledge measures are included in job descriptions:

1. Market Knowledge, 2. Knowledge of Culture and Customs, 3.Technology Expert Knowledge 4.Process Knowledge, 5.Professional Knowledge (degrees), 6 Technology User Knowledge 7.Foreign Language Knowledge, 8 .Team Work Knowledge Product Knowledge 9. Management Discipline Knowledge, 10 .Specialist Equipment Knowledge

The ADJUST system provides similar graphs as in the case of soft factors.

## 6  From Soft Factors and Knowledge to Pay Ranges

A pay range is provided per job title (manually computed by the user) depending on the overall Soft Factors and Knowledge Level as well as other targets related to critical success factors, process achievements and *experience*. The user can provide pay ranges and corresponding salary values. For Example:

| Pay Range | Min | Max | Factors |
|-----------|-------|-------|---------|
| 1 | 10000 | 15000 | List of Measures for the Employee to belong to this pay range (in Knowledge, Soft factors and other areas) |
| 2 | 13000 | 17000 | List of Measures for the Employee to belong to this pay range (in Knowledge, Soft factors and other areas) |

The system provides analysis graphs like job-titles and pay ranges as well as job titles and average salaries (planned and actual) as well as actual salary and number of people that correspond to this salary.

## 7  Capturing Declarative Knowledge through Workflow Transactions

In Adjust we integrate job descriptions, process and workflow elements with the aim of capturing the declarative-explicit knowledge of Agents in the organisation we have included the following fields in job descriptions.

| Main Job Title | This field can either be the sender or the recipient, depending on the value of the *Sender* field |
| --- | --- |
| Activity | This field holds the task that the sender carries out |
| Transaction | This field holds the transaction name that specifies an object that flows between agents (humans or software agents) in the organisation. |
| Other Job Title | This field can either be the sender or the recipient, depending on the value of the *Sender* field |
| Sender | If this field is checked, then the *Main Job Title* is the sender in this job description. If not, then the sender is the *Other Job Title* |
| Work / Information / Approval | This field indicates the purpose of the job description |

A transaction is an object that is exchanged between agents in the organisation. Agents are mainly human but can also be software agents. At the system level details about transactions are inserted . The declaration in the record explains that JOB TITLE 2 receives TRANSACTION1 in the context of ACTIVITY1 from JOB TITLE 1). The "Sender" check button remains unchecked in order to declare that JOB TILE 2 is a receiver of this transaction.

By repeating the same steps in various job descriptions for the same transaction we can produce the "transaction trace" as shown in the following figure:

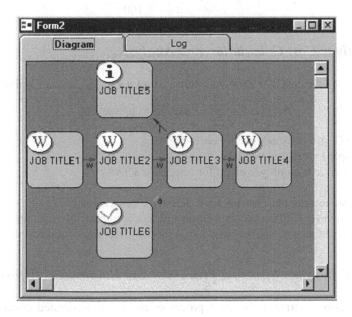

This transaction trace declares that there is a work-flow (rectangles marked with W) from job title (agent) 1 to 4. Additionally, job title 3 sends his result (object) for approval to job title 6 (rectangle marked with ✓) and when he receives the final approval he sends the final approved result to job title 5 for informational (rectangle marked with I) purposes.

A transaction trace is constructed for each transaction. The more approvals associated with the transaction the more control we impose on it. The less approval points the more flexible the transaction. Typical analysis results that will be produced by the system are the following: A similar kind of analysis will be performed at the level of a process, department and job title with more focus on the approval points. The calculation of the approval points per process or department will be performed on the basis of the job description where there is a declaration of approvals per transaction per activity.

## 8 Process Related Performance Measures and Links

In ADJUST only on the human resource related costs are measurement. The focus is on employee effort devoted to activities. Based on the effort per employee per activity the full time equivalents per activity is calculated. Which is the sum of employee's effort divided by 100.

$$\text{FTEs per Activity} = \frac{\Sigma(\text{EffperAct})}{100}$$

### 8.1 Activity Cost Calculation and Analysis

This analysis calculates the total activities cost for a specific level of activities.

$$ActCost = \sum ( EmpSal \times EffperAct )$$

Where ActCost = the total activity cost
        EmpSal = employee salary
        EffperAct= employee effort for a specific activity

The analysis returns the cost for each activity that belongs to a selected activity decomposition level. The decomposition level is automatically calculated during the creation of the analysis. The user can select any decomposition level.

### 8.2 Link between Strategy, Processes and Employees' Effort

The link between strategy, business processes and other associated concepts like technology fit, flexibility etc has been the focus for many researchers in recent years. Researchers argue that although business process management is a popular concept, it has not yet been properly theoretically grounded, lesding to problems in identifying critical success factors of BPM programs. Adjust proposes atheoretical framework with the utilization of three theories: contingency, dynamic capabilities and task–technology fit. The main premise is that the fit between the business environment and business processes is needed.

In adjust this identification is followed and an attempt to create a strong and sustained linkage between strategy and the way work is done is an enduring challenge in complex organizations. Business processes in BPR define how work is done and must be redesigned in a way that is consistent with organizational strategy.

In ADJUST a technique called Mission / non-Mission analysis is used, in order to analyse the extend to which agents' activities and organizational processes are secondary to the attainment of the organizations objectives or mission. The major benefit of this analysis is the ability to isolate and quantify non-missionary effort in selected work groups as well as the probable causes of it. An activity can be classified as non-mission according to the following criteria:

- The activity is organizationally misplaced.
- The activity is being duplicated because of a "do it by ourselves is better" attitude.
- The wrong skill level is being applied to the activity.

Mission-Non-Mission analysis is an extension to the critical success factors importance analysis.

The analysis on the above table is based on the degree of importance of each Activity and the level of achievement of the critical success factors via the activities. An Activity is considered as being very important (missionary) if it contributes to more than 50% of its total contribution potential. The total contribution potential is calculated by multiplying 10 (the highest grade in the scale) by the number of critical success factors. In the case we have 4 CSFs the contribution potential is 10 X 4 = 40. In this case, if the sum of the contribution in each CSF is more than 50 then the Activity is considered as very important.

The algorithm behind the analysis is as follows:

**For each Activity**
    Sum all CSFs values
    If Sum < Threshold then Activity is non-Mission
    If Sum > Threshold then Activity is Mission
**Loop**
**Threshold** = number of CSFs x highest CFS value x 0.5

The analysis will return results only if the user has associated activities with critical success factor values. Example reports produced by adjust are presented in the following diagram:

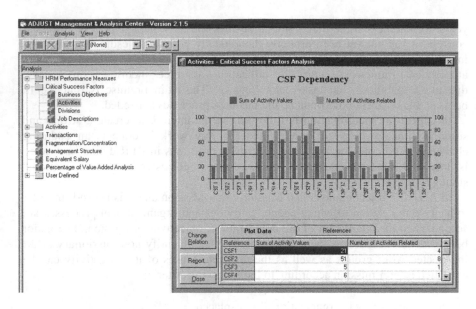

In a similar manner a CSF has a highest level of achievement if all Activities are associated with it and take a number of 10 in the scale. If the sum of the Activities is more than 50% of its achievement potential then the CSF is considered as multiply achieved.

## 8.3 Activity Value Added Analysis

In this analysis the focus is on the determination of each activity's (or set of activities) value to the "customers" and to the organization as a whole.

Activities can be classified into three primary types:

- **High value added activities.** These activities are critical to the achievement of existing organizational strategies and have high impact on product quality and customer satisfaction.
- **Low value added activities.** Activities, which provide low or negligible overall value from product, customer and organizational strategy perspective.
- **Business value added activities.** These activities are critical to the attainment of general business operations and usually support high value added activities.

The classification of each activity to each one of the above three categories is subjective and presented by the user in a scale from 1-10.

## 8.4  Fragmentation / Concentration Analysis

Fragmentation is the degree to which effort applied toward an activity is dispersed within a unit, a department, or the entire organization. It is expressed in terms of the number of employees it takes to generate one FTE (full time equivalent) and is calculated in the following ratio:

$$Frafmentation = \frac{\text{Number of Employees that have declared a \% of effort to the Activity}}{\text{FTE}}$$

the higher this ratio, the higher the degree of fragmentation.
  Where:

$$FullTimeEquivalent(FTE) = \sum \frac{effort\_of\_employees\_in\_activity}{100}$$

Concentration is the inverse of fragmentation. It is the average amount of time employees spend on a particular activity and is expressed in the following ratio:

$$Concetration = \frac{\text{FTE}}{\text{Number of Employees that have declared a \% of effort to the Activity}}$$

The lower this index, the lower the concentration or the higher the degree of fragmentation. This analysis calculates the fragmentation and concetration values of activity in a specific activity decomposition level. The calculation algorithm is shown below:

**For each Activity**
Full Time Equivalent (FTE) = Sum (effort of Employees in Activity)/100 = Full Time equivalent
  Concentration = FTE/Number of Employees that have declared a % of effort to the Activity
  Fragmentation = Number of Employees that have declared a % of effort to the Activity/ FTE
**Loop**
The analysis will return results only if the user has associated activities with employees and efforts.

## 8.5  Equivalent Salary Analysis

The purpose of equivalent salary analysis is to determine if work can be performed at a lower cost to the organization, either through internal re-allocation to less expensive employees or through outsourcing. It is expressed in the following ration:

$$Equivalent\_Salary = \frac{Total\_Activity\_Cost}{Activity\_FTEs}$$

(total activity cost divided by the total FTEs allocated to the activity) this kind of analysis is appropriate for:

- identifying trivial activities that are performed by highly qualified employees and vice versa
- evaluating whether compensation for activities, products or services is competitive with the marketplace, as in weighing a decision to outsource a particular product or service.

The analysis calculates the equivalent salary of a specific department for an activity decomposition level. The calculation algorithm is shown below:

**For each Activity**
  **For each employee that has declared % of effort to the Activity**
      Activity Total cost = (Salary x % of effort)
      Equivalent salary = Activity Total cost / FTE
  **Loop**
**Loop**

The analysis takes into consideration all the employee salaries that belong to a specific department as well as their declared effort for a specific activity decomposition level.

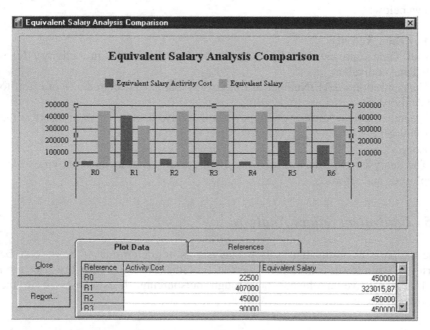

## 8.6 Process Cycle Time Analysis

This analysis identifies which steps in a business process are critical to the timely completion of the process. By mapping work step interdependencies the analysis identifies a (critical path), a set of work steps which, when shortened, will compress the process duration. This kind of analysis is often iterative.

When a work step is shortened, a resultant critical path may include a different set of work steps. This path is repeatedly refined as part of the overall redesign process. When the critical path in a process is identified the improvement of the efficiency of any individual work step in the critical path might result in big improvements on the efficiency of the process.

In ADJUST this kind of analysis is particularly useful when analysing a complex process that has multiple interrelated steps of sub-processes being executed concurrently.

# 9  Organizational Structure Related Processes

There are several key calculations that play a significant part in management structure analysis which evaluates structural efficiency and leveraging of management resources

## 9.1 Span of Control

This is how many employees a manager supervises excluding secretaries (or administrative support personnel).This analysis calculates the span of control in an organisational structure.

1.  For each manager find the manager/subordinate ratio. Each subordinate might be a manager.

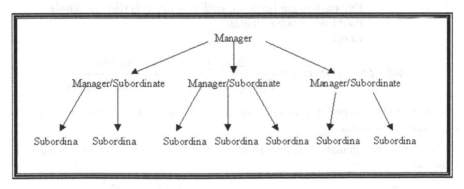

2.  Calculate the Average Ratio of manager/subordinate throughout the whole
    organisational structure
    This analysis calculates the span of control for a specific organizational
    structure (i.e.

$$AverageRatio = \frac{Sum(manager\,/\,subordinate\_ratios)}{Number\_of\_manager\,/\,subordinate\_occurances}$$

company, division, department, employee).

- Division span of control: The analysis calculates span of control for
  employees that belong to a specific division starting from the employee
  in that division that does not have any supervisors and working its way
  down to the lowest level of employees, which do not have subordinates.
- Company span of control: The analysis takes into consideration all the
  employees in the organisation starting from the employee that has no
  supervisors and working its way down to lowest level of the
  organisational structure (i.e. employees with no subordinates).
- Department span of control: The analysis calculates span of control for
  employees that belong to a specific department starting from the
  employee in that department that does not have any supervisors and
  working its way down to the lowest level of employees, which do not
  have subordinates.
- Employee span of control: The analysis calculates span of control for a
  specific employee starting from that employee and working its way down
  to the lowest level of employees which do not have subordinates.

## 9.2  Cost to Manage

This represents the manager's salary cost spread across all subordinate salary
costs.

**For each organisational entity or organisation as whole**
Find his/her subordinates
**Loop**

$$Cost\_to\_Manage = \frac{Salary\_of\_managers}{Sum(Salary\_of\_subordinates)}$$

Calculate the Average Ratio of manager/subordinate throughout the whole
organisational structure

This analysis calculates the cost to manage ratio for a specific organizational
structure

$$AverageRatio = \frac{Sum(Cost\_to\_Manage)}{Numbe\_of\_manager\,/\,subordinate\_occurances}$$

(i.e. company, division, department, employee).

## 9.3   *Layers of Management*

This represents the number of management layers in the organization. This analysis counts the depth of managers who report to other managers.

> For each organisational entity or organisation as a whole
> Repeat until there are no subordinates
> Find those who do not report to anybody
> Put them in Level 0
> For those who have been put in Level 0
> Find the subordinates and put them in Level 0+1
> Level 0 = Level 0 +1
> Loop
> Layers of management = number of Levels

This analysis calculates the layers of management ratio for a specific organizational structure (i.e. company, division, department, employee).

## 10   Conclusions

Existing performance measurement methodologies and tools have cost as the central topic while others the successful use of IT. Other means of business analysis techniques are mostly ignored.

ADJUST offers a state of the art analysis based on techniques from various disciplines that are applied in different perspectives. Organizational theories from both a holistic and an individualistic approach and existing work in process management have been the bases of this unique tool.

All analysis are based on existing applied theories.The analysis of the dimensions of transactions is influenced from the transaction costs economics theory. Mission / non-mission analysis and value added analysis from existing process management work.

One of the novelties of this system is the application of techniques like flexibility and dimensions of transactions. Other techniques like management structure analysis, concentration analysis, equivalent salary analysis and fractionalisation analyses are performed from a different viewpoint

Although management structure analysis is considered as unnecessary in process management, in ADJUST this technique is applied because it provides insight in the level of communication, coordination and control in the company before and after redesign.

The ability to analyse the resulting business model in a holistic way is compared compared to no other methodologies and tools.Users of Adjust have the tools to explore the relationship between organizational structure and processes, examine the alignment between business strategy, business processes and employee objectives. Furthermore they can analyse the organization from both the holistic and individualistic view and the level of effectiveness of activities and processes.The level of communication, coordination and control in existing structure and processes is examined and the impact of creating a market metaphor for organizational behaviour is analyzed. ADJUST has a broad range of applications for implementing sophisticated process controls which provide support at continuous improvement of processes, both within the company and across companies through the analysis of weak process points.It also offers a basis for benchmarking projects, both within the company and across companies, self - assessment by process participants in order to deside outsourcing.Early detection of weak points sets a basis for automated early-warning systems through collecting data for activity-based costing and simulation and calculating the efficiency and profitability of standard software implementations.

# Effects of Business Process Reengineering on Firm Performance: An Econometric Analysis

Yasin Ozcelik

Fairfield University
Dolan School of Business
Department of Information Systems & Operations Management
1073 North Benson Road, Fairfield, CT 06824, USA
yozcelik@fairfield.edu

**Abstract.** This chapter examines whether implementation of Business Process Reengineering (BPR) projects improve firm performance by analyzing a comprehensive data set on large firms in the United States. The performance measures utilized in the chapter are labor productivity, return on assets, and return on equity. We show that firm performance increases after the BPR projects are finalized, while it remains unaffected during execution. We also find that functionally focused BPR projects on average contribute more to performance than those with a broader cross-functional scope. This may be an indication that potential failure risk of BPR projects may increase beyond a certain level of scope.

**Keywords:** Business process reengineering, value and benefit, statistical analysis.

## 1 Introduction

Business Process Reengineering (BPR) is defined as a radical redesign of processes in order to gain significant improvements in cost, quality, and service. Firms have been reengineering various business functions for years, ranging from customer relationship management to order fulfilment, and from assembly lines to logistics. Anecdotal evidence suggests that many organizations gained benefits from BPR projects [30]. For instance, the CIGNA Corporation successfully completed a number of BPR projects and realized savings of $100 million by improving its customer service and reducing operating expenses [9]. Similarly, reengineering the accounts payable process at the Ford Motor Company increased the speed of payments and improved company relations with suppliers [17]. Arguably, some BPR projects fail to meet expectations. A survey conducted by the Arthur D. Little consulting firm found that 85% of executives surveyed were not satisfied with the outcome of their BPR projects [32]. Such poor outcomes may be attributed to several factors, including (i) expecting too much too soon [17], (ii) undertaking projects without a comprehensive cost-benefit analysis, (iii)

M. Glykas (Ed.): Business Process Management, SCI 444, pp. 99–110.
springerlink.com     © Springer-Verlag Berlin Heidelberg 2013

lack of expertise on redesigning a set of related activities [2], and (iv) lack of partnership between internal Information Technology (IT) department and other parts of firms [23].

BPR projects, by their nature, entail major changes in business processes that may lead to organizational instability and failure. Therefore, it is reasonable to expect BPR projects to have a significant and measurable effect on firm performance. In this chapter, we empirically investigate the performance effects of BPR projects both during and after the implementation periods using a new annual data set covering the period between 1984 and 2004. We utilize labor productivity, return on assets, and return on equity as firm-level performance variables. We use a panel-data regression model in order to take into account the cross-sectional and time-series nature of the data. We show that performance variables of firms remain unaffected during the implementation period of the BPR projects, which generally creates an initial turmoil in firm operations. The firm performance, however, significantly increases after the BPR projects are successfully completed. We also find that functionally focused BPR projects contribute more to performance than those with a broader cross-functional scope, suggesting that failure risk of BPR projects may increase beyond a certain scope.

The chapter is organized as follows. In the next section, we briefly survey previous studies on the topic, and then present our hypotheses in Section 3. We describe our data in Section 4 and regression variables in Section 5. We then describe our empirical methods in Section 6. Finally, we provide the regression results in Section 7 and conclude in Section 8.

## 2 Literature Review

The number of studies on the impact of BPR projects on firm performance is small but growing. The literature is rife with anecdotal evidence and short on empirical evidence of performance impacts of BPR projects. Most studies collectively suggest that there are substantial benefits for firms that successfully implement the structural changes associated with BPR projects [6]. Hunter et al. [19] and Murnane et al. [28] confirm this claim by analyzing data from the banking industry per se. Devaraj and Kohli [13] show that investments in IT can contribute to a higher level of revenue if they are supported by BPR initiatives. By studying the effect of three related innovations (IT, workplace reorganization, and new products and services) on demand for skilled labor, Bresnahan et al. [5] find that the demand for skilled labor is complementary with all the three innovations. Finally, Bertschek and Kaiser [4] find that workplace reorganization induces an increase in labor productivity that may be attributable to complementarities between IT and workplace reorganization.

The apparent lack of rigorous empirical research on the value of BPR projects indicates that there is a need to better measure BPR implementations through objective measures, and to relate them to organizational performance in the context of other variables that may also affect performance.

# 3 Hypotheses

BPR projects involve large investments in physical as well as human capital. The monetary costs of a BPR project include purchasing new equipment, hiring new personnel, and training employees to handle new roles. Indeed, organizations implementing BPR projects may need to increase their training budgets by 30 to 50 percent [1]. BPR projects may also have non-pecuniary costs due to problems encountered during implementation [16]. Such problems include (i) communications barriers between functional areas [24], (ii) lack of communication between top-level managers [25] as well as between BPR teams and other employees [14], (iii) resistance from employees [33], (iv) management reluctance to commit resources to BPR projects while expecting quick results [10], and (v) failing to address employee habits during implementation [15]. All of these factors suggest the following hypothesis:

**Hypothesis 1.** *Firms experience a drop in performance during BPR project implementation.*

Once BPR projects are finalized and implementation risks are resolved, employees are likely to become more comfortable with the new process design, and hence firms may be able to operate more efficiently. Thus, we expect firm performance to surpass its previous levels after the implementation, which leads to our second hypothesis:

**Hypothesis 2.** *Firm performance improves after the completion of BPR projects.*

A third issue of interest is the effect of project scope on firm performance. The scope of BPR projects vary; some projects focus on a single business function, such as order fulfilment or accounts payable, while others may be directed towards multiple functions. The scope of BPR projects may potentially affect the level of impact on firm performance. However, studies in the literature are far from providing consistent evidence on the direction of the impact. For example, Berry et al. [3] find that BPR projects with a large scope make the highest possible impact on firm performance. On the other hand, Dean [12] finds that the application of BPR across the entire firm may not produce as much benefit as a functionally-oriented project, such as switching to Just-in-Time (JIT) production system. In order to better investigate this issue empirically, we incorporate a scope variable into our analysis and suggest the following hypothesis:

**Hypothesis 3.** *The effect of BPR projects on firm performance increases with project scope.*

# 4 Description of Data

Our data covers BPR projects conducted by large U.S firms between 1985 and 2000. In order to avoid selection bias, we included in our sample all of the firms appearing in the 1998 edition of the Fortune 1000 list, regardless of whether they have implemented a BPR project. We used the ABI/INFORM and Lexis/Nexis online news resources to obtain press announcements for BPR projects, as well as the COMPUSTAT to extract our data. All of the monetary values in our data were inflation-adjusted by using the Consumer Price Index (CPI) values of the Federal Reserve Bank of Minneapolis, with the base year being chained at 1982-1984. We eliminated some of the firms due to missing data, reducing our sample size to 832 firms with a time span between 1984 and 2004. Of these firms, 93 have implemented a BPR project. We classified the BPR projects in to two groups with respect to project scope. Projects that likely affect a single business unit were classified as being functionally focused. Examples include reengineering of records management, sales force, and labor scheduling. Projects that potentially affect several departments were considered to have a cross-functional focus. Examples of such projects include restructuring and strategic rethinking of business for cost cutting or revenue growth purposes. Overall, 56 of the 93 projects in our data set were cross-functional and 37 were functionally oriented.

# 5 Description of Variables

## 5.1 Dependent and Independent Variables

We utilize labor productivity, Return on Assets (ROA), and Return on Equity (ROE) as dependent variables to measure firm performance from several dimensions. Table 1 below outlines the construction of these measures.

**Table 1** Construction of performance measures

| Performance Measure | Numerator | Denominator |
|---|---|---|
| Labor Productivity | Sales | Number of Employees |
| Return on Assets (ROA) | Income | Assets |
| Return on Equity (ROE) | Income | Equity |

Our key independent variables are the following. The implementation period is distinguished with a dummy that takes a value of one for all years during which a firm implements a BPR project, and zero otherwise. The post-BPR period is designated with another dummy that takes a value of one during all years after the implementation, and zero otherwise. These dummies allow us to take a longitudinal approach and are used for testing Hypotheses 1 and 2. Regarding Hypothesis 3, each of the above dummies is separated into two dummies, one for cross-functional BPR projects and the other for projects with a functional focus. For example, the dummy for the implementation period of cross-functional BPR projects is one for all years during which a firm implements the associated BPR project, and zero otherwise.

## 5.2 Control Variables

Our model includes four firm-level and two industry-level control variables. Past empirical studies have identified these controls as key determinants of firm performance [8, 36].

### 5.2.1 Firm-Level Controls

Our firm-level control variables are firm size, total IT budget, advertising expenditure, and market share. First, we use the natural logarithm of the number of employees as a proxy for firm size, as is standard in the literature. Second, in order to distinguish the effect of BPR projects on performance across firms with varying degrees of technical capability, we utilize the total IT budget as another firm-level control variable. Third, there is ample evidence in the literature supporting a positive relationship between advertising expenditure and firm performance [26, 29, 34]. Finally, market share is included as a control variable because both the efficiency theory [7, 11] and the market power theory [22, 35] provide evidence for a relationship between market share and firm performance. Market share can also serve as a proxy for other firm-specific assets not specifically captured in our study, such as managerial skills [20, 21].

### 5.2.2 Industry-Level Controls

The structure of an industry impacts the performance of firms within the industry [31]. We, therefore, utilize two variables frequently used in the literature to account for variation in firm performance due to idiosyncratic characteristics of different industries at the 2-digit SIC level: industry concentration and industry capital intensity. Consistent with the literature, industry concentration in our study is proxied by the four-firm concentration ratio, which is the total market share of the four largest firms in an industry. Industry capital intensity is included in the analysis to capture potential effects of entry barriers on firm performance. It is calculated as the sum of all capital expenditures divided by the sum of all sales in an industry.

### 5.2.3 Time Controls

We use separate dummy variables for each year to capture economy-wide shocks that may affect firm performance. The use of such dummies also helps us remove possible correlation between macroeconomic trends and firms' performances during the sample period.

## 6 Empirical Methods

We perform a panel data analysis to test our hypotheses as it accounts for both the time series and cross-sectional nature of our data. Panel data models have two estimation methods: fixed effects and random effects. The advantage of fixed effects estimation over random effects is that the former method allows the unobserved effect to correlate with the observed explanatory variables. The disadvantage is that fixed effects estimation produces less efficient estimators than random effects estimation can provide. The generally accepted way of choosing between fixed and random effects is running a Hausman test, and our regression results unanimously suggest using fixed effects estimation for all the regressions.

We use the logarithm of the numerator of each performance measure as a dependent variable, and the logarithm of its denominator as a control variable. This formulation relies on a property of the logarithm function, $\log(x/y) = \log(x) - \log(y)$. Such a specification has been used in past research as it provides flexibility in the relationship between the numerator and the denominator, while still retaining the interpretation as a performance measure [18]. Thus, the general form of the regression models used for testing Hypotheses 1 and 2 is:

$$\log (performance\ measure\ numerator)_{it} = intercept_i + \log (performance\ measure\ denominator)_{it}$$
$$+ implementation_{it} + post\text{-}implementation_{it}$$
$$+ firm\ controls_{it} + industry\ controls_{it} \qquad (1)$$
$$+ year\ dummies_t + \varepsilon_{it}$$

The implementation variable above is derived by interacting the dummy for the implementation period with firm size. This is equivalent to specifying that both the costs and benefits of BPR projects during the implementation period are proportional to firm size. We believe this is a more realistic specification than simply assuming identical costs and benefits across all firms, which would be the case if we were to include in the analysis the implementation period dummy per se. By the same token, the post-implementation variable is derived by interacting the dummy for the post-implementation period with firm size.

As for testing Hypothesis 3, we segment the implementation and post-implementation variables in to two types: functionally focused and cross-functional. Again, these variables are derived by interacting the associated dummies with firm size. Hence, the formulation of the fixed effects panel data models when investigating the effect of project scope becomes:

$$\log (performance\ measure\ numerator)_{it} = intercept_i + \log (performance\ measure\ denominator)_{it}$$
$$+ functional\ implementation_{it} + functional\ post\text{-}implementation_{it}$$
$$+ cross\text{-}functional\ implementation_{it} + cross\text{-}functional\ post\text{-}implementation_{it} \qquad (2)$$
$$+ firm\ controls_{it} + industry\ controls_{it} + year\ dummies_t + \varepsilon_{it}$$

# 7 Regression Results

The correlation matrix for all of our independent variables is presented in Table 2 below. All of the correlation entries among different variables in this table are significantly below 0.9, demonstrating that multicollinearity does not pose a serious problem to our analysis.

**Table 2** Correlation matrix for independent variables

| | Employees | Assets | Equity | IT budget | Advertising | Market share | Industry concentration | Industry cap. intensity |
|---|---|---|---|---|---|---|---|---|
| Employees | 1.000 | | | | | | | |
| Assets | 0.6040 *** | 1.000 | | | | | | |
| Equity | 0.6665 *** | 0.8950 *** | 1.000 | | | | | |
| IT budget | 0.7093 *** | 0.8617 *** | 0.8149 *** | 1.000 | | | | |
| Advertising | 0.7774 *** | 0.8463 *** | 0.8472 *** | 0.8179 *** | 1.000 | | | |
| Market share | 0.2426 *** | 0.2331 *** | 0.2232 *** | 0.2258 *** | 0.2163 *** | 1.000 | | |
| Industry concentration | -0.0649 *** | -0.2361 *** | -0.2288 *** | -0.1860 *** | -0.2244 *** | 0.4561 *** | 1.000 | |
| Industry cap. intensity | -0.0264 ** | -0.0198 * | 0.0676 *** | 0.0010 | -0.0215 ** | -0.0062 | -0.0166 * | 1.000 |

Note: *** $p<0.001$; ** $p<0.01$; * $p<0.05$.

The panel data regression results for the formulation described in Equation (1) above are presented in Table 3, where each column represents a different performance measure regression. According to Table 3, we do not find evidence supporting Hypothesis 1. None of the coefficients of the implementation variable is significantly different from zero. This may imply that potential negative impacts of BPR projects on firm performance variables during the implementation period are completely offset by their positive effects in the same period. Hypothesis 2 is uniformly supported as the coefficients of the post-implementation variable in Table 3 are significantly positive for all regressions. In other words, the firms in our sample have improved their performance in all three areas of interest after successfully implementing their BPR projects. Specifically, they have generated more sales and income per unit of input. In summary, we find a statistical association between improved firm performance and BPR projects during the post-implementation period without a significant drop in performance during the implementation period.

Recall that Hypothesis 3 is about the effect of project scope on firm performance. One would expect BPR-related benefits to increase with project scope, assuming that the risks of project implementation do not weigh in beyond a certain level of scope. The panel data regression results for the formulation described in Equation (2) are presented in Table 4.

**Table 3** Regression results for during and post-BPR implementation

| Dependent Variable | log (sales) | log (income) | log (income) |
|---|---|---|---|
| Interpretation | Labor productivity | Return on assets (ROA) | Return on equity (ROE) |
| During implementation | -0.00028 | -0.00105 | 0.00288 |
| Post-implementation | 0.00433 ** | 0.01038 ** | 0.01335 *** |
| log (employees) | 0.43507 *** | 0.15258 *** | 0.11808 *** |
| log (assets) | | 0.57193 *** | |
| log (equity) | | | 0.59733 *** |
| log (IT budget) | 0.17298 *** | 0.04928 *** | 0.05271 *** |
| log (advertising) | 0.07544 *** | 0.04186 *** | 0.02936 *** |
| Market share | 1.90082 *** | 0.09088 | 0.21720 |
| Industry concentration | -0.07500 ** | -0.22265 ** | -0.20055 ** |
| Industry capital intensity | 0.27603 ** | 0.77605 ** | 0.44557 * |
| $R^2$ | 0.8091 | 0.7044 | 0.7545 |

Note: *** $p<0.001$; ** $p<0.01$; * $p<0.05$.

**Table 4** Regression results for the effect of project scope

| Dependent Variable | log (sales) | log (income) | log (income) |
|---|---|---|---|
| Interpretation | Labor productivity | Return on assets (ROA) | Return on equity (ROE) |
| Functional during implementation | 0.00225 | 0.01600 | 0.01833 * |
| Functional post-implementation | 0.00596 ** | 0.02049 *** | 0.02209 *** |
| Cross-functional during implementation | -0.00173 | -0.01058 | -0.00588 |
| Cross-functional post-implementation | 0.00325 | 0.00399 | 0.00783 |
| log (employees) | 0.43500 *** | 0.15223 *** | 0.11758 *** |
| log (assets) | | 0.57113 *** | |
| log (equity) | | | 0.59664 *** |
| log (IT budget) | 0.17287 *** | 0.04881 *** | 0.05236 *** |
| log (advertising) | 0.07545 *** | 0.04208 *** | 0.02957 *** |
| Market share | 1.89565 *** | 0.05797 | 0.18845 |
| Industry concentration | -0.07531 ** | -0.22466 ** | -0.20338 ** |
| Industry capital intensity | 0.27894 ** | 0.79223 ** | 0.45955 |
| $R^2$ | 0.8091 | 0.7047 | 0.7548 |

Note: *** $p<0.001$; ** $p<0.01$; * $p<0.05$.

Similar to the results regarding Hypothesis 1, we do not find a significant association between functionally focused or cross-functional BPR projects and firm performance during the implementation period, with the exception that functionally focused BPR implementation is (weakly) associated with higher return on equity during the same period ($p$ value < 0.05). On the other hand, we do observe a statistically significant (and positive) relationship between functionally focused BPR implementation and all measures of firm performance after the implementation period. Compared to firms that have not engaged in BPR, firms that implemented a narrowly focused project have generated more sales and income per unit of input after the implementation. Interestingly, we do not find a statistically significant association between cross-functional BPR projects and firm performance after the implementation period. In fact, the coefficient for the cross-functional post-implementation variable is smaller than that for the functional post-implementation variable for all estimations. This may imply that firms implementing BPR projects with a larger focus make a higher level of investment in organizational capital and assets, which may not lead to comparable increases in sales and income after the implementation period. Another implication of this finding is that the failure risk of BPR projects may increase beyond a certain level of scope. In a sense, this result is parallel to the observations reported in a relevant case study [12].

The coefficients of our control variables in Tables 3 and 4 are in the expected direction. Firm size, IT budget, advertising expenditure, and market share are positively associated with all of our performance variables. Industry concentration is negatively associated with all of our performance variables. This implies that the average performance of a firm improves with the level of competitiveness in its industry, a finding that parallels those of [27]. Finally, we find a positive relationship between industry capital intensity and firm performance, which supports the view that incumbent firms in capital intensive industries could earn higher profits since they are likely to face fewer competitors [8].

# 8 Conclusion

This chapter contributes to the growing literature on the business value of BPR projects. We empirically investigated the effects of BPR projects on firm performance both during and after the implementation periods by considering a variety of measures, including labor productivity, return on assets, and return on equity. We utilized panel data regression models, and explicitly considered the scope of BPR projects in our empirical analysis. We used a comprehensive data set spanning the period between 1984 and 2004. We found that while overall performance of firms remains unaffected during the implementation of the BPR projects, it increases significantly after the implementation period. We also found that functionally focused BPR projects on average are associated more positively with firm performance than those with a cross-functional scope. This may indicate that potential failure risk of BPR projects may increase beyond a certain level of scope.

There are certain limitations to this study. First, our results capture the effect of BPR initiatives averaged over a wide variety of firms and their projects. Although we report a significant association between improved firm performance and BPR implementation at the functional level, it is conceivable that some of these projects could have actually failed. Hence, our results can only represent an average performance measurement across multiple firms and projects. Second, our observations are unavoidably limited to those BPR projects that are publicly announced. Therefore, we may have missed some of the projects that have not been announced, and consequently miscoded some companies as non-implementers when, in fact, they have undertaken a BPR project. Finally, our empirical results need to be interpreted as correlations rather than estimates of a causal model.

There are interesting avenues for future research on this subject. Arguably, effectiveness of BPR projects may not be uniform across all activities of a firm. Therefore, our model can be extended to analyze the effects of BPR at the strategic business units of firms, rather than at the organizational level. This would be possible by defining new performance measures for different business units and comparing the resulting differences across them. Such an analysis may provide more specific insights about the design and value of BPR initiatives to project managers.

# References

1. Al-Mashari, M., Zairi, M.: BPR implementation process: An analysis of key success and failure factors. Business Process Management Journal 5(1), 87–112 (1999)
2. Barua, A., Lee, B., Whinston, A.: The calculus of reengineering. Information Systems Research 7(4), 409–428 (1996)
3. Berry, D., Evans, G.N., Jones, R.M., Towill, D., The, B.P.R.: scope concept in leveraging improved supply chain performance. Business Process Management Journal 5(3), 254–275 (1999)
4. Bertschek, I., Kaiser, U.: Productivity effects of organizational change: Microeconometric evidence. Management Science 50(3), 394–404 (2004)
5. Bresnahan, T.F., Brynjolfsson, E., Hitt, L.: Information technology, workplace reorganization, and the demand for skilled labor: Firm-level evidence. Quarterly Journal of Economics 117(1), 339–376 (2002)
6. Brynjolfsson, E., Hitt, L.: Beyond computation: Information technology, organizational transformation and business performance. Journal of Economic Perspectives 14(4), 23–48 (2000)
7. Buzzell, R.D., Gale, B.T.: The PIMS principles: Linking strategy to performance. Free Press, New York (1987)
8. Capon, N., Farley, J.U., Hoenig, S.: Determinants of financial performance: A meta analysis. Management Science 36(10), 1143–1159 (1990)
9. Caron, J.R., Jarvenpaa, S.L., Stoddard, D.: Business reengineering at Cigna Corporation: Experiences and lessons learned from the first five years. MIS Quarterly 18, 233–250 (1994)
10. Cummings, J.: Reengineering is high on list, but little understood. Network World 27 (July 27, 1992)

11. Day, G.S., Montgomery, D.B.: Diagnosing the experience curve. Journal of Marketing 47(2), 44–58 (1983)
12. Dean, A.M.: Managing change initiatives: JIT delivers but BPR fails. Knowledge and Process Management 7(1), 11–19 (2000)
13. Devaraj, S., Kohli, R.: Information technology payoff in the health-care industry: A longitudinal study. Journal of Management Information Systems 16(4), 41–67 (2000)
14. Grover, V., Jeong, S.R., Kettinger, W.J., Teng, J.T.C.: The implementation of business process reengineering. Journal of Management Information Systems 12(1), 109–144 (1995)
15. Grover, V., Teng, J.T.C., Fiedler, K.D.: Information technology enabled business process redesign: An integrated planning framework. OMEGA 21(4), 433–447 (1993)
16. Guimaraes, T.: Field testing of the proposed predictors of BPR success in manufacturing firms. Journal of Manufacturing Systems 18(1), 53–65 (1999)
17. Hammer, M., Champy, J.: Reengineering the corporation: A Manifesto for business revolution. Harper Business Press, New York (1993)
18. Hitt, L.M., Wu, D.J., Zhou, X.: Investment in enterprise resource planning: Business impact and productivity measures. Journal of Management Information Systems 19(1), 71–98 (2002)
19. Hunter, L.W., Bernhardt, A., Hughes, K.L., Skuratowicz, E.: It's not just the ATMs: Firm strategies, work restructuring and workers' earnings in retail banking. Working Paper. University of Pennsylvania (2000)
20. Jacobson, R.: Unobservable effects and business performance. Marketing Science 9(1), 74–85 (1990)
21. Jacobson, R., Aaker, D.A.: Is market share all that it's cracked up to be? Journal of Marketing 49(4), 11–22 (1985)
22. Martin, S.: Market power and/or efficiency? Review of Economics and Statistics 70(2), 331–335 (1988)
23. Martinez, E.V.: Successful reengineering demands IS/business partnerships. Sloan Management Review 36(4), 51–60 (1995)
24. McKee, D.: An organizational learning approach to product innovation. Journal of Product Innovation Management 9(3), 232–245 (1992)
25. McPartlin, J.P.: Seeing eye to eye on reengineering. Information Week 74 (June 15, 1992)
26. Megna, P., Mueller, D.: Profit rates and intangible capital. Review of Economics and Statistics 73(4), 632–642 (1991)
27. Melville, N., Gurbaxani, V., Kraemer, K.: The productivity impact of information technology across competitive regimes: The role of industry concentration and dynamism. Decision Support Systems 43(1), 229–242 (2007)
28. Murnane, R.J., Levy, F., Autor, D.: Technological change, computers and skill demands: Evidence from the back office operations of a large bank. In: NBER Economic Research Labor Workshop (1999)
29. Nelson, P.B.: Advertising as information. Journal of Political Economy 82(4), 729–754 (1974)
30. Ozcelik, Y.: IT-enabled reengineering: Productivity impacts. Encyclopedia of Information Communication Technology 2, 498–502 (2008)
31. Porter, M.: Competitive strategy: Techniques for analyzing industries and competitors. Free Press, New York (1980)
32. Rock, D., Yu, D.: Improving business process reengineering. AI Expert 26(10), 27–34 (1994)

33. Ryan, H.W.: Managing change. Information Systems Management 9(3), 60–62 (1992)
34. Schmalensee, R.: A model of advertising and product quality. Journal of Political Economy 86(3), 485–503 (1978)
35. Smirlock, M., Gilligan, T., Marshall, W.: Tobin's q and the structure-performance relationship. American Economic Review 80(3), 618–623 (1984)
36. Syzmanski, D.M., Bharadwaj, S.G., Varadarajan, P.: An analysis of the market share-profitability relationship. Journal of Marketing 57(3), 1–18 (1993)

# Privacy-Aware Workflow Management

Bandar Alhaqbani, Michael Adams, Colin J. Fidge,
and Arthur H.M. ter Hofstede

Queensland University of Technology, Brisbane, Australia
b.alhaqbani@isi.qut.edu.au,
{mj.adams,c.fidge,a.terhofstede}@qut.edu.au

**Abstract.** Information security policies play an important role in achieving information security. Confidentiality, Integrity, and Availability are classic information security goals attained by enforcing appropriate security policies. Workflow Management Systems (WfMSs) also benefit from inclusion of these policies to maintain the security of business-critical data. However, in typical WfMSs these policies are designed to enforce the organisation's security requirements but do not consider those of other stakeholders. Privacy is an important security requirement that concerns the subject of data held by an organisation. WfMSs often process sensitive data about individuals and institutions who demand that their data is properly protected, but WfMSs fail to recognise and enforce privacy policies. In this paper, we illustrate existing WfMS privacy weaknesses and introduce WfMS extensions required to enforce data privacy. We have implemented these extensions in the YAWL system and present a case scenario to demonstrate how it can enforce a subject's privacy policy.

**Keywords:** Workflow Management Systems, Privacy, Authorisation, YAWL.

## 1 Introduction

Information security rests on confidentiality, integrity, and availability. These 'hard' security requirements are considered carefully in information system design and implementation. However, 'soft' security requirements, e.g. privacy and trust, are often neglected. In most cases, information security requirements are set to satisfy an organisation's security policy, but not those of other stakeholders, most notably the individuals and institutions who are the subjects of the data held by an organisation.

Privacy is a crucial security requirement that concerns users participating in e-business processes. In response to this concern, governments have set privacy laws, e.g. Australia's Privacy Act. According to Alan Westin [26],

M. Glykas (Ed.): Business Process Management, SCI 444, pp. 111–128.
springerlink.com

"Privacy is the claim of individuals, groups, or institutions to determine for themselves when, how, and to what extent information about them is communicated to others". However, there is a popular misconception that data confidentiality processes also satisfy data privacy requirements. In fact, traditional data confidentiality mechanisms aim to give the *owner* of data control over its accessibility, whereas privacy means giving the *subject* of data control over who accesses it.

In the information systems arena, Workflow Management Systems (WfMSs) are used to run day-to-day applications in numerous domains. A workflow separates the various activities of a given organisational process into a set of well-defined tasks. The tasks are executed according to the *organisation*'s policies to achieve certain objectives. Among these policies, security policies are crucial for ensuring that the organisation adheres to its own security objectives. However, many workflows deal with different types of data that originate from various sources. Once the data is retrieved for a particular workflow case, the organisation, through its WfMS, is responsible for maintaining data confidentiality as per the organisation's confidentiality policy. Well-crafted workflow access control mechanisms help the organisation to achieve such security objectives by assigning tasks' execution to authorised (human) resources only.

However, the workflow system might also hold descriptive data about a user which, in the user's opinion, would cause a privacy violation if it was accessed by certain workflow authorised resources. In order to execute the workflow case securely, and satisfy the user's privacy wishes, we need to consider the user's privacy policy, which states the user's access authorisation, in the workflow access control mechanism. Current WfMS structures do not provide a way to capture a user's privacy wishes because they fail to recognise and respect the wishes of the workflow's *subject*.

The subject's privacy policy impacts workflow execution in two ways. On the one hand, it affects the resource allocation process. Usually, this process is implemented according to the organisation's rules. However, the subject's privacy policy acts as a filter that should exclude workers not authorised by the subject from the workflow's allocatable resources. On the other hand, it affects data presentation when rendering data forms. Sensitive data should not be revealed to users not authorised by the data's subject.

In this paper, we introduce the notion of *subject* and its implications into a workflow system's security state, especially the *privacy* filtering aspect. This is demonstrated with three distinct examples. In addition, we present a conceptual data model which introduces the *subject* notion into the workflow authorisation model. In order to validate this model, we have implemented it in the YAWL environment [23], producing a novel secure work-resource allocation strategy with auxiliary data properties which are used to control access to private data. Finally, we use a healthcare case scenario to demonstrate the effectiveness of our implemented approach.

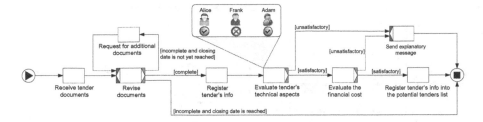

**Fig. 1** Tender evaluation workflow model

## 2  Motivating Examples

Workflow subjects and their privacy requirements have not gained sufficient attention when designing and executing workflow models. In this section, we use the YAWL notation to present three distinct examples which highlight the privacy and conflict-of-interest implications that result from neglecting the subject of a workflow. From these examples, we derive the workflow extensions that are described in more detail in Sect. 3.

### 2.1  Avoiding Conflict of Interest: Contract Tender Evaluations

In business, contract tenders are evaluated in several steps as illustrated in Fig. 1. The process starts by receiving the tenderer's documents and putting them through technical and financial evaluations. These tasks are allocated to the organisation's available resources to carry out as per the organisation's security authorisation policy.

However, we can identify a security threat in this example that results from not considering the subject of the workflow in the authorisation process. Let's assume that the ACME company has submitted a tender document to a government agency. As per the agency's authorisation policy, either of *Alice*, *Frank*, or *Adam* can perform the technical evaluation for any submitted tender. Let's further assume that *Frank* is a shareholder in ACME and is allocated the technical evaluation task for ACME's tender. This allocation creates a *conflict-of-interest* which might compromise *Frank*'s actions in a way that does not serve the organisation's best interests.

This problem occurs because the organisation cannot express an authorisation constraint that excludes those human resources that are in a conflict-of-interest with the tendering company. Instead, the company's identity should be used as the workflow system's 'subject' so that the organisation can easily create an authorisation constraint to protect against any conflicts of interest.

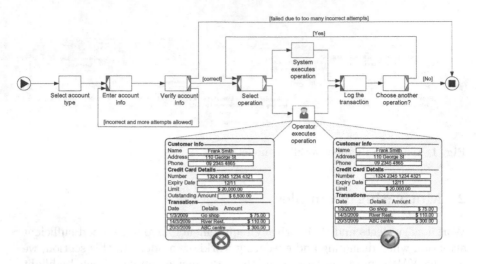

**Fig. 2** Phone banking workflow model

## 2.2   Hiding Personal Data: Phone Banking

In the banking sector, phone banking is a useful service that provides substantial benefits. Fig. 2 illustrates a phone banking workflow model that receives and processes customer requests. Requests are processed automatically by the system or manually by an operator. In this particular case, the customer has no control over what can be seen of his bank account's data by the workflow-authorised operator. This is due to data access control being managed by the bank's security policy without consideration of the customer's privacy wishes.

For instance, a customer's privacy desire to *hide* his credit card balance from a bank operator cannot be enforced in current WfMSs. However, extending the WfMS to recognise the subject of the workflow would allow a WfMS to retrieve and enforce the customer's privacy policy. This could be achieved by concealing the customer's private credit card balance from forms visible to unauthorised resources, from the customer perspective. In Fig. 2 different information is revealed to the operator depending on the customer's privacy settings.

## 2.3   Generalising Data: Social Networking

Social networks hold collections of data with different privacy levels. The workflow model example in Fig. 3 aims to produce a 'friends' album from the Facebook network. In this process, the user selects friends that he wants to include in his album and then selects the information that he wants to

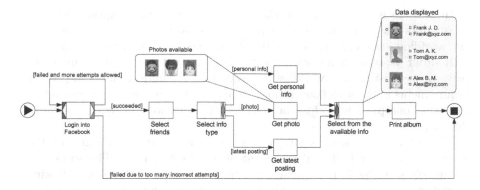

**Fig. 3** 'Create friends album' workflow model

retrieve about them. Before producing the album, the user examines the retrieved information and selects the information that he wants to save.

In a current WfMS, this process would be executed by retrieving the user's friends', i.e. the workflow's subjects, information by presenting them to the user without considering the subjects' privacy filters. Instead, the WfMS should consider the workflow subjects' privacy wishes when executing this process. For instance, if *Tom* does not want anyone else to use his photo in any of their pages, the workflow should understand and enforce *Tom*'s privacy desire by, e.g. *generalising the data* made available to John, in this instance by substituting *Tom*'s photo with a generic male image as shown in Fig. 3.

## 3   Workflow Implications

The examples in Sect. 2 illustrated the inability of current WfMS authorisation policies to preserve a workflow subject's privacy and avoid conflicts of interest. In this section, we introduce four technical requirements needed to overcome these security problems and to enhance workflow authorisation constraints. Also, we present a practical approach for adding these technical requirements to a workflow management system.

### 3.1   Adding the Subject to Workflow Designs

The workflow design phase defines a workflow specification required to achieve a certain objective. A specification comprises tasks, resources, control flow, and work allocation strategies, including a workflow authorisation policy that consists of authorisation constraints defined by the workflow administrator (e.g. separation of duties).

However, such a workflow authorisation policy cannot capture security constraints related to how a user's or an organisation's data is employed by the workflow case, e.g. the bank customer in the phone banking workflow.

In order to strengthen workflow authorisation to protect against threats to privacy, we must use the workflow subject's relevant information and privacy requirements in the workflow authorisation process. Therefore, we must introduce the concept of *subject* to the workflow specification during the workflow design phase.

We define a workflow's subject as an entity that owns some of the workflow's data or is described and identified by this data. The subject is a uniquely identifiable individual, e.g. a bank customer in the phone banking workflow, or an institution, e.g. the tendering company in the tender evaluation workflow. Also, it can be a single entity, e.g. a bank customer, or several, e.g. all of the user's friends in the 'create friends album' workflow model. In the workflow design phase, we can then create the subject's related authorisation constraints by using the workflow's subject reference which identifies the entity that a particular data item describes. This reference can then be used by the workflow engine while executing a workflow case to retrieve the subject's relevant information, e.g. the subject's privacy policy, and employ this in its access authorisation mechanism.

## 3.2 Auxiliary Data Properties for Privacy Requirements

In Workflow Management Systems, the workflow data perspective is developed during the design phase where it describes the data that will be manipulated by the workflow case. In order to include privacy properties, we need to capture their definition in the same data perspective. That is, we need a way to introduce data properties in an ad-hoc fashion without editing the primary workflow data.

To do this, we use auxiliary data properties that are metadata descriptors (attribute-value pairs) associated with workflow data elements. Each workflow data element may have a number of auxiliary properties that at runtime may influence certain actions, or change the presentation of data when it is rendered. We can use this data at any stage of the workflow design phase and link it to its associated workflow data. This predefined data is used for various functions while executing a workflow case, e.g. to define the way messages are displayed or forms are rendered.

For privacy purposes, we can use such auxiliary data properties in our access control mechanism so that, based on their values, they may influence the workflow engine to protect private data. The privacy-related auxiliary data property values are set by considering the privacy rules in the subject's privacy policy and the resource(s) that will execute a task. To preserve a subject's privacy we need two auxiliary data properties:

1. **The *hide* property**   is an auxiliary data property that is assigned for each data element. It serves to direct the form rendering engine to hide the existence of a private field and is set dynamically according to the subject's privacy policy. This is accomplished by not showing the private data field at all when rendering the form that results from executing a workflow task. Hiding the credit card balance field in Sect. 2.2 is an example of this.
2. **The *generalise* property**   functions similarly except that it instructs the workflow form engine to display the field but generalise the data in such a way that the observer's knowledge of the private data is minimised. Substituting a generic male image in Sect. 2.3 is an example.

## 3.3   Privacy-Preserving Work Allocation

Several work allocation patterns have been introduced to accommodate work-resource allocation requirements in workflows [19]. However, these work allocation strategies did not consider the subject of the workflow and thus do not take into account privacy requirements and potential conflicts-of-interest. As illustrated earlier, the subject's privacy requirements direct the work-resource allocation process to allocate work items to non-restricted resources (from the subject's perspective). Let's recall the phone banking example from Sect. 2.2 and assume that *John* has called the system and there are two employees who can take *John*'s request, *Tom* and *Matt*. *Tom* is restricted by *John* from accessing *John*'s credit card balance whereas *Matt* is not. In this case, the WfMS should consider *John*'s privacy policy and preferentially offer the task to *Matt*.

However, in some cases assigning the task to a non-restricted resource cannot be achieved. For example, let's assume that *John* also restricts *Matt* from accessing his credit card balance. As a result, *Tom* and *Matt* are both restricted by *John*. To solve this problem, the workflow management should allocate tasks to the resources that have the lowest restriction level.

## 3.4   Data Patterns for Private Information

Several workflow data patterns were introduced by Russell et al. [18]. Among these patterns, both workflow data pull and push patterns are required to enhance workflow privacy awareness. Usually the subject's data, which includes his privacy policy, would be stored in an external database. In order for the workflow system to obey the subject's privacy policy, it needs to retrieve this data from the external database. The workflow data pull pattern is defined as the ability of a workflow to request data elements from resources (e.g. external databases) or services in the operational environment. This pattern can make the necessary connection between the WfMS and the subject's privacy rules and thus enhances the WfMS's privacy awareness.

In addition, the workflow data push pattern allows the workflow to initiate the passing of data elements to a resource (e.g. an external database) or service in the operational environment. This pattern thus allows the workflow to update the external database and to include any new information, for instance, to alert subjects to attempts to access their private data.

## 4  Objects and Roles for Privacy-Aware Workflow Management

To capture these requirements, we developed a conceptual Object Role (OR) model [13] that addresses the meta data requirements for subjects and their privacy policies for use by Workflow Management Systems. For clarity, we partitioned the conceptual model into five parts, where each part concerns a specific concept.

Fig. 4 depicts the OR model for our resource concept in a WfMS. Resources come in different types, e.g. `subject` and `employee`. A user can be either a subject, e.g. a bank customer, an employee, e.g. a bank teller, or both. In the organisational structure, an employee occupies one or more job positions that are uniquely identified by an ID. For administrative supervision purposes, the holder of a job position may report to a higher administrative job position. Each job position belongs to a unique organisational unit that is identified by an ID. Similar to job supervision, each organisational unit administratively may belong to another organisational unit. Within an organisation, each workflow process specification must be owned by an organisational unit. A workflow process specification might be about one or more subjects. An employee may possess some capabilities that can be used by the WfMS to determine suitable resources to execute a task. In some business cases, there is a requirement to build a team that is responsible for handling a specific task. The team members might have additional access privileges that are allowed by the team's security.

The role-task concept in our model, shown in Fig. 5, uses some of the entity types from the resource model. These are shaded to indicate that these entity types are external and have been retrieved from other models. Fig. 5 shows the relation between tasks and roles, and role assignments. One or more roles can be assigned to a job position, and also a job position can be assigned to many roles. In addition, a resource can be assigned directly to an additional role that is not part of the resource's job position's roles. In order to comply with the role inheritance feature that is needed for Role-Based Access Control [22], in our model a role may belong to one or more parent roles.

Delegation of authority is a useful feature that allows a user to temporarily transfer his privileges to another user to carry out a specific task on his behalf.

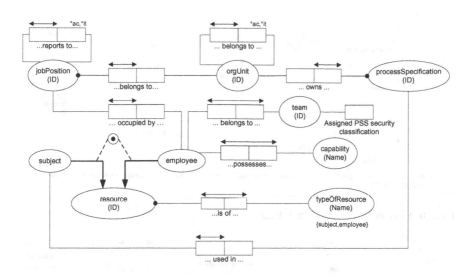

**Fig. 4** Conceptual model - resources

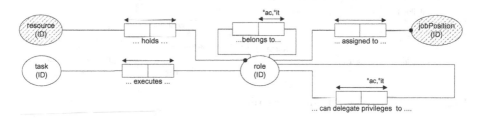

**Fig. 5** Conceptual model - roles and tasks

In our model, we allow for this feature by specifying which role can be delegated and to which roles this delegation should be given. Each role can be used in more than one task and a task can be executed by more than one role.

In Fig. 6 we show the OR model for the privilege concept. A privilege is identified by a unique privilege identifier and has a unique combination of an action and a data record which implies that the permitted action is allowed on the data record. The privilege is assigned to a resource by either linking it to a task or through a referral process. In the referral process, an employee (delegator) refers a subject's case to another employee (delegatee) but with certain privileges. In order to complete this referral successfully, the delegator's role must be allowed to delegate to the delegatee's role, which can be checked by looking into the role delegation relation in Fig. 5.

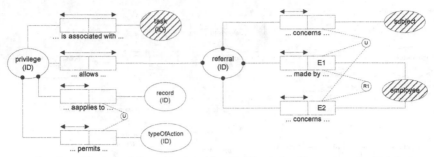

R1 **If** employee E1's role permits him to delegate a privilege P to E2's role, **then** E1 can make a referral that allows privilege P to E2.

**Fig. 6** Conceptual model - privileges

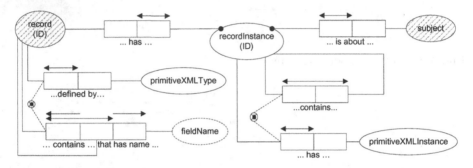

**Fig. 7** Conceptual model - data structures

The data concept of our model is captured by the OR model in Fig. 7. Each record in the OR model is identified by a unique ID. Each record is either a parent of other records or a child. If it is a parent, we capture the record ID and its children's IDs and names. Each child is a record that has a name and unique ID. If a record does not have a child, we capture the record's primitive XML type. With regard to the data value part, we use a record instance to capture the data value characteristics. Each record instance has a unique ID and must relate to a record and be owned by a subject. The record instance is either a parent of other record instance(s) or a leaf (i.e. childless). The data value is captured by the leaf record instance that corresponds to a leaf record.

Fig. 8 represents the last part of our conceptual OR model. It shows the authorisation structure that captures the subject's privacy requirements. The subject's privacy policy consists of access authorisations that are modelled by an entity access policy. Each access policy has a unique ID and must be set by a subject to authorise or restrict the capabilities of certain employees. The access policy can be applied either on the record level, which affects its instances that are owned by the subject, or on a particular record instance.

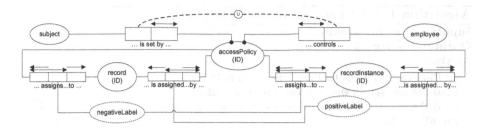

**Fig. 8** Conceptual model - authorisation

We use positive and negative authorisation approaches to express the subject's required authorisation [8]. In positive authorisation, we use a positive label to flag a certain record or record instance and assign it to an employee. As a result, any employee who has the required positive label of a record or a record instance can access its data, otherwise the employee is disallowed. In negative authorisation, we use negative labels instead of positive labels to restrict certain employees from accessing a certain record or record instance.

## 5   A Privacy-Aware Workflow Engine

To implement all of the aforementioned concepts we extended the YAWL environment [23]. Our implementation began by converting the conceptual schema above to a relational schema. The resulting privacy database tables were then created in the PostgreSQL server. YAWL's Java work resource allocation framework allowed us to implement a new Java work allocator class that performs the secure allocation strategy.

This class interacts with YAWL's database and our privacy database to retrieve information according to a privacy-aware work allocation process, Algorithm 1. This algorithm evaluates a restriction weight for each potential participant who can execute a given task that pertains to a particular workflow subject, to be used in deciding which resource should be given a task.

To implement our auxiliary data properties (hide and generalise), we took advantage of YAWL's form rendering framework and implemented a new Java class to preprocess the form generated by YAWL's form engine. This class uses the subject ID and participant ID to determine those fields that the resource processing the task is not authorised to access, by getting information from our privacy database. The restricted field's attributes are set accordingly to either hide or generalise sensitive data. When the form is returned to the form rendering engine, it can either totally hide the existence of the restricted field or replace the field's content with generic text.

**Algorithm 1.** Least-restricted resource allocation

**Input:** `subjectId, taskId`
**Output:** `resourceId`
**Method:**

{:: Find the resources that can execute `taskId` ::}
Find the role set `RO` that can execute task `taskId`
Find the resource set `RE` that can play any role `r ∈ RO`
**for all** `s ∈ RE` **do**
    `s.Weight ← 0`
**end for**
{:: Calculate the restriction weight at the record level ::}
Find the record set `RC` that is accessed by `taskId`
`SR ←` number of positive labels that are set in `RC` by the `subjectId`
**for all** `s ∈ RE` **do**
    `SPR ←` number of positive labels that are set by `subjectId` for `s` in `RC`
    `SNR ←` negative labels that are set by `subjectId` for `s` in `RC`
    `s.Weight ← (SR − SPR) + SNR`
**end for**
{:: Calculate the restriction weight at the record instance level ::}
Find the instance set `IN` that are accessed by `taskId`
`SI ←` number of positive labels that are set in `IN` by the `subjectId`
**for all** `s ∈ RE` **do**
    `SIP ←` number of positive labels that are set by `subjectId` for `s` in `IN`
    `SIN ←` number of negative labels that are set by `subjectId` for `s` in `IN`
    `s.Weight ← s.Weight + (SI − SIP) + SIN`
**end for**
{:: Find the least restricted resource ::}
`resourceId ← s`, where `∀x ∈ RE • s.Weight ≤ x.Weight`

# 6 Healthcare Case Scenario

There is growing privacy concern in the healthcare domain over the use
of patients' Electronic Health Records that is obstructing their introduction [12, 16]. To overcome this concern, patients should be given a certain degree of control over non-safety-critical private data within their EHRs [5, 20].
Here we use a healthcare case scenario to demonstrate how our extended
workflow engine can satisfy a patient's privacy desires in a healthcare workflow system by controlling its work assignments and manipulating data presentation. For the scenario, we consider a patient's visit to a hospital's
emergency department. We modelled the emergency room process using the
YAWL editor (Fig. 9), defining the tasks and their associated roles. In this
model, data are retrieved from the patient's Electronic Health Record (EHR)
that resides in an external database to be used in this process.

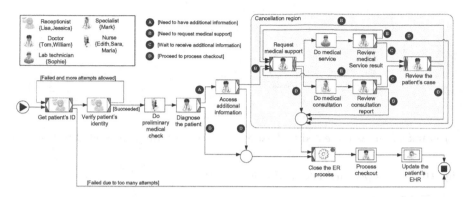

**Fig. 9** The hospital's emergency process model

In order to execute the emergency room workflow model, we populated our privacy database with data samples. We loaded several patients' EHR data into our database, and set our database tables to reflect the hospital's emergency room employees list and their roles (top left of Fig. 9). As an example we assumed that a patient Frank has expressed his privacy wishes by setting his access control policies to deny access to his birth date using negative authorisation labels for Jessica, Edith and Sara, and has granted access to his Chlamydia diagnosis only to William by assigning him a positive authorisation label.

Now assume that Frank has appeared at the reception desk requiring treatment. Since there is no prior information about the workflow's subject, the task *get patient's ID* is assigned randomly to one of the receptionists. If the task is executed by Jessica then she will enter Frank's ID and then execute the *verify patient's identity* task. The form that is rendered by our privacy-aware workflow engine to Jessica is shown in Fig. 10(a). The `DateOfBirth` field is entirely hidden by setting the hide auxiliary data property, because Frank disallowed Jessica from seeing this field. By contrast, if the two tasks were assigned to Lisa, the form will show Frank's `DateOfBirth` field as depicted in Fig. 10(b) because Frank did not set any restrictions for Lisa.

Once the receptionist has verified Frank's identity, the workflow engine will allocate the task *do preliminary medical check* to an appropriate nurse. YAWL's workflow engine uses our privacy-aware work allocation strategy (Algorithm 1) to choose a nurse. In Frank's case, the workflow engine allocates the task to Maria because Frank has not placed any privacy restriction on her, unlike the other nurses Sara and Edith.

The same allocation strategy is used by the workflow engine to determine a suitable doctor to execute the diagnosis task. This task is allocated by the workflow engine to William because his privacy restriction weight is lower

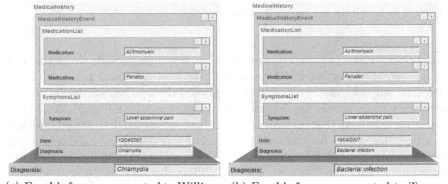

(a) Frank's form as presented to Jessica    (b) Frank's form as presented to Lisa

**Fig. 10** Frank's personal information form

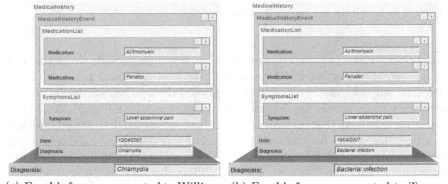

(a) Frank's form as presented to William    (b) Frank's form as presented to Tom

**Fig. 11** Frank's medical history form

than other doctors such as Tom. William, as per Frank's privacy policy, is able to access Frank's recorded diagnosis of Chlamydia whereas Tom cannot. The form that is generated to William is depicted in Fig. 11(a). However, now let's assume that William is busy and cannot take Frank's case. In this situation, the task will be reallocated to Tom assuming there are no other doctors available. However, the workflow engine knows that Tom is not authorised to know about Frank's Chlamydia diagnosis. Therefore, the workflow engine renders the form so that the Chlamydia diagnosis is replaced by the generic term `Bacterial infection` so that Tom does not know the specifics of Frank's sensitive diagnosis as shown in the form produced by YAWL in Fig. 11(b).

When generalising medical data to protect patient privacy we must, of course, still provide as much information as is permissable to medical staff. This can be done, for instance, by choosing the most specific generalisation from the hierarchical *International Statistical Classification of Diseases* [21].

# 7   Related Work

Authorisation is an important workflow security requirement [7]. It refers to enforcing access control to ensure that only authorised resources are allowed to execute a workflow task. Sandhu introduced a Role-Based Access Control (RBAC) model [22] that breaks the traditional authorisation link between subjects and permissions and inserts a role notion in the middle to ease the authorisation management process. However, the RBAC model is role-centric and does not consider the task notion in WfMSs. Conceptual foundations for Task-Based authorisations are presented by Thomas and Sandhu [24] where privileges for assignment and revocation are discussed in order to provide active access control enforcement. Oh and Park [17] introduced a Task-Role-Based Access Control (T-RBAC) model that is built on top of the RBAC model. A task notion is inserted between roles and permissions, allowing task execution to be assigned to role(s). This development results in better authorisation management from a workflow perspective. In order to have further access control, authorisation constraints are introduced as additional filters to be applied on subject-role, role-task, and subject-task relations [9]. However, the T-RBAC model and the authorisation constraints that are introduced do not consider the subject of the workflow in the authorisation policy, hence they fail to address privacy requirements.

The T-RBAC model can be extended further by adding a new organisational element 'the functional level' and by using event-condition-action rules to present an authorisation constraint [11]. The event part denotes when an authorisation may need to be modified. The condition part verifies that the event actually requires modifications of authorisations, and determines the involved agents, roles, tasks and processes. However, this authorisation model fails to address the subject's privacy due to not considering the subject's authorisation policy in its condition part. Similarly, access control models that are developed to satisfy the workflow separation-of-duties security requirement [15] fail to obey the workflow subject's privacy rules.

Research has been done on introducing new security constraints to guard workflow executions [29]. However, none has recognised the workflow subject's privacy requirements and therefore neither privacy-constraints nor secure work-resource allocations have been introduced. Cao et al. [10] addressed the importance of human involvement in workflow applications and noted that poor design of work-resource assignment strategies is one of the critical issues in workflow projects. They introduced four authorisation models, they are staff-authorisation, role-authorisation, team-authorisation, and department-authorisation. These are used to provide a dynamic task authorisation policy that is expressed by a task authorisation policy language. However, this policy does not consider the workflow subject's privacy concerns because they discuss only authorisation requirements.

Wolter et al. [27] argue that current process modelling standards are incapable of capturing security goals such as confidentiality, integrity, or dynamic authorisation. Therefore, they proposed a security policy model that contains a set of security constraint models. In the authorisation constraint model, permissions are inserted between a subject (a resource in our case) to a target (a task) but they did not introduce an owner (a subject) for the tasks that are used in the process model. As a result, these authorisation models do not satisfy the privacy requirement for the subject of the workflow.

Xu et al. [28] proposed algorithms to optimise resource allocation in order to execute a business process within time and cost constraints. They take into account the structural characteristics of a business process such as task dependencies. However, security constraints have not been considered in their optimised work allocation strategy, so their work fail to address not only the privacy requirements but security constraints in general.

Open source workflow management systems such as jBPM [2], ruote (OpenWFEru) [3], and Enhydra Shark [1]; and commercial WfMSs such as IBM WebSphere [14], TIBCO [4], and FLOWer [25] similarly do not capture the subject of the workflow process in their workflow specifications which means that their workflow engine is incapable of retrieving the subject's privacy rules in order to integrate them into their work-resource allocation models.

Privacy policy languages, e.g. XACML and EPAL, are used to translate access control requirements into processable XML files [6]. The languages can implement several access control models, e.g. RBAC and Mandatory Access Control. However, the security policy languages have not yet been incorporated in WfMSs. In our work, we used a simple privacy policy where a patient states which data is authorised/unauthorised to be accessed by a certain user. As future work, it is interesting to consider the use of privacy policy languages to capture privacy requirements and to enhance the YAWL environment with capabilities to interpret and enforce privacy policies specified in such languages during workflow execution.

# 8  Conclusion

Workflow Management Systems (WfMSs) enforce an organisation's security policies while executing a workflow process. However, they fail to incorporate the security concerns, privacy, of other stakeholders. In this paper, we explained the importance of privacy requirements and presented their implications for workflow functions. We introduced the notion of data *subject* to workflows and defined it as part of a workflow specification. This extension allows a WfMS to be aware of the data subject's identity and consequently to retrieve the subject's privacy policy using a workflow data pull pattern. In addition, we presented a new secure work allocation strategy that uses the subject's privacy policy to assign workflow tasks to the least-restricted resource from a privacy perspective, and the concept of a privacy-aware work-flow form rendering engine that uses auxiliary data properties to enforce the appropriate concealment actions. We then implemented these extensions in the YAWL system.

We then showed through a case scenario that the extended WfMS is capable of capturing and enforcing a data subject's privacy policy.

# References

1. Enhydra Shark: Open source workflow,
   http://shark.ow2.org/doc/1.1/index.html (accessed August 20, 2009)
2. jBPM user guide, http://docs.jboss.com/jbpm/v4.0/userguide (accessed August 20, 2009)
3. ruote: Open source Ruby workflow engine,
   http://openwferu.rubyforge.org/documentation.html (accessed August 20, 2009)
4. TIBCO BPM resource center,
   http://www.tibco.com/solutions/bpm/default.jsp (accessed August 28, 2009)
5. Alhaqbani, B., Fidge, C.J.: Access Control Requirements for Processing Electronic Health Records. In: ter Hofstede, A.H.M., Benatallah, B., Paik, H.-Y. (eds.) BPM Workshops 2007. LNCS, vol. 4928, pp. 371–382. Springer, Heidelberg (2008)
6. Anderson, A.H.: A comparison of two privacy policy languages: EPAL and XACML. In: Proceedings of the 3rd ACM Workshop on Secure Web Services, Alexandria, USA, November 3, pp. 53–60. ACM Press, New York (2006)
7. Atluri, V., Warner, J.: Security for workflow systems. In: Gertz, M., Jajodia, S. (eds.) Handbook of Database Security: Application and Trends, pp. 213–230. Springer (2008)
8. Bertino, E., Buccafurri, F., Rullo, P.: An Authorization Model and Its Formal Semantics. In: Quisquater, J.-J., Deswarte, Y., Meadows, C., Gollmann, D. (eds.) ESORICS 1998. LNCS, vol. 1485, pp. 127–142. Springer, Heidelberg (1998)
9. Bertino, E., Ferrari, E., Alturi, V.: The specification and enforcement of authorization constraints in workflow management systems. ACM Transactions on Information and System Security 2(3), 65–104 (1999)
10. Cao, J., Chen, J., Zhao, H., Li, M.: A policy-based authorization model for workflow-enabled dynamic process management. Journal of Network and Computer Applications 32(2), 412–422 (2009)
11. Casati, F., Casanto, S., Fugini, M.: Managing workflow authorization constraints through active database technology. Information Systems Frontiers 3(3), 319–338 (2001)
12. Chhanabhai, P., Holt, A.: Consumers are ready to accept the transition to online and electronic records if they can be assured of the security measures. Medscape General Medicine 9(1), 8 (2007)
13. Haplin, T.: Information Modeling and Relational Databases: From Conceptual Analysis to Logical Design. Morgan Kaufmann Publishers (2001)

14. IBM. WebSphere business modeler, version 6.2.0,
    `http://publib.boulder.ibm.com/infocenter/dmndhelp/v6r2mx/index.`
    `jsp?topic=/com.ibm.btools.modeler.advanced.help.doc/doc/concepts/`
    `modelelements/processdiagram.html` (accessed August 25, 2009)
15. Jiang, H., Lu, S.: Access Control for Workflow Environment: The RTFW Model.
    In: Shen, W., Luo, J., Lin, Z., Barthès, J.-P.A., Hao, Q. (eds.) CSCWD. LNCS,
    vol. 4402, pp. 619–626. Springer, Heidelberg (2007)
16. Meier, E.: Medical privacy and its value for patients. Seminars in Oncology
    Nursing 18(2), 105–108 (2002)
17. Oh, S., Park, S.: Task-role-based access control model. Information Sys-
    tems 28(6), 533–562 (2003)
18. Russell, N., ter Hofstede, A.H.M., Edmond, D., van der Aalst, W.M.P.: Work-
    flow Data Patterns: Identification, Representation and Tool Support. In: Del-
    cambre, L.M.L., Kop, C., Mayr, H.C., Mylopoulos, J., Pastor, Ó. (eds.) ER
    2005. LNCS, vol. 3716, pp. 353–368. Springer, Heidelberg (2005)
19. Russell, N., van der Aalst, W.M.P., ter Hofstede, A.H.M., Edmond, D.: Work-
    flow Resource Patterns: Identification, Representation and Tool Support. In:
    Pastor, Ó., Falcão e Cunha, J. (eds.) CAiSE 2005. LNCS, vol. 3520, pp. 216–
    232. Springer, Heidelberg (2005)
20. Sadan, B.: Patient data confidentiality and patient rights. International Journal
    of Medical Informatics 62, 41–49 (2001)
21. Safian, S.C.: The Complete Diagnosis Coding Book. McGraw Hill Higher Ed-
    ucation (2009)
22. Sandhu, R.S., Samarati, P.: Access control: Principles and practice. IEEE Com-
    munications Magazine 32(9), 40–48 (1994)
23. ter Hofstede, A.H.M., van der Aalst, W.M.P., Adams, M., Russell, N. (eds.):
    Modern Business Process Automation: YAWL and its Support Environment.
    Springer, Heidelberg (2009)
24. Thomas, R., Sandhu, R.: Task-based authorization controls (TBAC): A family
    of models for active and enterprise-oriented autorization management. In: Lin,
    T.Y., Qian, S. (eds.) Proceedings of the IFIP TC11 WG11.3 11th International
    Conference on Database Security XI: Status and Prospects (DBSec 1997), Lake
    Tahoe, California, USA, August 10-13. IFIP, vol. 113, pp. 166–181. Chapman
    & Hall (1997)
25. Wave-Front. FLOWer 3: Designers guide (2004)
26. Westin, A.: Privacy and Freedom. The Bodley Head Ltd. (1970)
27. Wolter, C., Menzel, M., Schaad, A., Miseldine, P., Meinel, C.: Model-driven
    business process security requirements specification. Journal of Systems Archi-
    tecture 55, 211–223 (2009)
28. Xu, J., Liu, C., Zhao, X.: Resource Allocation vs. Business Process Improve-
    ment: How They Impact on Each Other. In: Dumas, M., Reichert, M., Shan,
    M.-C. (eds.) BPM 2008. LNCS, vol. 5240, pp. 228–243. Springer, Heidelberg
    (2008)
29. Yao, L., Kong, X., Xu, Z.: A task-role based access control model with multi-
    constraints. In: Kim, J., et al. (eds.) Proceedings of 4th International Confer-
    ence on Networked Computing and Advanced Information Management (NCM
    2008), Gyeongju, Korea, September 2-4, vol. 1, pp. 137–143. IEEE (2008)

# Performance Measurement in Business Process, Workflow and Human Resource Management

Apostolia Plakoutsi[1], Georgia Papadogianni[1], and Michael Glykas[1,2,3]

[1] Department of Financial and Management Engineering, University of the Aegean, Greece
[2] Aegean Technopolis, The Technology Park of the Aegean Area, Chios, Greece
[3] Faculty of Business and Management, University of Wallangong. Dubai

## 1 Introduction

In today's turbulent business environment companies aim to become more and more competitive in their effort to increase their profits. They concentrate on business processes efficiency and effectiveness .

Most business process modelling notations concentrate on modelling the holistic process model and assume that the micro-individualistic employee level follows process execution. Work-flow modelling notations on the other hand concentrate on the flow of documents and work amongst job description holders and pay more attention to automation of document-process flows (Oinn et. al.), (Freefluo), (Pillai et. al.), (Michalickova et. al.), (Taylor et. al. (A)), (Taylor (2007) et. al.), (Winder et. al.). Human resource management (HRM) systems concentrate on the management of a firms most valuable element the employee (Rowe et. al.), (Ghanem et. al. (B)), (Curcin et. al.), (Richards et. al.), (Guo et. al.), (Boudreau et. al.). They target performance measurement at the micro employee level, they allow the development of systems like rewards and incentives, training and education, salary and pay ranges, career pathing etc (Balasubramanian et. al.), (Allen et. al.), (Taylor et. al. (B)).

In this chapter, we present a process monitoring methodology and tool (called ADJUST) that incorporates concepts and tools emerging from process modeling and analysis tools, human resources management and work-flow management tools. The resulting methodology allows performance measurement to be performed not only at the individualistic (employee) level but also at holistic (process-organizational unit).

## 2 The Function and Characteristics of Business Process Modeling Notations (BPMN)

The Business Process Modelling Notation (BPMN) (Nakava et. al.) provides a standard business process modelling notation for business stakeholders, including business analysts, technical developers who manage and monitor these processes. Modelling and execution can be inter-connected by the implementation of

M. Glykas (Ed.): Business Process Management, SCI 444, pp. 129–156.

business processes due to the generation of execution definitions. (Brooks et. al.), (Lee and Xiong), (Nagatani), (Goderis et. al.), (Chen and van Aalst), (van der Aalst et. al.).BPMN allows the creation of Business Process Diagrams that represent graphically the activities and the order of the business processes and the flow controls that are performed.

The Business Process Modelling Language (BPML) (Nakava et. al.) provides a model for expressing abstract and executable processes, managing the whole of business processes (e.g. data management, operational semantics). The persistence and interchange of definitions across different systems and modelling tools is allowed due to the provision of a grammar in an XML Schema format.

Its formalism with state transitions is illustrated in the figure below:

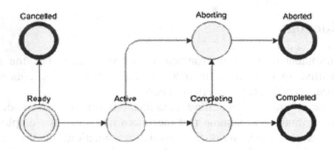

The Discovery Process Markup Language (DPML) (Nakava et. al. (A))is a declarative XML processing language, which   allows conditional processing, loops, sub-routines, and exception handling. XML documents are considered ass the basic data quantum and URIs as the method of managing them. The XML processing instructions and resolution of URI schemes are abstracted from the language so as to be fully extensible.

The Grid Job Definition Language (GjobDL)  is part of a group of definition languages called Grid Application Definition Language (GADL), developed within the Fraunhofer Resource Grid (Hoheisel and Der), (Hoheisel), (Hoheisel (A)).

The Generic Work-flow Description Language (GWorkflowDL) (Hoheisel), (Hoheisel (A)) is an extension of the existing Grid Job Definition Language, an XML-based language which is based on the formalism of Petri nets in order to reflect the dynamic behaviour of distributed Grid jobs.

The basic characteristics of the advanced GWorkflowDL are the following: definition of transitions, including sub nets, edges link transition to both input and output places, placement of tokens containing high    level control or data related to the parameters of the Web Service operations.

One implementation of a Workflow Engine for the GWorkflowDL is named as GWES (Generic Work-flow Execution Service) (Hoheisel).

The Figure below represents the XML Schema of the GWorkflowDL version 0.4.

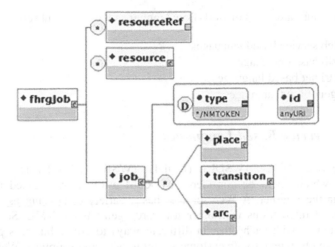

The Grid Services (AGWL), (Hoheisel and Der (A) ), (Chen and van Aalst), (Richards et. al.) Flow Language is an XML based language that enables the specification of work-flow descriptions for Grid services. Its major features are the following:

- *Service Providers*, which are the list of services taking part in the work-flow
- *Activity Model*, which outlines the list of fundamental activities in the work-flow
- *Composition Model*, which outlines the interactions between the individual services
- *Lifecycle Model,* which outlines the lifecycle for the various activities and the services which are part of the work-flow

The Service Work-flow Language (SWFL) (Hoheisel (A)), (W3C), is an XML-based metalanguage for the development of scientific work-flows. Its role is the support of a new set of conditional operators and loop constructs

The SWFL devopment is attributed to Cardiff University during the period of 2002-2003. SWFL supports a new set of conditional operators and loop constructs as well as arrays and objects.

# 3  What Is XML Process Definition Language

XPDL provides a framework for implementing business process management and workflow engines, and for designing, analyzing, and exchanging business processes. One of the key elements of the XPDL is its extensibility to handle information used by a variety of different work-flow products. XPDL thus supports a number of differing approaches. The specification is intended for use by software vendors, system integrators, consultants and any other individual or organization concerned with the design, implementation, and analysis of business process management systems as well as with interoperability among workflow systems.

There are four categories of work-flow languages with their subsets:

- Web service based languages
- Grib based languages
- Petri net based languages
- Agent based languages.

## 3.1  Web Service Based Languages

A Web service (W3C), (WSDL), (UDDI), (WSIL), (WSFL) is a software application which could be considered as XLM artefact. Automated resources accessed via the Internet. A Web service has a variety of advantages. Firstly it supports direct interactions with other software agents using XML. Secondly, it can be combined with each other in different ways to create business processes that enable you to interact with customers, employees, and suppliers. Web service also controls relationships between interactions and allows the actions of a conversation between participants.

Web service based languages have 3 subsets:

a)  Business Process Execution Language For Web Services (BPEL4WS)
Business Process Execution Language for Web Services (BPEL4WS) provides an XML-based process definition and execution language that enables the description of rich business processes capable of consuming and providing Web services in a reliable and dependable manner. BPEL4WS enables portability and interoperability by defining constructs to implement executable business processes and message exchange protocols, thereby supporting both executable and abstract business processes.

b)  Web services flow language (WSFL)
The web services flow language is useful in order to describe Web Services compositions. There are two types of Web Services compositions:

1 the suitable usage pattern of a collection of Web Services in order the result to show how to achieve a particular business goal.
2 The interaction pattern of a collection of Web Services.

Finally, WSFL uses Web Services Description Language (WSDL) for the exposition of service interfaces and their protocol links.

c)  XWEL(XLM-Based Work-Flow Language
It is a simple language which explains work flows with tasks, parameters as well as data dependency links.

## 3.2  GRIB Based Languages

A Grid is a layer of networked services that allow users single sign-on access to a distributed collection of information technology data and application resources.

The Grid services permit the entire collection to be seen as an information processing system that the user can access from any location.

GBL have 3 subsets:

a) Abstract Grid Work-flow language( AGWL)

AGWL describes Grib workflow application at a high level of abstraction. It is a language which is based on XML and permit a programmer to define and interconnect activities that mention to computational tasks or user interactions. As a result, activities are connected by control and data flow links. Moreover, using the AGWL, the user can focus on specifying scientific applications without dealing with either the complexity of the Grib or any specific implementation technology. It is worth noting that AGWL supports a total high level access mechanism to data repositories and is the main interface to the ASKALON Grid application development environment and has been applied to many real world applications.

b) Data Grib Language(DGL)

The Data Grid Language (DGL) is used to interact with the data grids and dataflow environments. A simple example of the use of dataflow systems is the management of the ingestion of a collection into a data grid. The SCEC project implemented the collection ingestion as a dataflow using the data grid language and executed the dataflow using a SDSC Matrix Grid workflow engine.

c) Simple Conceptual Unified Flow Language(SCUFL)

It is used in the myGrid project. It defines a high-level work-flow description language that allows the user to map a conceptual task to a single entity with a minimal amount of implementation specific information.

## 3.3 Petri Net Based Languages

The use of Petri Net Based Languages is as follows:

1  They control the run of the workflow.
2  They model the interaction between software resources, which are represented by software transitions, and data resources, which are represented by data places.
3  They are capable of describing the consequent and parallel execution of tasks with or without synchronization.
4  It is possible to define loops on the conditional execution of tasks.

### 3.3.1 Yawl: Yet Another Work-Flow Language

YAWL is a BPM/Workflow system, based on a concise and powerful modelling language, that handles complex data transformations, and full integration with organizational resources and external Web Services. It is built upon two main concepts: work-flow patterns and Petri net.

YAWL offers:

1  native data handling using XML Schema, XPath and XQuery.
2  a service-oriented architecture that provides an environment that can easily be tuned to specific needs.
3  Tasks in YAWL can be mapped to human participants, Web Services, external applications or to Java classes.
4  YAWL aims to be straightforward to deploy. It offers a number of automatic installers and an intuitive graphical design environment.
5  Cancellation of a part of a work-flow by a particular task.
6  OR, AND, and XOR splits/joins.

## 3.4  Agent Based Languages

Recently, efforts are made to integrate upgraded management systems in the field of software agents. The characteristics of these agents are that they are very basic , used to implement autonomous activities, and are multi systems. Applications are divided into two forms:

1) agent- enhanced work-flow management
2) agent-based work-flow management

Several ideas have appeared  for distributed work-flow enactment mechanisms based on the Agent example. However, there are differences in approaches in the following areas:

1  In the supported dimensions of distribution
2  In the adopted coordination model
3  In the exploited MAS organizational structure

Some authors propose  an agent-based work-flow engine, centred on a hierarchical structure in which a  Process Agent executes a work-flow instance by requesting the execution of the tasks composing the work-flow to a set of Resource Agents.

Others suggest an agent-based approach for enacting work-flows specified in BPEL4WS. The novelty of the approach is that the enactment of the work-flows is carried out by peer agents that can be associated with web services. The final peer-to-peer agent-based enactment  approach is presented in (Yan et. al.). In this approach the work-flow to be enacte, is  decomposed into a set of interrelated task partitions. Each task partition represents a service and its position, i.e., the interaction and dependency with the other services in the process.

## XLANG

The main example of agent based work-flow is XLANG which model business processes as Autonomous Agents. XLANG provides its users the following capabilities:

- Transaction support
- Custom correlation of messages
- Flexible handling of exceptions
- Dynamic service referral
- Contracts to agglomerate services.

# 4   The Work-flow Management Coalition (WfMC) Standards

The Workflow Management Coalition (WfMC) is a global organization of adopters, developers, consultants, analysts, as well as university and research groups engaged in workflow and BPM. The WfMC creates and contributes to process related standards, educates the market on related issues, and is the only standards organization that concentrates purely on process. The WfMC created Wf-XML and XPDL, the leading process definition language used today in over 80 known solutions to store and exchange process models. XPDL is a process design format for storing the visual diagram and all design time attributes.product attributes.

The WfMC has developed a Work-flow Reference Model for work-flow management systems, which describes a generic model for the construction of work-flow systems and identifies how it may be related to various alternative implementation approaches.

## *4.1   The WfMC Workflow Reference Model*

All work-flow systems include a number of general components that interact in a defined set of ways. To achieve interoperability between work-flow products a standardized set of interfaces and data interchange formats between such components is necessary. The model is shown in the figure below:

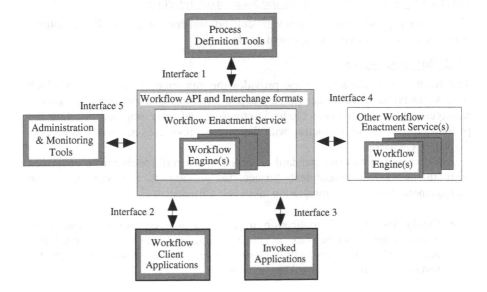

There are five main categories of conformance which are described as follows:

Interface 1.Process Definition Interchange Interface

Goal: to allow the use of different work-flow process definition tools to produce process descriptions to be used by several different work-flow engines.

This interface is termed the process definition import/export interface that would provide a common interchange format for the following types of information:

- Process start and termination conditions
- Identification of activities within the process, including associated applications and work-flow relevant data
- Identification of data types and access paths
- Information for resource allocation decisions
- Definition of transition conditions and flow rules

Interface 2. Work-flow Client Application Interface

Goal: to enable work-flow client applications to perform, submit and monitor work. Not only it defines standards for the construction of a common work-list handler to provide work-list management for one or more work-flow systems, but also it allows the interaction between several work-flow services controlled from the desktop environment.

Interface 3. Invoked Application Interface

Goal: to permit both application agents and applications that have been designed to be "workflow-enabled" to interact directly with the work-flow engine.

Interface 4.Work-flow Interoperability Interface

Goal: to allow interoperability between heterogeneous work-flow systems.

Interface 5. System Administration and Monitoring Interface

Goal : to support common management, administration and audit operations across several work-flow management products.

Work-flow enactment Service

The work-flow enactment service provides the run time environment in which process instantiation and activation occurs, utilising one or more work-flow management engines, responsible for interpreting and activating part or all, of the process definition and interacting with the external resources necessary to process the various activities.

A wide range of industry standard or application specific tools can be integrated with the work-flow enactment service to provide a complete work-flow management system. This integration takes two formats:

- Firstly, the invoked application interface, which enables the work-flow, engine directly to activate a specific application to tackle a particular activity. This would typically be server based and require no user action, for instance to conjure up email application.

- Secondly, the work-flow client application interface through which the work-flow engine interacts with a separate work-flow client application responsible for organising work on behalf of particular user.

# 5 Real Time Organisational Performance Measurement in Work-Flow Engines

The aim of ADJUST is not to develop yet another work-flow engine-tool but rather a platform that will get information from languages, work-flow engines or tools (WFMS, BPR) that comply to the WfMC standards in order to perform a holistic analysis of business models in real time.

## 5.1 WfMC Interface 1: Process Definition Interchange Process Model

Interface 1 of the WfMC reference model is implemented in ADJUST as a process definition tool that has the capability of interfacing with other tools that are using the same standard.

To provide a common method to access and describe work-flow definitions, a work-flow process definition meta-data model has been established. This meta-data model identifies entities that are used commonly in the two work-flow engines or languages.

The WPDL exporters and importers are capable of "translating" a limited number of entities that compose the "Minimum Meta Model" necessary for the calculations and ratios used in the ADJUST Business Analysis Engine (ABAE).

One of the most important elements of the WPDL is a generic construct structure that captures work-flow engine-tool specific attributes in a generic form. Both batch and API based interface architectures of ADJUST contain a "filter" mechanism to support the information exchange. Via this "WfMC interface/1 layer" input or output of real-time information can be translated to the generic WPDL form and from this to engine or tool specific representations if needed. The transfer mechanism could be either API-based or batch oriented (via files or memory transfers). The transfer mechanism that has been selected in ADJUST for common use is the batch oriented approach. The "WfMC interface/1 layer" is implemented as a client-server application as required by the WfMC Reference Model.

### 5.1.1 ADJUST Vendor Tool Exporters

There are two possible approaches to achieve proper interchange of a work-flow process definition:

1  Define build-time APIs for the creation of the objects
2  Define a common language for describing work-flow processes that can be transferred as part of a textual file.

The WPDL grammar is directly related to these objects and attributes. This approach needs two two types of interfaces to be provided by a vendor:

1 Import a work-flow definition from a character stream of definitions according to the common process definition language into the vendor's internal representation.

2 Export a work-flow definition from the vendor's internal representation to a character stream according to the common process definition language.

### 5.1.2 Process Definition Interchange

The requirement to transfer process definitions, in whole or part, may occur for a number of business reasons:

1 the use of a vendor specific tool to define, model, analyse, simulate or document a business process prior to its enactment on a (different) work-flow execution service (or services) enables separate selection of work-flow and process definition/modelling tools, allowing the most suitable tool to be used for each part of the overall system

2 the transfer and retrieval of process definitions to/from a common database repository, accessed by a number of different tools or run-time systems, may be desirable to provide a single consistent-controlled process definition repository to all different tools and run-time systems and thus identifying, presenting and bridging conceptual or functional inconsistencies amongst them.

3 the modification of existing process definitions by authorised users during enactment, either on a one-off or persistent basis need to be broadcast in all systems in a consistent and controlled manner.

4 the transfer of a process definition (in whole or part) from one work-flow engine to another may be necessary to facilitate interoperability of that process between two or more work-flow engines during process enactment. Such transfer may take place prior to, or in some cases during, enactment.

## 5.2 The ADJUST Meta-model in WPDL

The ADJUST Meta-Model describes the top-level entities contained within a Work-flow Process Definition, their relationships and attributes (including some, which may be defined for simulation purposes rather than work-flow enactment). It also defines various conventions for grouping process definitions into related process models and the use of common definition data across a number of different process definitions or models as shown in the figure below.

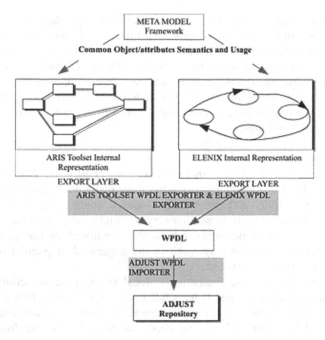

## Entities Overview

The meta-model identifies the basic set of entities and attributes used in the exchange of process definitions. In this section we will also give an overview of the proposed WPDL (WfMC) grammar for each entity. Grammar description will present WPDL output from the Exporter tools. The top-level entities are as follows:

### Work-flow Process Definition

This describes the process itself, i.e. ID and textual description, and provides other optional information associated with administration of the process definition (creation date, author, etc.) or to be used at process level during process execution (initiation parameters to be used, execution priority, time limits to be checked, person to be notified, simulation attributes, etc.). The Work-flow Process Definition entity thus provides header information for the process definition and is therefore related to all other entities in that process.

### Work-flow Process Activity

A process definition consists of one or more activities, each comprising a logical, self-contained unit of work within the process definition. An activity represents work which will be processed by a combination of resources and/or computer applications (specified by application assignment). Other optional information may be associated with the activity such as whether it can be started / finished automatically by the work-flow management system or its priority level relative to other activities when congestion for resource or system services occurs. Usage of specific work-flow relevant data items by the activity may also be specified.

The scope of an activity is local to a specific process definition. An activity may be atomic and in this case is the smallest unit of self contained work which is specified within the process. It can also generate several individual work items for presentation to a user invoking, for example, different IT tools. An activity may also be specified as a loop, which acts as controlling activity for repeated execution of a set of activities within the same process definition. In this case the set of looping activities is connected to the "controlling (loop) activity" by special loop begin/end transitions.

*Transition Information*

Activities are related to one another via flow control conditions (transition information). Each individual transition has three elementary properties, the from-activity, the to-activity and the condition under which the transition is made. Transition from one activity to another may be conditional (involving expressions which are evaluated to permit or inhibit the transition) or unconditional. The transitions within a process may result in the sequential or parallel operation of individual activities within the process.

The scope of a particular transition is local to the process definition, which contains the activity itself and its associated activities. More complex transitions, which cannot be expressed using the simple elementary transition (ROUTE) attributes and the split and join functions associated with the from- and to-activities, are formed using dummy activities, which can be specified as intermediate steps between real activities allowing additional combinations of split and/or join operations. Using the basic transition entity plus dummy activities, routing structures of arbitrary complexity can be specified. Since several different approaches to transition control exist in different work-flow tools and engines, several conformance classes need to be specified within the ADJUST WPDL grammar.

Transitions were not used in ADJUST. In order though to be compliant with standards we translated transitions that appear in Satellite tools to WPDL. The transition part of a WPDL file was not transferred into ADJUST Repository.

*Work-flow Participant Declaration*

This provides descriptions of resources that can act as performer of activities. The resource, which can be assigned to perform an activity, is specified in the *participant assignment* attribute, which links the activity to the set of resources (within the work-flow participant declaration) which may be allocated to it. The work-flow participant declaration does not necessarily refer to a single person, user or a job description, but rather it may refer to a set of people of appropriate skill or responsibility, or machine automata resources or even other software programs. The meta-model includes four simple types of resources for the work-flow participant attribute.

*The ADJUST Organisational Model*

In ADJUST there exists an extensive use of the [<extended attribute list>] // function for the participant specific attribute in order to construct the organisational model. WPDL does not provide any semantics in order to construct an Organisational Model (OM). The use of extended attributes in ARIS Exporter has solved this problem.

In more sophisticated scenarios the participant declaration may refer to an Organisational Model, external to the work-flow process definitions, which enables the evaluation of more complex expressions, including reference to business functions, organisational entities and relationships. Reference to separate resource assignment expressions may also be required under various circumstances. This first version of specification does not include a fully standardised specification for either OM or process history functions; such functions has been realised for the ADJUST project by use of the "extended library" function.

# 6 Adjust Repository API, Its Classes and Their Methods

The ADJUST repository API is a set of functions put in a Dynamic Linked Library (DLL) and allows an authorised user to build interface bridges with ADJUST Repository. The API has been tested in ADJUST by:

- **The WPDL Importer** which is using the API functions in order to import WPDL files into temporary ADJUST Repository WfMS compliant data sets and to a Temporary ADJUST Repository business process data sets.
- **Performance now interface** which is using API in order to import information that has been extracted from Descriptions/Performance Now (INSPERITY) into the temporary ADJUST Repository human resource related data sets.
- **Adjust Engine** which is using API functions in order to add, delete, and update data that are stored into the final ADJUST repository.

Regarding the classes, There are thirteen in the Adjust Engine API, one for each of the twelve tables, and another one, which has general purpose, database–oriented methods. All table classes are named after the "cls"<tablename> convention, i.e. clsActivites for the Activities table. These classes have some methods in common:

1 Public Function Fetch (ID as String) As ADODB. Record set:
   These methods return a *recordset* object for the table which the current class is referred to. The *recordset* object has all the fields of the record whose ID was entered as a parameter.

2 Public Function FetchGrid(ID As Integer, Index As Integer) As ADODB. Record set:
   These methods act like the previous ones, but they fetch data from tables related by foreign keys to the class owner table. The parameter *index* exists only when there are more than one such tables and identifies the one to be read. (Michael Glykas, Performance Measurement in Business Process, Workflow and Human Resource Management, Volume 18, Issue 4, pages 241–265, October/December 2011).

## 7 Aris Toolset - WPDL Exporter

This interface is based on mechanism that converts information that is kept in ARIS Toolset from its internal representation to WPDL. The exporter has been build using ARIS toolset's own scripting language (ARIS BASIC) which enables users to create their own reports. The reports can be produced in different formats such as Rich Text Format (RTF) Hypertext Mark-up Language (HTML) and of course as a standard ASCII text file. In ADJUST we use the latter format in accordance with WfMC's Interface 1.

With the use of the WPDL exporter pieces of information and data in ARIS repository are converted to WPDL. The WPDL text file, once produced, is then read from the WPDL Importer, which performs syntax checking and importing of the information into the BPR temporary repository.

The WPDL ARIS exporter file generated supports a sub-set of WPDL grammar rules in addition to the extended attributes. The WPDL Importer on the other hand can interpret and translate the full WPDL grammar or any sub-set of it.

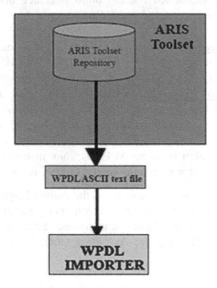

Below we present a list of WPDL ARIS exporter attributes and their respective representations in the ADJUST Repository:

- PARTICIPANTS: depending whether they are of HUMAN, ROLE, ORGANISATIONAL_UNIT they are going to be mapped into EMPLOYEE, JOB TITLE, and DEPARTMENT or DIVISION RESPECTIVELY.
- ACTIVITIES: they will be mapped into Activities as well
- BELONGS_TO extended attribute: it is used to create the following relationships:

~Division-Department
~Department - Sub-Department
~Job Title - Employee
~Employee - Division or Department

- PERFORMER attribute : it is used to create the relationship between Activity Employee
- EXECUTED_BY extended attribute: same as the above
- IMPLEMENTATION attribute: it is used to create the Activity - Sub-Activity relationship.

# 8 WPDL Exporter

The work-flow system WPDL exporter is an interface tool that converts work-flow concepts to WPDL. It uses some of the WPDL concepts. In Elenix for example, (which is a work-flow system) these concepts are Activities, Routes, Participants, Transitions, Data, Applications and Work-flow.

Activities represent some kind of work to be performed by an application or by a Work-flow.
Applications represent some kind of program/application that executes a specific amount of work.
Transactions represent links from one object to another. These links can have logical conditions associated with them that guarantee that the link is only followed when the logical conditions are fulfilled (become true).
Routes objects are kinds of Activities that obey "if then else" statements.
Data objects contain types and variables.
Extended Attributes are used to insert new concepts into WPDL.

When exporting the Elenix Floway to WPDL, there is a mapping between the languages. Bellow there is a table illustrating this mapping.

| Elenix Floway | WPDL |
|---|---|
| Stage | Activity |
| Condition | Route |
| Link | Transition |
| Groups | Participant |
| Variable | Data |
| Tasks | Application |

Almost every concept in Elenix has a direct mapping into WPDL, but that is not entirely true. In some cases further translation is necessary, for example when converting a Condition, we need to put its logical conditions to the Transitions outgoing of the Route, and not on the Route itself.

# 9 WPDL Importer

## 9.1 Objective

The main function of the WPDL Importer is to convert a WPDL file into ADJUST API calls. This is done by a parser written in *Flex, Bison* and *C++*. This parser verifies the WPDL syntax and it semantics, and finally calls the ADJUST DLL to insert all the data in the ADJUST database.

## 9.2 Parts of WPDL Importer

Syntatic Parser
The Syntactic Parser was built using the *Bison* grammar given by *WfMC*. This grammar was modified in order to return the tokens values, and consistent error messages. The modifications talked above where implemented by inserting rules in the grammar in order to save the WPDL data to memory.

Semantic Parser
The second part is the semantic parser, this, reads all the WPDL data from memory, constructs all the Activities trees, Participants trees, and all other important data. Also in this part is written an Output file with all the list of the API functions to be called and their respective parameters.

For last, the third part has the job to read the output file generated by the Semantic Parser and execute its instructions in order to fill the ADJUST database.

## 9.3 Mapping between WPDL and ADJUST Data Base

WPDL mapping to the ADJUST database is implemented in two phases. In the first phase all organisational concepts are imported and saved in the database. Four tables have been created in the ADJUST database to host these concepts, namely: *JobTitles, Employees, Departments, Divisions* and *EmpAct (Employee/Activity)*. The *JobTile* table is saving all roles. The Employees table is filled with all Employees (names). The *Departments* and *Divisions* tables host all organisational elements. They also contain associated organisational hierarchies by utilising the Extended Attributes presented before and thus save data about superordinate-subordinate relationships. In the table bellow we present the WPDL mapping in ADJUST.

| WPDL | ADJUST |
|------|--------|
| Participant – ORG_UNIT level 0 | Division |
| Participant – ORG_UNIT level 1 | Division |
| Participant – ORG_UNIT level $2 \Rightarrow n$ | Department |
| Participant – ROLE | **JobTitle** |
| Participant – HUMAN | Employee |

The second phase consists in inserting the Activities and connecting them with the Employees. For each Activity inserted we have a list of Employees that executed this Activity. With this list, a search is performed in internal tables to find the database Key for each of these employees. Having this key and the Activity database Key returned when it was inserted we can connect an Employee with an Activity, using the *EmpAct (Employee/Activity)* table.

## 9.4 API Usage

The ADJUST DLL application is used as a COM Component. All methods in the DLL application are encapsulated by C++ classes. Available classes are:

> _clsActivities
> _clsBusinessObjectives
> _clsCriticalSuccessFactors
> _clsDBTools
> _clsDepartments
> _clsDivisions
> _clsEmployees
> _clsFactorCatgegories
> _clsFactors
> _clsJobDescriptions
> _clsJobTitles
> _clsPayRanges
> _clsTransactions

From this list, the ones currently used are: *_clsActivities, _clsDBTools, _clsDepartments, _clsDivisions, _clsEmployees* and *_clsJobTitles*.

The class *_clsDBTools* is only used for the deletion of data in the database before inserting any new data in it and is used for data consistency in the database.

# 10  HRM Interface

This is the interface bridge between Descriptions/Performance Now and the temporary ADJUST Repository for HRM. The interface application reads the data that has been extracted from Descriptions/Performance Now and after doing some necessary transformation imports them into the temporary ADJUST Repository via the ADJUST Repository API (Figure 1). Thus information has been extracted from Descriptions/Performance Now is located in an Access database file.

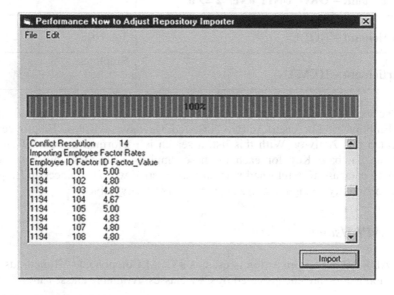

**Fig. 1** Adjust Repository HRM Interface

## 10.1   The HRM Tool Functions

A user can perform the following functions with the tool:

- Open a database file that contains Descriptions/Performance Now data
- Import the data from the Access database to temporary ADJUST repository

The data transformation function is important in order to achieve the compatibility between the data accepted by ADJUST repositories and the data extracted from the Access database. The tool automatically performs the following transformations:

**Employee Name**

ADJUST repositories internal representation: <Full Name> as string
Exported Access database internal representation: <LastName>,<First Name> as string

**Employee Salary**

ADJUST repositories internal representation: <Salary> as integer
Exported Access database internal representation: <Salary><Currency> as string

This is a very important transformation because ADJUST is using salaries for calculations.

**Performance Rate**

ADJUST repositories internal representation: An integer number from 1 to 10
Exported Access database internal representation: A real number from 1 to 5
Example Transformation: 4,6 to 9

This is also important in order to achieve compatibility between the tools' rates.

Figure 2 shows the interface tool after the transformed data have imported into temporary ADJUST repository.

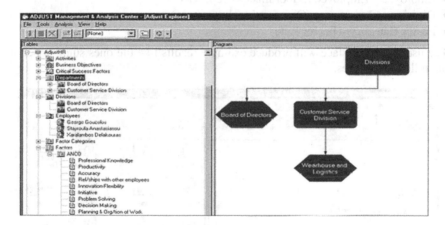

**Fig. 2** Adjust Tool HRM Departments Interface

The tool imports the following information in the temporary ADJUST Repository:

• Employee Information (Name, Salary, Address)
• Job Titles Information (Name)
• Division Information (Name)
• Department Information (Name)
• Factor Category Information (Name)
• Factors (Name)

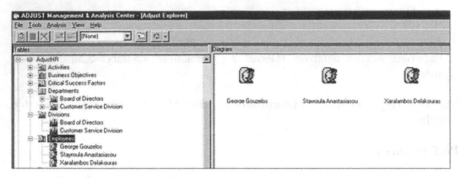

**Fig. 3** Adjust Tool Employees Interface

The tool also creates the following relationships:

- Employee - Job Title (Organisational Structure)
- Employee - Division or Department (Organisational Structure)
- Employee – Factor Rate
- Employee – Employee (Subordinate – Supervisor)
- Division – Department (Organisational Structure)
- Factor Category – Factor Rate

The use of the interface will produce example results like the ones summarised in the following figures.

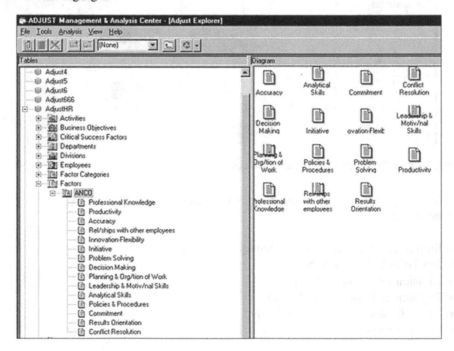

**Fig. 4** Adjust Tool Performance Factors Interface

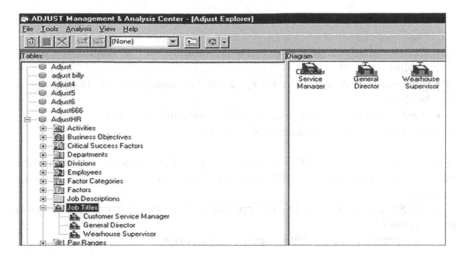

**Fig. 5** Adjust Tool Job Titles Interface

# 11 Conclusions and Further Research

We presented the Adjust Toolset that aims to integrate existing business modelling, human resource management and work-flow systems via interfaces that operate not only during organisational design but also in real time. All interfaces were developed using the WPDL language developed by the WfMC. These interfaces are composed of importing and exporting facilities from and to the three types of systems, namely: hrm, work-flow and business modelling. When the Adjust database is populated with data from the three types of associated systems the Adjust engine can perform business analysis. This is achieved via a series of metrics and techniques both quantitative and qualitative mainly based on effort based costing.

Different analysis techniques from various disciplines are applied in different perspectives. Most of these techniques are influenced form organizational theories, strategic planning, accounting, information system development, transaction cost economics and existing work in process management.

The application a smaller subset of techniques like flexibility and dimensions of transactions are novel to performance measurement. Other techniques like management structure analysis, concentration analysis, equivalent salary analysis and fractionalization analysis are used from a different viewpoint.

The major contribution of ADJUST in business analysis is the ability to analyse the resulting business model much more holistically compared to other methodologies and tools.

Business analysis in ADJUST allows users to:

- Analyse the organization from both the holistic and individualistic view
- Explore the relationship between organizational structure and processes

- Examine the alignment between business strategy, business processes and employee objectives
- Examine the level of communication, coordination and control in existing structure and processes
- Analyse the impact of creating a market metaphor for organizational behaviour
- Analyse the level of effectiveness of activities and processes
- Analyse the level of efficiency of activities and processes

Applications of the ADJUST toolset in a variety of case studies help companies measure and evaluate the efficiency of their standard software-supported processes.

ADJUST has a broad range of applications for implementing sophisticated process controls.

The information and evaluations provided support at:

- Continuous improvement of processes, both within the company and across companies
- Weak point analysis during process engineering
- A basis for benchmarking projects, both within the company and across companies
- Outsourcing decisions
- Self-assessment by process participants
- A basis for automated early-warning systems
- Collecting data for activity-based costing and simulation
- Calculating the efficiency and profitability of standard software implementations.

## 12 Further Research

Adjust has been applied in various business sectors like banking, production and manufacturing, telecoms, information technology. The results of case studies as well as user's views on the techniques and ratios used will be integrated into the methodology and will help in adding new techniques and ratios and even eliminate others that are not considered as useful.

Currently there is an effort undertaken for further research in the area of ratio and business concept interlinking which will help ADJUST users of benefit from predefined links amongst the ratios used. For example in organisational analysis the lower the span of control ratio or percentage the lower the cost to manage and vice versa. We have introduced the theory of fuzzy cognitive maps and created a tool for performance measures interlinking and scenario planning at different organisational levels (Glykas and Valiris (C)), (IDS-SCHEER), (SMD), (Valiris and Glykas), (Valiris and Glykas (A)), (Valiris and Glykas (B)), (Xirogiannis and Glykas), (Xirogiannis and Glykas (A)), (Glykas and Xirogiannis), (Xirogiannis and Glykas (B)), (Xirogiannis et. al.).

# References

(Addis et. al.) Addis, M., Ferris, J., Greenwood, M., Li, P., Marvin, D., Oinn, T., Wipat, A.: Experiences with e-science workflow specification and enactment in bioinformatics. In: Cox, S. (ed.) e-Science All Hands Meeting, pp. 459–466 (2003)

(AGWL) http://dps.uibk.ac.at/projects/agwl/

(Allen et. al.) Allen, G., Davis, K., Dramlitsch, T., Goodale, T., Kelley, I., Lanfermann, G., Novotny, J., Radke, T., Rasul, K., Russell, M., Seidel, E., Wehrens, O.: The gridlab grid application toolkit. In: HPDC 2002: Proceedings of the 11th IEEE International Symposium on High Performance Distributed Computing HPDC-11 20002 (HPDC 2002), p. 411. IEEE Computer Society, Washington, DC (2002)

(Andrews et. al.) Andrews, T., Curbera, F., Dholakia, H., Goland, Y., Klein, J., Leymann, F., Liu, K., Roller, D., Smith, D., Thatte, S., Trickovic, I., Weerawarana, S.: Business process execution language for web services version 1.1. BEA Systems, IBM, Microsoft, SAP, Siebel Systems, Tech. Rep. (2003), http://www-128.ibm.com/developerworks/library/specification/wsbpel/

(ASKALON) http://www.dps.uibk.ac.at/projects/askalon/

(Balasubramanian et. al.) Balasubramanian, R., Babak, S., Churches, D., Cokelaer, T.: Geo600 online detector characterization system. Classical and Quantum Gravity 22, 4973 (2005), http://www.citebase.org/abstract?id=oai:arXiv.org:gr-qc/0504140

(biztalk) http://www.cs.unibo.it/~laneve/papers/biztalk.pdf

(Boudreau et. al.) Boudreau, T., Glick, J., Spurlin, V.: NetBeans: The Definitive Guide. O'Reilly & Associates, Inc., Sebastopol (2002)

(Bradley et. al.) Bradley, J., Brown, C., Carpenter, B., Chang, V., Crisp, J., Crouch, S., Roure, D.D., Newhouse, S., Li, G., Papay, J., Walker, C., Wookey, A.: The omii software distribution. In: All Hands Meeting 2006 (2006), pp. 748–753, http://eprints.ecs.soton.ac.uk/13407/

(Brooks et. al.) Brooks, C., Lee, E., Liu, X., Neuendorffer, S., Zhao, H.Z. e. Y.: Heterogeneous concurrent modeling and design in java (volume 2: Ptolemy ii software architecture). EECS Dept., UC Berkeley, Tech. Rep. (July 22, 2005), http://chess.eecs.berkeley.edu/pubs/63.html

(Buhler and Vidal) Buhler, P., Vidal, J.M.: Towards Adaptive Workflow Enactment Using Multiagent Systems. Journal of Information Technology and Management 6, 61–87 (2005)

(Buhler and Vidal (A)) Buhler, P., Vidal, J.M.: Enacting BPEL4WS specified workflows with multiagent systems. In: Proceedings of the Workshop on Web Services and Agent-Based Engineering (2004)

(Buhler et. al.) Buhler, P., Vidal, J.M., Verhagen, H.: Adaptive Workflow = web services + agents. In: Proceedings of the First International Conference on Web Services, pp. 131–137 (2003)

(Chen and van Aalst) Chen, Y., van Aalst: On scientific workflows. IEEE Computer Society's Technical Committee for Scalable Computing (2007), http://citeseer.ist.psu.edu/plotkin03origins.html

(Curcin et. al.) Curcin, V., Ghanem, M., Guo, Y., Rowe, A., He, W., Pei, H., Qiang, L., Li, Y.: It service infrastructure for integrative systems biology. In: SCC 2004: Proceedings of the 2004 IEEE International Conference on Services Computing, pp. 123–131. IEEE Computer Society, Washington, DC (2004)

(Ehler et. al.) Ehrler, L., Fleurke, M., Purvis, M.A., Savarimuthu, B.T.R.: AgentBased Workflow Management Systems(Wfmss): JBees - A Distributed and Adaptive WFMS with Monitoring and Controlling Capabilities. Journal of Information Systems and e-Business Management 4(1), 5–23 (2006)

(Freefluo) Freefluo enactment engine (2008), http://freefluo.sourceforge.net

(Ghanem et. al.) Ghanem, M., Curcin, V., Wendel, P., Guo, Y.: Building and using analytical workflows in discovery net. In: Dubitzky, W. (ed.) Data Mining on the Grid. John Wiley and Sons (2008)

(Ghanem et. al. (A)) Ghanem, M., Guo, Y., Lodhi, H., Zhang, Y.: Automatic scientific text classification using local patterns: KDD CUP 2002 (task 1). SIGKDD Explor. Newsl. 4(2), 95–96 (2002)

(Ghanem et. al. (B)) Ghanem, M., Chortaras, A., Guo, Y.: Web Service programming for biomedical text mining. In: SIGIR Workshop on Search and Discovery in Bioinformatics Held in Conjunction with the 27th Annual International ACM SIGIR Conference, Sheffield, UK (July 2004), http://pubs.doc.ic.ac.uk/web-service-text-mining/

(Glykas) Glykas, M.: Effort based performance measurement in business process management. Knowledge and Process Management 18(1), 10–33 (2011)

(Glykas and Valiris (A)) Gykas, M., Valiris, G.: Formal Methods in Object Oriented Business Modelling. The Journal of Systems and Software 48, 27–41 (1999)

(Glykas and Valiris (B)) Glykas, M., Valiris, G.: Management Science Semantics for Object Oriented Business Modelling in BPR. Information and Software Technology 40(8), 417–433 (1998)

(Glykas and Valiris (C)) Glykas, M., Valiris, G.: ARMA: A Multidisciplinary Approach to Business Process Redesign. Knowledge and Process Management 6(4), 213–226 (1999)

(Glykas and Xirogiannis) Glykas, M., Xirogiannis, G.: A soft knowledge modeling approach for geographically dispersed financial organizations. Soft Computing 9(8), 593 (2004)

(Goderis et. al.) Goderis, A., Brooks, C., Altintas, I., Lee, E.A., Goble, C.: Heterogeneous composition of models of computation. EECS Department. University of California, Berkeley, Tech. Rep. UCB/EECS- 2007-139 (November 2007)

(Goderis et. al. (A)) Goderis, A., Brooks, C.H., Altintas, I., Lee, E.A., Goble, C.A.: Composing Different Models of Computation in Kepler and Ptolemy II. In: Shi, Y., van Albada, G.D., Dongarra, J., Sloot, P.M.A. (eds.) ICCS 2007. LNCS, vol. 4489, pp. 182–190. Springer, Heidelberg (2007)

(Guo et. al.) Guo, Y., Liu, J.G., Ghanem, M., Mish, K., Curcin, V., Haselwimmer, C., Sotiriou, D., Muraleetharan, K.K., Taylor, L.: Bridging the macro and micro: A computing intensive earthquake study using discovery net. In: SC 2005: Proceedings of the 2005 ACM/IEEE Conference on Supercomputing, p. 68. IEEE Computer Society, Washington, DC (2005)

(Guo et. al.) Guo, L., Robertson, D., Chen-Burger, Y.-H.: Enacting the Distributed Business Workflows Using BPEL4WS on the Multi-agent Platform. In: Eymann, T., Klügl, F., Lamersdorf, W., Klusch, M., Huhns, M.N. (eds.) MATES 2005. LNCS (LNAI), vol. 3550, pp. 35–46. Springer, Heidelberg (2005)

(Hoheisel) Hoheisel, A.: Workflow Management for Loosely Coupled Simulations. In: Proceedings of the Biennial meeting of the International Environmental Modelling and Software Society (iEMSs), Osnabrck (2004)

(Hoheisel (A)) Hoheisel, A., Der, U.: Dynamic Workflows for Grid Applications. In: Proceedings of the Cracow Grid Workshop 2003, Krakau, Polen (2003)

(Hoheisel (B)) Hoheisel, A.: Grid Application Definition Language? GADL 0.2. Technischer Report, Fraunhofer FIRST (2002)

(Hoheisel and Der) Hoheisel, A., Der, U.: An XML-Based Framework for Loosely Coupled Applications on Grid Environments. In: Sloot, P.M.A., Abramson, D., Bogdanov, A.V., Gorbachev, Y.E., Dongarra, J., Zomaya, A.Y. (eds.) ICCS 2003, Part I. LNCS, vol. 2657, pp. 245–254. Springer, Heidelberg (2003)

(Hoheisel and Der (A)) Hoheisel, A., Der, U.: An XML-Based Framework for Loosely Coupled Applications on Grid Environments. In: Sloot, P.M.A., Abramson, D., Bogdanov, A.V., Gorbachev, Y.E., Dongarra, J., Zomaya, A.Y. (eds.) ICCS 2003, Part I. LNCS, vol. 2657, pp. 245–254. Springer, Heidelberg (2003)

(Hoheisel and Der (B)) Hoheisel, A., Der, U.: Dynamic workflows for Grid applications. In: Proceedings of the Cracow Grid Workshop 2003, Cracow, Poland (2003)

(Hull et. al.) Hull, D., Wolstencroft, K., Stevens, R., Goble, C., Pocock, M.R., Li, P., Oinn, T.: Taverna: A tool for building and running workflows of services. Nucleic Acids Research 34, W729–W732 (2006), web Server Issue

(IDS-SCHEER) http://www.ids-scheer.de

(Inforsense) Inforsense ltd (2008), http://www.inforsense.com/

(INSPERITY) http://www.insperity.com

(ISO-15909-1) ISO 15909-1, High-level Petri nets - Part 1: Concepts, definitions and graphical notation (2004)

(ISO-15909-2) ISO 15909-2, High-level Petri nets - Part 2: Transfer Format, Working Draft (2005)

(Jensen) Jensen, K.: An Introduction to the Theoretical Aspects of Coloured Petrinets. In: de Bakker, J.W., de Roever, W.-P., Rozenberg, G. (eds.) REX 1993. LNCS, vol. 803, pp. 230–272. Springer, Heidelberg (1994)

(Lee and Xiong) Lee, E.A., Xiong, Y.: A behavioral type system and its application in ptolemy ii. Form. Asp. Comput. 16(3), 210–237 (2004)

(Ludascher et. al.) Ludascher, B., Altintas, I., Berkley, C., Higgins, D., Jaeger, E., Jones, M., Lee, E.A., Tao, J., Zhao, Y.: Scientific workflow management and the kepler system: Research articles. Concurr. Comput.: Pract. Exper. 18(10), 1039–1065 (2006)

(Michalickova et. al.) Michalickova, K., Bader, G.D., Dumontier, M., Lieu, H., Betel, D., Isserlin, R., Hogue, C.W.V.: Seqhound: biological sequence and structure database as a platform for bioinformatics research. BMC Bioinformatics 3, 32 (2002)

(Minglu et. al.) Li, M., Sun, X.-H., Deng, Q.: Grid and cooperative computing. In: Second International Workshop, GCC 2003, Springer, Heidelberg (2003) ISBN 032-9743

(Mulyar and van der Aalst) Mulyar, N.A., van der Aalst, W.M.P.: Patterns in Colored Petri Nets, BETA Working Paper Series, WP 139. Eindhoven University of Technology, Eindhoven (2005)

(Nagatani) Nagatani, T.: Interaction between buses and passengers on a bus route. Physica A Statistical Mechanics and its Applications 296, 320–330 (2001)

(Nakava et. al.) Harmon, P., Davenport, T.: Business Process Modeling Notation BPMN CORE NOTATION Business Process Change, pp. 513–516 (2007)

(Nakava et. al. (A)) Nakaya, J., Kimura, M., Hiroi, K., Ido, K., Yang, W., Tanaka, H.: Genomic Sequence Variation Markup Language (GSVML). International Journal of Medical Informatics 79(2), 130–142 (2010)

(Oinn et. al.) Oinn, T., Addis, M., Ferris, J., Marvin, D., Greenwood, M., Goble, C., Wipat, A., Li, P., Carver, T.: Delivering web service coordination capability to users. In: WWW Alt. 2004: Proceedings of the 13th International World Wide Web Conference on Alternate Track Papers & Posters, pp. 438–439. ACM, New York (2004)

(Pillai et. al.) Pillai, S., Silventoinen, V., Kallio, K., Senger, M., Sobhany, S., Tate, J.G., Velankar, S.S., Golovin, A., Henrick, K., Rice, P., Stoehr, P., Lopez, R.: Soap-based services provided by the european bioinformatics institute. Nucleic Acids Research 33(web-server-issue), 25–28 (2005)

(Purvis et. al.) Purvis, M. A., Savarimuthu, B.T.R., Purvis, M.K.: A Multi-agent Based Workflow System Embedded with Web Services. In: Proceedings of the Second International Workshop on Collaboration Agents: Autonomous Agents for Collaborative Environments (COLA 2004), Beijing, China, pp. 55–62. IEEE/WIC Press (September 2004)

(Repetto et. al.) Repetto, M., Paolucci, M., Boccalatte, A.: A Design Tool to Develop Agent-Based Workflow Management Systems. In: Proc. of the 4th AI*IA/TABOO Joint Workshop "From Objects to Agents": Intelligent Systems and Pervasive Computing, Villasimius, CA, Italy, September 10-11, pp. 100–107. Pitagora Editrice Bologna (2003)

(Richards et. al.) Richards, M., Ghanem, M., Osmond, M., Guo, Y., Hassard, J.: Gridbased analysis of air pollution data. Ecological Modelling 194, 274–286 (2006)

(Rowe et. al.) Rowe, A., Kalaitzopoulos, D., Osmond, M., Ghanem, M., Guo, Y.: The discovery net system for high throughput bioinformatics. Bioinformatics 19(90001), 225–2231 (2003)

(Savarimuthu et. al.) Savarimuthu, B.T.R., Purvis, M.A., Purvis, M.K., Cranefield, S.: Agent-Based Integration of Web Services with Workflow Management Systems. Information Science Discussion Paper, vol. (05). University of Otago, Dunedin (2005) ISSN 1172-6024

(SCUFL) http://www.mygrid.org.uk/usermanual1.7 /scufl_language_wb_features.html

(Slominsky et. al.) Slominsky, A.: Adapting BPEL to Scientific Workflows. In: Taylor, I., Deelman, E., Gannon, D., Shields, M. (eds.) Workflows for e-Science, pp. 208–226. Springer, New York (2007)

(SMD) http://www.smd.com.pt

(Syed et. al.) Syed, J., Ghanem, M., Guo, Y.: Discovery processes: Representation and reuse. In: Proceedings of First UK e-Science All-hands Conference, Sheffield, UK (2002)

(Taverna) http://www.taverna.org.uk

(Taylor et. al.) Taylor, I.J., Deelman, E., Gannon, D.B.: Workflows for e-Science: Scientific Workflows for Grids. Springer (December 2006)

(Taylor et. al. (A)) Taylor, I.J., Shields, M., Wang, I., Harrison, A.: Visual Grid Workflow in Triana. Journal of Grid Computing 3(3-4), 153–169 (2005), http://www.springerlink.com/openurl.asp? genre=article&issn=15707873&volume=3&issue=3&spage=153

(Taylor et. al. (B)) Taylor, I.J., Shields, M., Wang, I., Rana, O.: Triana Applications within Grid Computing and Peer to Peer Environments. Journal of Grid Computing 1(2), 199–217 (2003), http://journals.kluweronline.com/article.asp?PIPS=5269002

(Taylor (2007) et. al.) Taylor, I.J., Shields, M., Wang, I., Harrison, A.: The Triana Workflow Environment: Architecture and Applications. In: Taylor, I., Deelman, E., Gannon, D., Shields, M. (eds.) Workflows for e-Science, pp. 320–339. Springer, New York (2007), http://www.uddi.org

(UDDI) UDDI: Universal Description, Discover and Integration of Business for the Web, http://www.uddi.org

(Upstill and Boniface) Upstill, C., Boniface, M.: Simdat. CTWatch Quarterly, vol. 1(4), pp. 16–20 (November 2005), http://eprints.ecs.soton.ac.uk/11622/

(van der Aalst et. al.) van der Aalst, W.M.P., ter Hofstede, A.H.M., Kiepuszewski, B., Barros, A.P.: Workflow patterns. Distributed and Parallel Databases 14, 5–51 (2003)

(Valiris and Glykas) Valiris, G., Glykas, M.: Critical Review of existing BPR Methodologies: The Need for a Holistic Approach. Business Process Management Journal 5(1), 65–86 (1999)

(Valiris and Glykas (A)) Valiris, G., Glykas, M.: A Case Study on Reengineering Manufacturing Processes and Structures. Knowledge and Process Management 7(2), 20–28 (2000)

(Valiris and Glykas (B)) Valiris, G., Glykas, M.: Developing Solutions for Redesign: A Case Study in Tobacco Industry. In: Zupančič, J., Wojtkowski, G., Wojtkowski, W., Wrycza, S. (eds.) Evolution and Challenges in System Development. Plenum Press (1998)

(Van der Aalst and A. Hofstede) van der Aalst, W., Hofstede, A.: Yawl: Yet another workflow language (2002),
http://citeseer.ist.psu.edu/vanderaalst03yawl.html

(van der Aalst and Kumar) van der Aalst, W.M.P., Kumar, A.: XML based schema definition for support of inter-organizational workflow. University of Colorado and University of Eindhoven report (2000)

(van der Aalst (B)) van der Aalst, W.M.P.: The Application of Petri Nets to Workflow Management. Eindhoven University of Technology, Eindhoven

(Wassermann et. al.) Wassermann, B., Emmerich, W., Butchart, B., Cameron, N., Chen, L., Patel, J.: Sedna: A BPEL-based environment for visual scientific workflow modelling. In: Taylor, I., Deelman, E., Gannon, D., Shields, M. (eds.) Workflows for eScience - Scientific Workflows for Grids. Springer (2006)

(Wassermann (2006) et. al.) Wassermann, B., Emmerich, W., Butchart, B., Cameron, N., Chen, L., Patel, J.: Sedna: A BPEL-based environment for visual scientific workflow modelling. In: Taylor, I., Deelman, E., Gannon, D., Shields, M. (eds.) Workflows for eScience - Scientific Workflows for Grids. Springer (2006)

(Web Services) Introducing the Web Services Flow Language,
http://www.ibm.com/developerworks/library/ws-ref4/

(Weber and Kindler) Weber, M., Kindler, E.: The Petri Net Markup Language. In: Ehrig, H., Reisig, W., Rozenberg, G., Weber, H. (eds.) Petri Net Technology for Communication-Based Systems. LNCS, vol. 2472, pp. 124–144. Springer, Heidelberg (2003)

(WfMC) WfMC document number WfMC TC-1016-P

(WFMC (A)) http://www.wfmc.org

(Winder et. al.) Winder, R.: Hello groovy! CVU 18(3), 3–7 (2006)

(WS-BPEL) http://www-106.ibm.com/developerworks/webservices/library/ws-bpel/

(WSDL) Web Services Description Language (WSDL) 1.1, W3C,
  http://www.w3.org/TR/wsdl
(WSFL) Web Services Flow Language (WSFL), http://www-4.ibm.com/
  software/solutions/webservices/pdf/WSFL.pdf
(WSIL) Web Services Inspection Language (WSIL),
  http://xml.coverpages.org/IBM-WSInspection-Overview.pdf
(W3C) Web Services Glossary, W3C Working Group Note (February 11, 2004),
  http://www.w3.org/TR/ws-gloss/
(Xirogiannis and Glykas) Xirogiannis, G., Glykas, M.: Fuzzy Cognitive Maps in Business
  Analysis and Performance Driven Change. Journal of IEEE Transactions in Engineering
  Management 13(17) (2004)
(Xirogiannis and Glykas (A)) Xirogiannis, G., Glykas, M.: Fuzzy Cognitive Maps as a
  Back-End to Knowledge-Based Systems in Geographically Dispersed Financial
  Organizations. Knowledge and Process Management Journal 11(1) (2004)
(Xirogiannis and Glykas (B)) Xirogiannis, G., Glykas, M.: Intelligent Modeling of
  e-Business Maturity. Expert Systems with Applications 32(2), 687–702 (2006)
(Xirogiannis et. al.) Xirogiannis, G., Chytas, P., Glykas, M.: George Valiris. Intelligent
  impact assessment of HRM to the shareholder value
(Yan et. al.) Yan, J., Yang, Y., Kowalczyk, R., Nguyen, X.T.: A service workflow
  management framework based on peer-to-peer and agent technologies. In: Proc. of the
  International Workshop on Grid and Peer-to-Peer based Workflows co-hosted with the
  5th International Conference on Quality Software, Melbourne, Australia, September 19 -
  21 (2005)
(YAWL) http://www.yawlfoundation.org/
(Yuhong et. al.) Yan, Y., Maamar, Z., Shen, W.: Integration of Workflow and Agent
  Technology for Business Process Management. In: Proceeding of The Sixth
  International Conference on CSCW in Design, London, Ontario, Canada, July 12-14
  (2001)

# Understanding the Costs of Business Process Management Technology

Bela Mutschler[1] and Manfred Reichert[2]

[1] Business Informatics Group, University of Applied Sciences Ravensburg-Weingarten, Germany
bela.mutschler@hs-weingarten.de
[2] Institute of Databases and Information Systems, University of Ulm, Germany
manfred.reichert@uni-ulm.de

**Abstract.** Providing effective IT support for business processes has become crucial for enterprises to stay competitive in their market. Business processes must be defined, configured, implemented, enacted, monitored and continuously adapted to changing situations. Process life cycle support and continuous process improvement have therefore become critical success factors in enterprise computing. In response to this need, a variety of process support paradigms, process specification standards, process management tools, and supporting methods have emerged. Summarized under the term Business Process Management (BPM), they have become a success-critical instrument for improving overall business performance. However, introducing BPM approaches in enterprises is associated with significant costs. Though existing economic-driven IT evaluation and software cost estimation approaches have received considerable attention during the last decades, it is difficult to apply them to BPM projects. In particular, they are unable to take into account the dynamic evolution of BPM projects caused by the numerous technological, organizational and project-specific factors influencing them. The latter, in turn, often lead to complex and unexpected cost effects in BPM projects making even rough cost estimations a challenge. What is needed is a comprehensive approach enabling BPM professionals to systematically investigate the costs of BPM projects. This chapter takes a look at both known and often unknown cost factors in BPM projects, shortly discusses existing IT evaluation and software cost estimation approaches with respect to their suitability for BPM projects, and finally introduces the EcoPOST framework. EcoPOST utilizes evaluation models to describe the interplay of technological, organizational, and project-specific BPM cost factors as well as simulation concepts to unfold the dynamic behavior and costs of BPM projects.

## 1 Introduction

### 1.1 Motivation

While the benefits of *Process-Aware Information Systems* (PAISs) and BPM technology are usually justified by improved process performance [53, 67, 69], there

M. Glykas (Ed.): Business Process Management, SCI 444, pp. 157–194.
springerlink.com      © Springer-Verlag Berlin Heidelberg 2013

exist no approaches for systematically analyzing related cost factors and their dependencies. Though software cost estimation [4] has received considerable attention during the last decades and has become an essential task in information systems engineering, it is difficult to apply existing approaches to BPM projects. This difficulty stems from the inability of these approaches to cope with the numerous technological, organizational and project-driven cost factors to be considered in the context of BPM projects [40]. As example consider the significant costs for redesigning business processes. Another challenge deals with the many dependencies existing between the different cost factors. Activities for *business process redesign*, for example, can be influenced by intangible impact factors like available *process knowledge* or *end user fears*. These dependencies, in turn, result in dynamic effects which influence the overall costs of BPM projects. Existing evaluation techniques [45] are usually unable to deal with such dynamic effects as they rely on rather static models based upon snapshots of the considered project context.

What is needed is an approach that enables organizations to investigate the complex interplay between the many cost and impact factors that arise in the context of PAIS introduction and BPM projects [52]. This chapter presents the EcoPOST methodology, a sophisticated and practically validated, model-based methodology to better understand and systematically investigate the complex cost structures of such BPM projects.

## 1.2   IT Evaluation - Challenges and Approaches

Generally, the adoption of *information technology* (IT) can be described by means of an *S curve* (cf. Fig. 1A) [10, 11, 62]. When new IT emerges at the market, it is unproven, expensive and difficult to use. Standards have not been established yet and best practices still have to established. At this point, only "first movers" start projects based on the emerging IT. They assume that the high costs and risks for being an innovator will be later compensated by gaining competitive advantage [7].

Picking up an emerging IT at a later stage, by contrast, allows to wait until it becomes more mature and standardized, resulting in lower introduction costs and risks. However, once the value of IT has become clear, both vendors and users rush to invest in it. Consequently, technical standards emerge and license costs decrease. Soon, the IT is widely spread, with only few enterprises having not made respective investment decisions. The S curve is then complete. Factors that typically push a new IT up the S curve include standardization, price deflation, best practice diffusion, and consolidation of the vendor base. All these factors also erode the ability of IT as enabler for differentiation and competitive advantage. In fact, when dissemination of IT increases, its strategic potential shrinks at the same time. Finally, once the IT has become part of the general infrastructure, it is difficult to achieve further strategic benefits (though rapid technological innovation often continues). This can be illustrated by a *Z curve* (cf. Fig. 1B).

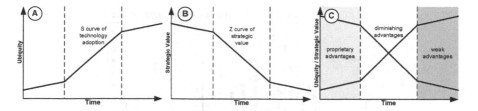

**Fig. 1** The Curves of Technology Adoption

Considering the different curves of IT adoption, decisions about IT investments (and the right point in time to realize them) constitute a difficult task [5, 18] (cf. Fig. 1C). Respective decisions are influenced by numerous factors [30, 36, 39, 37]. Hence, policy makers often demand for a business case [51] summarizing the key parameters of an IT investment. Thereby, different evaluation dimensions are taken into account [31]. As examples consider the costs of an investment, its assumed profit, its impact on work performance, business process performance, and the achievement of enterprise goals.

To cope with different evaluation goals, numerous evaluation approaches have been introduced (e.g. [48, 57, 58]). Fig. 2 shows results of an evaluation of 19 IT-evaluation approaches we presented in [45]. Today's policy makers usually rely on simple and static decision models as well as on intuition and experiences rather than on a profound analysis of an IT investment decision. Further, rules of thumb such as "invest to keep up with the technology" or "invest if the competitors have been successful" are often applied. In many cases, there is an asymmetric consideration of costs and benefits. For example, many financial calculations (cf. Section 4) over-estimate benefits in the first years in order to realize a positive ROI. Besides, many standard evaluation approaches are often not suitable to be used at early planning stages of IT investments. In fact, many projects (especially those utilizing innovative information technology) - despite their potential strategic importance - have a negative economic valuation result at an early stage. This situation results in a high risk of false rejection. This means that enterprises with independently operating business units, targeting at maximizing the equity of a company in short term, have to overcome the problem to not routinely reject truly important IT investments due to the use of insufficient evaluation techniques.

## 1.3 Chapter Overview

Section 2 summarizes the EcoPOST cost analysis method. Section 3 introduces evaluation patterns as an important means to simplify the application of the EcoPOST cost analysis approach. Section 4 introduces rules for designing EcoPOST evaluation models, while Section 5 describes general modeling guidelines. Section 6 introduces a case study to illustrate the use of EcoPOST and its practical benefits. Finally, Sections 7 and 8 conclude with a discussion of our approach and a summary.

**The Big Picture**

Caption: x: supported  o: optional  ↑: positive  ↓: negative  →: neutral

| The Big Picture | Return On Investment (ROI) | Payback Period | Accounting Rate of Return | Breakeven Analysis | Net Present Value (NPV) | Internal Rate of Return (IRR) | Zero Base Budgeting | Cost Effectiveness Analysis | Total Cost of Ownership (TCO) | Target Costing | Times Savings Times Salary Model | Hedonic Wage Model | Activity-based Costing | Business Process Intelligence | Nolan's Approach | Porter's Competitive Forces Model | Parson's Approach | Approach of McFarlan and McKenney | Real Option Theory |
|---|---|---|---|---|---|---|---|---|---|---|---|---|---|---|---|---|---|---|---|
| **I. Classification Criteria** | | | | | | | | | | | | | | | | | | | |
| **Evaluation Viewpoint** | | | | | | | | | | | | | | | | | | | |
| Financial Viewpoint | x | x | x | x | x | x | x | x | x | x | | | | | o | o | | | x |
| Work Performance Viewpoint | | | | | | | | | | | x | x | x | x | | | | | |
| Strategic Viewpoint | | | | | | | | | | | | | | | x | x | x | x | x |
| **Evaluation Dimension** | | | | | | | | | | | | | | | | | | | |
| Evaluation of Benefits | x | x | x | x | x | x | | x | | x | x | x | x | x | x | x | x | x | x |
| Evaluation of Costs | x | x | x | x | x | x | x | x | x | x | | | x | x | | x | | | x |
| Evaluation of Risks | o | o | o | o | o | o | o | | | | | | | | | x | | | |
| Evaluation of Work Performance | | | | | | | | | | | | x | | x | x | | | | |
| **Evaluation Scope** | | | | | | | | | | | | | | | | | | | |
| ex-ante | x | x | x | x | x | x | x | x | x | x | x | x | x | x | x | x | x | x | x |
| ex-post | x | x | x | x | | | | | | | | | x | x | x | | | | |
| **Evaluation Outcome** | | | | | | | | | | | | | | | | | | | |
| Quantification \| Absolute Measures | x | x | | x | x | | | x | | | x | x | x | | | | | | x |
| Quantification \| Relative Measures | | | x | | x | | | | | | | | x | | | | | | x |
| Visualization of Outcome | o | | o | | | | o | | o | | | x | x | x | | | | | x |
| Qualitative Conclusions | | | | | | | | x | | x | | | | | x | x | x | x | x |
| **II. Evaluation Criteria** | | | | | | | | | | | | | | | | | | | |
| **Plausibility** | | | | | | | | | | | | | | | | | | | |
| Interpretability of Evaluation Results | → | ↑ | → | ↑ | ↑ | ↑ | ↑ | ↑ | ↑ | ↑ | ↑ | ↑ | ↑ | ↑ | → | → | → | → | → |
| Transparency of Result Generation | ↑ | ↑ | ↑ | ↑ | ↑ | ↑ | ↑ | → | ↑ | → | ↑ | ↑ | ↑ | ↑ | → | ↓ | → | → | → |
| **Objectiveness** | | | | | | | | | | | | | | | | | | | |
| Degree of Objectiveness | ↑ | → | ↑ | → | → | → | → | ↑ | ↑ | → | ↑ | ↑ | ↑ | ↑ | ↓ | ↓ | ↓ | ↑ | → |
| **Sensitivity** | | | | | | | | | | | | | | | | | | | |
| Error-Proneness of Evaluations | → | → | → | → | ↓ | ↓ | ↑ | → | ↑ | → | ↓ | ↓ | ↓ | ↑ | ↓ | ↓ | ↓ | → | → |
| Resistance against Manipulation | → | ↓ | → | ↓ | ↓ | ↓ | → | → | → | → | → | → | → | → | ↓ | ↓ | ↓ | → | → |
| **Practical Applicability** | | | | | | | | | | | | | | | | | | | |
| Flexibility | ↑ | → | ↑ | → | ↑ | → | ↑ | ↑ | ↑ | ↑ | ↑ | ↑ | ↑ | ↑ | → | → | → | → | ↑ |
| Efficiency (Effort) – Data Collection & Preparation – Availability | → | → | → | → | → | → | ↓ | ↓ | ↑ | ↓ | ↑ | ↑ | → | ↑ | ↓ | → | → | → | → |
| Efficiency (Effort) – Data Collection & Preparation – Heterogeneity | → | → | → | → | → | → | ↑ | ↑ | ↑ | ↓ | ↑ | ↑ | ↑ | → | ↓ | → | → | → | → |
| Efficiency (Effort) – Data Collection & Preparation – Data Quality | → | → | → | → | → | → | → | ↓ | → | ↓ | ↓ | ↓ | ↓ | ↓ | → | → | ↑ | ↑ | → |
| Efficiency (Effort) – Overall Effort | ↑ | ↑ | ↑ | ↑ | ↑ | ↑ | ↑ | ↓ | → | ↓ | ↑ | → | ↑ | → | ↓ | ↑ | ↑ | ↑ | ↑ |
| Effectiveness – Comparability | ↑ | ↑ | ↑ | ↑ | ↑ | ↑ | → | ↑ | → | ↑ | ↑ | ↑ | ↑ | ↑ | ↓ | ↓ | ↓ | ↓ | ↑ |
| Effectiveness – Correctness of Conclusions | → | → | → | → | → | → | ↑ | ↑ | → | ↑ | → | → | → | ↑ | ↑ | ↑ | ↑ | → | → |
| **Theoretical Foundation** | | | | | | | | | | | | | | | | | | | |
| Degree of Formalization | ↑ | ↑ | ↑ | ↑ | ↑ | ↑ | → | → | → | → | ↑ | ↑ | → | ↑ | ↓ | ↓ | ↓ | ↓ | ↑ |
| **Tool Support** | | | | | | | | | | | | | | | | | | | |
| Simplicity of providing Tool Support | ↑ | ↑ | ↑ | ↑ | ↑ | ↑ | ↑ | ↑ | ↑ | ↑ | ↑ | → | → | ↑ | ↓ | → | → | → | → |

**Fig. 2** IT Evaluation Approaches: The Big Picture

## 2   The EcoPOST Cost Analysis Methodology

Our EcoPOST methodology was designed to support the introduction of Process-aware Information Systems (PAISs) [14, 50, 49] and BPM technology [13, 22, 70, 55] in the automotive industry (and was, consequently, also validated and piloted in several BPM projects in this domain). The EcoPOST methodology comprises seven steps (cf. Fig. 3). *Step 1* concerns the comprehension of an evaluation scenario. This is crucial for developing problem-specific evaluation models. The following two steps (Steps 2 and 3) deal with the identification of two different kinds of *Cost Factors* representing costs that can be quantified in terms of money (cf. Table 1): *Static Cost Factors* (SCFs) and *Dynamic Cost Factors* (DCFs).

**Table 1**  Cost Factors

| | Description |
|---|---|
| SCF | *Static Cost Factors* (SCFs) represent costs whose values do not change during an BPM project (except for their time value, which is not further considered in the following). Typical examples: software license costs, hardware costs and costs for external consultants. |
| DCF | *Dynamic Cost Factors* (DCFs), in turn, represent costs that are determined by activities related to an BPM project. The (re)design of business processes prior to the introduction of PAIS, for example, constitutes such an activity. As another example consider the performance of interview-based process analysis. These activities cause measurable efforts which, in turn, vary due to the influence of intangible *impact factors*. The DCF "Costs for Business Process Redesign" may be influenced, for instance, by an intangible factor "Willingness of Staff Members to Support Process (Re)Design Activities". Obviously, if staff members do not contribute to a (re)design project by providing needed information (e.g., about process details), any redesign effort will be ineffective and result in increasing (re)design costs. If staff willingness is additionally varying during the (re)design activity (e.g., due to a changing communication policy), the DCF will be subject to even more complex effects. In the EcoPOST framework, intangible factors like the one described are represented by *impact factors*. |

*Step 4* deals with the identification of *Impact Factors* (ImFs), i.e., intangible factors that influence DCFs and other ImFs. We distinguish between organizational, project-specific and technological ImFs. ImFs cause the value of DCFs (and other ImFs) to change, making their evaluation a difficult task to accomplish. As examples consider factors such as "End User Fears", "Availability of Process Knowledge" and "Ability to (Re-)design Business Processes". Finally, ImFs can be *static* or *dynamic* (cf. Table 2).

**Table 2**  Impact Factors

| | Description |
|---|---|
| *Static ImF* | Static ImFs do not change, i.e., they are assumed to be constant during an BPM project; e.g., when there is a fixed degree of user fears, process complexity, or work profile change. |
| *Dynamic ImF* | Dynamic ImFs may change during an BPM project, e.g., due to interference with other ImFs. As examples consider process and domain knowledge which is typically varying during an BPM project (or a subsidiary activity). |

Unlike SCFs and DCFs the values of ImFs are not quantified in monetary terms. Instead, they are "quantified" by experts[1] using qualitative scales describing the degree of an ImF. As known from software cost estimation models, such as COCOMO [4], the qualitative scales we use comprise different "values" (typically ranging from "very low" to "very high"). These values are used to express the strength of an ImF on a given cost factor (just like in COCOMO).

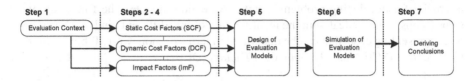

**Fig. 3** Main Steps of the EcoPOST Methodology

Generally, dynamic evaluation factors (i.e., DCFs and dynamic ImFs) are difficult to comprehend. In particular, intangible ImFs (i.e., their appearance and impact in BPM projects) are not easy to follow. When evaluating the costs of BPM projects, therefore, DCFs and dynamic ImFs constitute a major source of misinterpretation and ambiguity. To better understand and to investigate the dynamic behavior of DCFs and dynamic ImFs, we introduce the notion of *evaluation models* as basic pillar of the EcoPOST methodology (*Step 5*; cf. Section 3). These evaluation models can be simulated (*Step 6*) to gain insights into the dynamic behavior (i.e., evolution) of DCFs and dynamic ImFs (*Step 7*). This is important to effectively control the design and implementation of PAIS as well as the costs of respective BPM projects.

## 2.1  Evaluation Models

In EcoPOST, dynamic cost/impact factors are captured and analyzed by evaluation models which are specified using the System Dynamics [54] notation (cf. Fig. 4). An evaluation model comprises SCFs, DCFs and ImFs corresponding to model variables.

Different types of variables exist. *State variables* can be used to represent dynamic factors, i.e., to capture changing values of DCFs (e.g., the "Business Process Redesign Costs"; cf. Fig. 4A) and dynamic ImFs (e.g., "Process Knowledge"). A state variable is graphically denoted as rectangle (cf. Fig. 4A), and its value at time $t$ is determined by the accumulated changes of this variable from starting point $t_0$ to present moment $t$ ($t > t_0$); similar to a bathtub which accumulates – at a defined moment $t$ – the amount of water poured into it in the past. Typically, state variables are connected to at least one *source* or *sink* which are graphically denoted as cloud-like symbols (except for state variables connected to other ones) (cf. Fig. 4A). Values of

---

[1] The efforts of these experts for making that quantification is not explicitly taken into account in EcoPOST, though this effort also increases information system development costs.

state variables change through inflows and outflows. Graphically, both flow types are depicted by twin-arrows which either point to (in the case of an inflow) or out of (in the case of an outflow) the state variable (cf. Fig. 4A). Picking up again the bathtub image, an *inflow* is a pipe that adds water to the bathtub, i.e., inflows increase the value of state variables. An *outflow*, by contrast, is a pipe that purges water from the bathtub, i.e., outflows decrease the value of state variables. The DCF "Business Process Redesign Costs" shown in Fig. 4A, for example, increases through its inflow ("Cost Increase") and decreases through its outflow ("Cost Decrease"). Returning to the bathtub image, we further need "water taps" to control the amount of water flowing into the bathtub, and "drains" to specify the amount of water flowing out. For this purpose, a *rate variable* is assigned to each flow (graphically depicted by a valve; cf. Fig. 4A). In particular, a rate variable controls the inflow/outflow it is assigned to based on those SCFs, DCFs and ImFs which influence it. It can be considered as an interface which is able to merge SCFs, DCFs and ImFs.

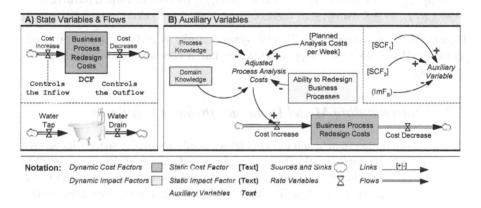

**Fig. 4** Evaluation Model Notation and initial Examples

Besides state variables, evaluation models may comprise *constants* and *auxiliary variables*. Constants are used to represent static evaluation factors, i.e., SCFs and static ImFs. Auxiliary variables, in turn, represent intermediate variables and typically bring together – like rate variables – cost and impact factors, i.e., they merge SCFs, DCFs and ImFs. As example consider the auxiliary variable "Adjusted Process Analysis Costs" in Fig. 4B, which merges the three dynamic ImFs "Process Knowledge", "Domain Knowledge", and "Ability to Redesign Business Processes" and the SCF "Planned Analysis Costs per Week". Both constants and auxiliary variables are integrated into an evaluation model with *links* (not flows), i.e., labeled arrows. A *positive link* (labeled with "+") between x and y (with y as dependent variable) indicates that y will tend in the same direction if a change occurs in x. A *negative link* (labeled with "-") expresses that the dependent variable y will tend in the opposite direction if the value of x changes. Altogether, we define:

***Definition 2.1 (Evaluation Model)*** *A graph EM = (V, F, L) is denotes as evaluation model, if the following holds:*

- $V := S \overset{.}{\cup} X \overset{.}{\cup} R \overset{.}{\cup} C \overset{.}{\cup} A$ *is a set of model variables with*

  - *S is a set of state variables,*
  - *X is a set of sources and sinks,*
  - *R is a set of rate variables,*
  - *C is a set of constants,*
  - *A is a set of auxiliary variables,*

- $F \subseteq ((S \times S) \cup (S \times X) \cup (X \times S))$ *is a set of edges representing flows,*
- $L \subseteq ((S \times A \times Lab) \cup (S \times R \times Lab) \cup (A \times A \times Lab) \cup (A \times R \times Lab) \cup$
  $(C \times A \times Lab) \cup (C \times R \times Lab))$ *is a set of edges representing links with*
  $Lab := \{+, -\}$ *being the set of link labels:*

  - $(q_i, q_j, +) \in L$ *with* $q_i \in (S \overset{.}{\cup} A \overset{.}{\cup} C)$ *and* $q_j \in (A \overset{.}{\cup} R)$ *denotes a positive link,*
  - $(q_i, q_j, -) \in L$ *with* $q_i \in (S \overset{.}{\cup} A \overset{.}{\cup} C)$ *and* $q_j \in (A \overset{.}{\cup} R)$ *denotes a negative link.*

Generally, the evolution of DCFs and dynamic ImFs is difficult to comprehend. Thus, EcoPOST additionally provides a simulation component for capturing and analyzing this evolution (cf. Step 6 in Fig. 3).

## 2.2 Understanding Model Dynamics through Simulation

To enable simulation of an evaluation model we need to formally specify its behavior. EcoPOST accomplishes this by means of a *simulation model* based on *mathematical equations*. Thereby, the behavior of each model variable is specified by one equation (cf. Fig. 5), which describes how a variable is changing over time during simulation.

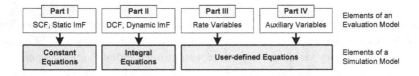

**Fig. 5** Elements of a Simulation Model

Fig. 6A shows a simple evaluation model.[2] Assume that the evolution of the DCF "Business Process Redesign Costs" (triggered by the dynamic ImF "End User Fears") shall be analyzed. End user fears can lead to emotional resistance of users and, in turn, to a lack of user support when redesigning business processes (e.g., during an interview-based process analysis). For model variables representing an

---

[2] It is the basic goal of this toy example to illustrate how evaluation models are simulated. Generally, evaluation models are much more complex. Due to lack of space we do not provide a more extensive example here.

SCF or static ImF the equation specifies a constant value for the model variable; i.e., SCFs and static ImFs are specified by single numerical values in *constant equations*. As example consider EQUATION A in Fig. 6B. For model variables representing DCFs, dynamic ImFs or rate/auxiliary variables, the corresponding equation describes how the value of the model variable evolves over time (i.e., during simulation). Thereby, the evolution of DCFs and dynamic ImFs is characterized by *integral equations* [16]. This allows us to capture the accumulation of DCFs and dynamic ImFs from the start of a simulation run ($t_0$) to its end ($t$):

**Definition 2.2 (Integral Equation)** *Let EM be an evaluation model (cf. Definition 2.1) and S be the set of all DCFs and dynamic ImFs defined by EM. An integral equation for a dynamic factor $v \in S$ is defined as follows:*

$$v(t) = \int_{t_0}^{t}[inflow(s) - outflow(s)]ds + v(t_0) \text{ where}$$

- *$t_0$ denotes the starting time of the simulation run,*
- *$t$ represents the end time of the simulation run,*
- *$v(t_0)$ represents the value of $v$ at $t_0$,*
- *$inflow(s)$ represents the value of the inflow at any time $s$ between $t_0$ and $t$,*
- *$outflow(s)$ represents the value of the outflow at any time $s$ between $t_0$ and $t$.*

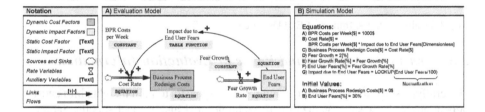

**Fig. 6** Dealing with the Impact of End User Fears

As example consider EQUATION C in Fig. 6B which specifies the increase of the DCF "Business Process Redesign Costs" (based on only one inflow). Note that in Fig. 6B the equations of the DCF "Business Process Redesign Costs" and the dynamic ImF "End User Fears" are presented in the way they are specified in Vensim [65], the simulation tool used by EcoPOST, and not as real integral equations.

Rate and auxiliary variables are both specified in the same way, i.e., as user-defined functions defined over the variables preceding them in the evaluation model. In other words, rate as well as auxiliary variables are used to merge static and dynamic cost/impact factors. During simulation, values of rate and auxiliary variables are dynamic, i.e., they change along the course of time. Reason is that they are not only influenced by SCFs and static ImFs, but also by evolving DCFs and dynamic ImFs. The behavior of rate and auxiliary variables is specified in the same way:

**Definition 2.3 (User-defined Equation)** *Let EM be an evaluation model (cf. Def.
2.1) and X be the set of rate/auxiliary variables defined by EM. An equation for
$v \in X$ is a user-defined function $f(v_1, ..., v_n)$ with $v_1, ..., v_n$ being the predecessors of
v in EM.*

As example consider EQUATION B in Fig. 6B. The equation for rate variable "Cost
Rate" merges the SCF "BPR Costs per Week" with the auxiliary variable "Impact
due to End User Fears". Assuming that activities for business process redesign are
scheduled for 32 weeks, Fig. 7A shows the values of all dynamic evaluation factors
of the evaluation model over time when performing a simulation. Fig. 7B shows the
outcome of the simulation. As can be seen there is a significant negative impact of
end user fears on the costs of business process redesign.

| TIME | Change ($) | BPR Costs ($) | Cost Rate ($) | Change (%) | User Fears (%) |
|------|------------|---------------|---------------|------------|----------------|
| 00   | -          | 0             | 1000          | -          | 30             |
| 01   | 1000       | 1000          | 1010          | 2          | 32             |
| 02   | 1010       | 2010          | 1020          | 2          | 34             |
| 03   | 1020       | 3030          | 1030          | 2          | 36             |
| 04   | 1030       | 4060          | 1040          | 2          | 38             |
| 05   | 1040       | 5100          | 1050          | 2          | 40             |
| 06   | 1050       | 6150          | 1060          | 2          | 42             |
| ...  | ...        | ...           | ...           | ...        | ...            |
| 30   | 1840       | 38300         | 1900          | 2          | 90             |
| 31   | 1900       | 40200         | 2020          | 2          | 92             |
| 32   | 2020       | 42220         | 2140          | 2          | 94             |

**Fig. 7** Dealing with the Impact of End User Fears

## 2.3 Sensitivity Analysis and Reuse of Evaluation Information

Generally, results of a simulation enable PAIS engineers to gain insights into causal
dependencies between organizational, technological and project-specific factors.
This helps them to better understand resulting effects and to develop a concrete
"feeling" for the dynamic implications of EcoPOST evaluation models. To investigate how a given evaluation model "works" and what might change its behavior,
we simulate the dynamic implications described by it – a task which is typically too
complex for the human mind. In particular, we conduct "behavioral experiments"
based on series of simulation runs. During these simulation runs selected simulation
parameters are manipulated in a controlled manner to systematically investigate the
effects of these manipulations, i.e., to investigate how the output of a simulation will
vary if its initial condition is changed. This procedure is also known as *sensitivity
analysis*. Simulation outcomes can be further analyzed using graphical charts.

Designing evaluation models can be a complicated and time-consuming task.
Evaluation models can become complex due to the high number of potential cost
and impact factors as well as the many causal dependencies that exist between
them. Taking the approach described so far, each evaluation and simulation model
has to be designed from scratch. Besides additional efforts, this results in an exclusion of existing modeling experience and prevents the reuse of both evaluation and

simulation models. In response to this problem, Section 3 introduces a set of reusable *evaluation patterns* (EP). EPs do not only ease the design and simulation of evaluation models, but also enable the reuse of evaluation information. This is crucial to foster the practical applicability of the EcoPOST framework.

## 2.4 Other Approaches

To enable cost evaluation of BPM technology, numerous evaluation approaches exist – the most relevant ones are depicted in Fig. 2. All these approaches can be used in the context of BPM projects. However, only few of them address the specific challenges tackled by EcoPOST.

*Activity-based costing* (ABC) belongs to this category. It does not constitute an approach to unfold the dynamic effects triggered by causal dependencies in BPM projects. Instead, ABC provides a method for allocating costs to products and services. ABC helps to identify areas of high overhead costs per unit and therewith to find ways to reduce costs. Doing so, the scope of the business activities to be analyzed has to be identified in a first step (e.g., based on activity decomposition). Identified activities are then classified. Typically, one distinguishes between *value adding* or *non-value adding* activities, between *primary* or *secondary* activities, and between *required* or *non-required* activities. An activity will be considered as value-adding (compared to a non-value adding one) if its output is directly related to customer requirements, services or products (as opposed to administrative or logistical outcomes). Primary activities directly support the goals of an organization, whereas secondary activities support primary ones. Required (unlike non-required) activities are those that must always be performed. For each activity creating the products or services of an organization, costs are gathered. These costs can be determined based on salaries and expenditures for research, machinery, or office furniture. Afterwards, activities and costs are combined and the total input cost for each activity is derived. This allows for calculating the total costs consumed by an activity. However, at this stage, only costs are calculated. It is not yet determined where the costs originate from. Following this, the "activity unit cost" is calculated. Though activities may have multiple outputs, one output is identified as the primary one. The "activity unit cost" is calculated by dividing the total input cost (including assigned costs from secondary activities) by the primary activity output. Note that the primary output must be measurable and its volume or quantity be obtainable. From this, a "bill of activities" is derived which contains a set of activities and the amount of costs consumed by each activity. Then, the amount of each consumed activity is extended by the activity unit cost and is added up as a total cost for the bill of activity. Finally, the calculated activity unit costs and bills of activity are used for identifying candidates for business process improvement. In total, ABC is an approach to make costs related to business activities (e.g., business process steps) transparent. Therefore, aplying the method can be an accompanying step of EcoPOST in order to make certain cost and impact factors transparent. However, the correct accomplishment of an

ABC analysis causes significant efforts and requires a lot of experience. Often, it may be not transparent, for example, which costs are caused by which activity.

Besides, there are formalisms (no full evaluation approaches!) that can be applied to unfold the dynamic effects triggered by causal dependencies in BPM projects.

Causal *Bayesian Networks* (BN) [23], for example, promise to be a useful approach. BN deal with (un)certainty and focus on determining probabilities of events. A BN is a directed acyclic graph which represents interdependencies embodied in a given joint probability distribution over a set of variables. This allows to investigate the interplay of the components of a system and the effects resulting from this. BN do not allow to model feedback loops since cycles in BN would allow infinite feedbacks and oscillations that prevent stable parameters of the probability distribution.

*Agent-based modeling* provides another interesting approach. Resulting models comprise a set of reactive, intentional, or social agents encapsulating the behavior of the various variables that make up a system [6]. During simulation, the behavior of these agents is emulated according to defined rules [59]. System-level information (e.g., about intangible factors being effective in a BPM project) is thereby not further considered. However, as system-level information is an important aspect in our approach, we have not further considered the use of agent-based modeling.

Another approach is provided by *fuzzy cognitive maps* (FCM) [17]. An FCM is a cognitive map appyling relationships between the objects of a "mental landscape" (e.g., concepts, factors, or other resources) in order to compute the "strength of impact" of these objects. To accomplish the latter task fuzzy logic is used. Most important, an FCM can be used to support different kinds of planning activities. As example consider (BPM) projects; for them, an FCM alows to analyze the mutual dependencies between respective project resources.

## 3   EcoPOST Evaluation Patterns

BPM projects often exhibit similarities, e.g., regarding the appearance of certain cost and impact factors. We pick up these similarities by introducing customizable patterns. This shall increase model reuse and facilitate practical use of the EcoPOST framework. *Evaluation patterns* (EPs) do not only ease the design and simulation of evaluation models, but also enable the reuse of evaluation information [42, 35, 38]. This is crucial to foster practical applicability of the EcoPOST framework.

Specifically, we introduce an *evaluation pattern* (EP) as a predefined, but customizable EcoPOST model, i.e., an EP can be built based on the elements introduced in Section 2. An EP consists of an *evaluation model* and a corresponding *simulation model*. More precisely, each EP constitutes a template for a specific DCF or ImF as it typically exists in many BPM projects. Moreover, we distinguish between *primary* EPs (cf. Section 3.2) and *secondary* ones (cf. Section 3.3). A primary EP describes a DCF whereas a secondary EP represents an ImF. We denote an EP representing an ImF as secondary as it has a supporting role regarding the design of EcoPOST cost models based on primary EPs.

The decision whether to represent cost/impact factors as static or dynamic factors in EPs also depends on the model designer. Many cost and impact factors can be modeled both as static or dynamic factors. Consequently, EPs can be modeled in alternative ways. This is valid for all EPs discussed in the following.

Patterns were first introduced to describe best practices in architecture [1]. However, they have also a long tradition in computer science, e.g., in the fields of software architecture (*conceptual patterns*), design (*design patterns*), and programming (*XML schema patterns*, *J2EE patterns*, etc.). Recently, the idea of using patterns has been also applied to more specific domains like workflow management [66, 68, 56, 26] or inter-organizational control [24]. Generally, patterns describe solutions to recurring problems. They aim at supporting others in learning from available solutions and allow for the application of these solutions to similar situations. Often, patterns have a generative character. Generative patterns (like the ones we introduce) tell us how to create something and can be observed in the environments they helped to shape. Non-generative patterns, in turn, describe recurring phenomena without saying how to reproduce them.

Reusing System Dynamics models has been discussed before as well. Senge [61], Eberlein and Hines [15], Liehr [29], and Myrtveit [46] introduce generic structures (with slightly different semantics) satisfying the capability of defining "components". Winch [72], in turn, proposes a more restrictive approach based on the parameterization of generic structures (without providing standardized modeling components). Our approach picks up ideas from both directions, i.e. we address both the definition of generic components and customization.

## 3.1   Research Methodology and Pattern Identification

As sources of our patterns (cf. Tables 3 and 4) we consider results from surveys [41], case studies [35, 43], software experiments [44], and profound experiences we gathered in BPM projects. These projects addressed a variety of typical settings in enterprise computing which allows us to generalize our experiences.

**Table 3** Overview of primary Evaluation Patterns and their Data Sources

| Pattern Name | Discussed in chapter | Survey | Case Study | Literature | Experiment | Experiences |
|---|---|---|---|---|---|---|
| Business Process Redesign Costs | yes | x | x | x | - | x |
| Process Modeling Costs | yes | - | - | x | x | x |
| Requirements Definition Costs | yes | - | x | x | - | x |
| Process Implementation Costs | yes | x | x | x | x | x |
| Process Adaptation Costs | no | x | x | x | x | x |

To ground our patterns on a solid basis we first create a list of candidate patterns. For generating this initial list we conduct a detailed literature review and rely on our experience with PAIS-enabling technologies, mainly in the automotive industry (e.g., [3, 34, 22]. Next we thoroughly analyze the above mentioned material to find empirical evidence for our candidate patterns. We then map the identified evaluation data to our candidate patterns and – if necessary – extend the list of candidate patterns.

**Table 4** Overview of Secondary Evaluation Patterns and their Data Sources

| Pattern Name | Discussed in chapter | Survey | Case Study | Literature | Experiment | Experiences |
|---|---|---|---|---|---|---|
| Process Knowledge | yes | x | - | x | x | x |
| Domain Knowledge | yes | x | - | x | x | x |
| Process Evolution | yes | x | - | x | - | x |
| Process Complexity | yes | - | - | x | - | - |
| Process Maturity | no | - | - | x | - | x |
| Work Profile Change | no | x | - | x | x | x |
| End User Fears | no | x | x | x | - | x |

A pattern is defined as reusable solution to a commonly occurring problem. We require each evaluation pattern to be observed at least three times in literature and our empirical research. Only those patterns, for which such empirical evidence exists, are included in the final list of patterns presented in the following.

## 3.2 Primary Evaluation Patterns

**Business Process Redesign Costs.** The EP shown in Fig. 8 deals with the costs of business process redesign activities. Prior to PAIS development such activities become necessary for several reasons. As examples consider the need to optimize business process performance or the goal of realizing a higher degree of process automation. This EP is based on our experiences (from several process redesign projects) that business process redesign costs are primarily determined by two SCFs: "Planned Costs for Process Analysis" and "Planned Costs for Process Modeling". While the former SCF represents planned costs for accomplishing interviews with process participants and costs for evaluating existing process documentation, the latter SCF concerns costs for transforming gathered process information into a new process design. Process redesign costs are thereby assumed to be varying, i.e., they are represented as DCF.

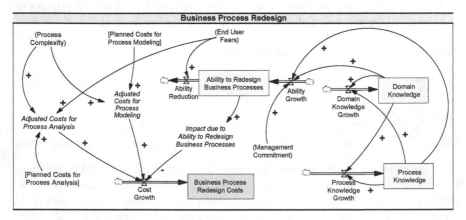

**Fig. 8** Primary Evaluation Pattern: Business Process Redesign Costs

This EP reflects our experience that six additional ImFs are of particular importance when investigating the costs of process redesign activities: "Process Complexity", "Management Commitment", "End User Fears", "Process Knowledge", "Domain Knowledge", and "Ability to Redesign Business Processes" (cf. Fig. 8). The importance of these factors is confirmed by results from one of our surveys. While process analysis costs (i.e., the respective SCF) are influenced by "Process Complexity" and "End User Fears" (merged in the auxiliary variable "Adjusted Costs for Process Analysis"), process modeling costs are only influenced by "Process Complexity" (as end users are typically not participating in the modeling process). Business process redesign costs are further influenced by a dynamic ImF "Ability to Redesign Business Processes", which, in turn, is influenced – according to our practical experiences – by four ImFs (causing the ImF "Ability to Redesign Business Processes" to change): "Management Commitment", "End User Fears", "Process Knowledge", and "Domain Knowledge". Note that – if desired – the effects of the latter three ImFs can be further detailed based on available secondary EPs.

**Process Modeling Costs.** The EP depicted in Fig. 9 deals with the costs of process modeling activities in BPM projects. Such activities are typically accomplished to prepare the information gathered during process analysis, to assist software developers in implementing the PAIS, and to serve as guideline for implementing the new process design (in the organization). Generally, there exist many notations that can be used to specify process models. Our EP, for example, assumes that process models are expressed as *event-driven process chains* (EPC).

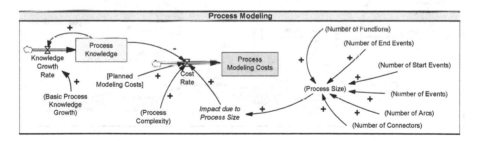

**Fig. 9** Primary Evaluation Pattern: Process Modeling Costs

Basically, this EP (cf. Fig. 9) reflects our experiences that "Process Modeling Costs" are influenced by three ImFs: the two static ImFs "Process Complexity" and "Process Size" (whereas the impact of process size is specified based on a table function transforming a given process size into an EcoPOST impact rating [35]) and the dynamic ImF "Process Knowledge" (which has been also confirmed by our survey described in [35]). The ImF "Process Complexity" is not further discussed here. Instead, we refer to [35] where this ImF has been introduced in detail. The ImF "Process Size", in turn, is characterized based on (estimated) attributes of the process model to be developed. These attributes depend on the used modeling formalism. As aforementioned, the EP from Fig. 9 builds on the assumption that the EPC formalism is used for process modeling. Taking this formalism, we specify process size based on the "Number of Functions", "Number of Events", "Number of Arcs", "Number of Connectors", "Number of Start Events", and "Number of "End Events". Finally, the DCF "Process Modeling Costs" is also influenced by the dynamic ImF "Process Knowledge" (assuming that an increasing amount of process knowledge results in decreasing modeling costs).

**Requirements Definition Costs.** The EP from Fig. 10 deals with costs for defining and eliciting requirements [35]. It is based on the two DCFs "Requirement Definition Costs" and "Requirement Test Costs" as well as on the ImF "Requirements to be Documented". This EP reflects the observation we made in practice that the DCF "Requirements Definition Costs" is determined by three main cost factors: costs of a requirements management tool, process analysis costs, and requirements documentation costs. Costs of a requirements management tool are constant and are therefore represented as SCF. The auxiliary variable "Adjusted Process Analysis Costs", in turn, merges the SCF "Planned Process Analysis Costs" with four process-related ImFs: "Process Complexity", "Process Fragmentation", "Process Knowledge", and "Emotional Resistance of End Users" (whereas only process knowledge is represented as dynamic ImF).

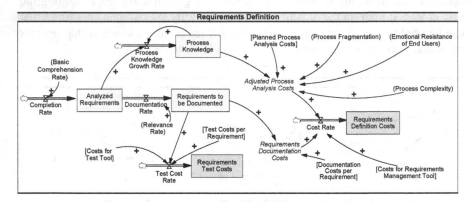

**Fig. 10** Primary Evaluation Pattern: Requirements Definition Costs

Costs for documenting requirements (represented by the auxiliary variable "Requirements Documentation Costs") are determined by the SCF "Documentation Costs per Requirement" and by the dynamic ImF "Requirements to be Documented". The latter ImF also influences the dynamic ImF "Process Knowledge" (resulting in a positive link from "Analyzed Requirements" to the rate variable "Process Knowledge Growth Rate"). "Requirements Test Costs" are determined by two SCFs ("Costs for Test Tool" and "Test Costs per Requirement") and the dynamic ImF "Requirements to be documented" (as only documented requirements need to be tested). Costs for a test tool and test costs per requirement are assumed to be constant (and are represented as SCFs).

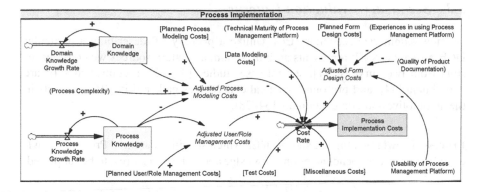

**Fig. 11** Primary Evaluation Pattern: Process Implementation Costs

**Process Implementation Costs.** The EP depicted in Fig. 11 deals with costs for implementing a process and their interference through impact factors [35]. An additional EP (not shown here) deals with the costs caused by adapting the process(es) supported by an PAIS. This additional EP is identical to the previous EP "Process Implementation Costs" – except for the additional ImF "Process Evolution".

Basic to this EP is our observation that the implementation of a process is determined by six main cost factors (cf. Fig. 11): "Adjusted Process Modeling Costs", "Adjusted User/Role Management Costs", "Adjusted Form Design Costs", "Data Modeling Costs", "Test Costs", and "Miscellaneous Costs". The first three cost factors are characterized as "adjusted" as they are influenced – according to interviews with software developers and process engineers – by additional process-related ImFs. Therefore, they are represented by auxiliary variables which merge the SCFs with ImFs. Process modeling costs, for example, are influenced by "Process Knowledge", "Domain Knowledge", and "Process Complexity". User/role management costs are only biased by "Process Knowledge". Form design costs are influenced by some technology-specific ImFs (cf. Chapter 4.7.4): "Technical Maturity of Process Management Platform", "Experiences in using Process Management Platform", "Usability of Process Management Platform", and "Quality of Product Documentation". Note that this EP strongly simplifies the issue of process implementation. In particular, we assume that the identified six main cost factors aggregate all other potential cost factors. The SCF "Data Modeling Costs", for example, may include costs for providing database management functionality and for configuring a database management system. However, it is thereby not further distinguished between subsidiary cost factors, i.e., other cost factors are not made explicit. If this is considered as necessary, additional SCFs or DCFs can be introduced in order to make specific cost factors more explicit. Note that we have analyzed this EP in more detail in a controlled software experiment [44].

### 3.3 Secondary Evaluation Patterns

We now summarize secondary EPs. Unlike a primary EP describing a particular DCF, a secondary EP represents an ImF. Again, as pattern sources we consider results from two surveys [41], several case studies [35, 43], a controlled software experiment [44], and profound practical experiences gathered in BPM projects in the automotive and clinical domain [34, 28].

**Process Knowledge.** Fig. 12 shows an EP which specifies the ImF "Process Knowledge", i.e., causal dependencies on knowledge about the process(es) to be supported.

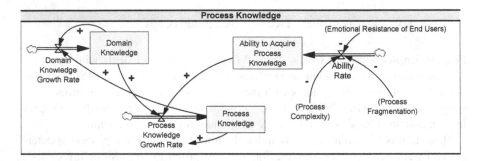

**Fig. 12** Secondary Evaluation Pattern: Process Knowledge

Process knowledge includes, for example, knowledge about process participants and their roles as well as knowledge about the flow of data. Acquiring process knowledge necessitates the ability to acquire process knowledge. This ability, however, strongly depends on three ImFs: "Emotional Resistance of End Users", "Process Complexity" and "Process Fragmentation". Besides, process knowledge is also influenced by the dynamic ImF "Domain Knowledge".

**Domain Knowledge.** The EP from Fig. 13 deals with the evolution of domain knowledge along the course of an BPM project. Our practical experiences allow for the conclusion that "Domain Knowledge" is a dynamic ImF influenced by three other ImFs: the period an PAIS engineer is working in a specific domain (captured by the dynamic ImF "Experience"), the dynamic ImF "Process Knowledge" and the complexity of the considered domain (represented by the static ImF "Domain Complexity"). Besides, the dynamic ImF "Domain Knowledge" is additionally influenced by the static ImF "Basic Domain Knowledge Growth". This static ImF reflects the situation that domain knowledge is continuously increasing during a BPM project (or a subsidiary activity).

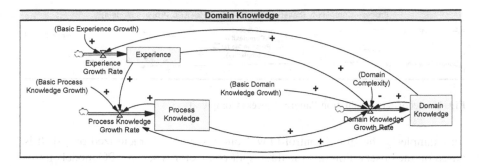

**Fig. 13** Secondary Evaluation Pattern: Domain Knowledge

**Process Evolution.** The EP depicted in Fig. 14 covers the static ImF "Process Evolution". Specifically, it describes drivers of process evolution. Basically, this EP reflects the assumption that business process evolution is caused by various drivers. Note that arbitrary drivers of evolution may be included in the EP.

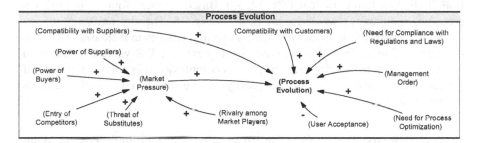

**Fig. 14** Secondary Evaluation Pattern: Business Process Evolution

**Process Complexity.** The EP from Fig. 15 deals with the ImF "Process Complexity". Note that this EP does not specify process complexity itself, but defines it based on an easier manageable replacement factor. In our context, this replacement factor corresponds to the complexity of the process model describing the business process to be supported [8]. Thus, we extend process complexity to "Process Complexity / Process Model Complexity". The EP from Fig. 15 further aligns with the assumption that respective process models are formulated using EPC notation. According to the depicted EP, the static ImF "Process Complexity/Process Model Complexity" is determined by four other static ImFs: "Cycle Complexity", "Join Complexity" (JC), "Control-Flow Complexity" (CFC), and "Split-Join-Ratio" (SJR) (whereas the latter ImF is derived from the SCFs "Join Complexity" and "Control-Flow Complexity"). The complexity driver "Cycle Complexity" has been motivated in [9, 27]. Arbitrary cycles, for example, can lead to EPC models without clear semantics (cf. [25]

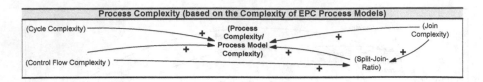

**Fig. 15** Secondary Evaluation Pattern: Process Complexity

for examples). The ImF "Control-Flow Complexity" is characterized by [8]. It is based on the observation that the three split connector types in EPC models introduce a different degree of complexity. According to the number of potential post-states an AND-split is weighted with 1, an XOR-split is weighted with the number of successors $n$, and an OR-split is weighted with $2n - 1$. The sum of all connector weights of an EPC model is then denoted as "Control-Flow Complexity" [19]. The ImF "Join Complexity" can be defined as the sum of weighted join connectors based on the number of potential pre-states in EPC models [32, 33]. Finally, the mismatch between potential post-states of splits and pre-states of joins in EPC models is included as another driver of complexity. This mismatch is expressed by the static ImF "Split-Join-Ratio" (= JC/CFC) [32, 33]. Based on these four static ImFs (or drivers of complexity), we derive the EP from Fig. 15. Thereby, an increasing cycle complexity results in higher process complexity. Also, both increasing CFC and increasing JC result in increasing process complexity. A JSR value different from 1 increases error probability and thus process complexity. It is noteworthy that – if desired – other drivers of process complexity may be considered as well. Examples can be found in [27, 33].

**Process Maturity.** The EP from Fig. 16 specifies the static ImF "Process Maturity". This EP is based on the assumption that increasing process maturity results in lower costs. This static ImF builds upon the 22 process areas of the *capability maturity model integration* (CMMI) [12], a process improvement approach providing organizations with elements of effective processes. The overall ImF "Process Maturity" is determined by the maturity of the four categories of the CMMI continuous representation: "Process Management", "Engineering", "Project Management", and "Support". Each category is further detailed by CMMI process areas.

**Work Profile Change.** This EP (not shown here, but discussed in [35]) deals with the change of end user work profiles (and the effects of work profile changes). More specifically, it relates the perceived work profile change to changes emerging in the five job dimensions of Hackman's *job characteristics model* [20, 21]: (1) *skill variety*, (2) *task identity*, (3) *task significance*, (4) *autonomy*, and (5) *feedback from the job*. For each of these five core job dimensions, the emerging change is designated based on the level before and after PAIS introduction.

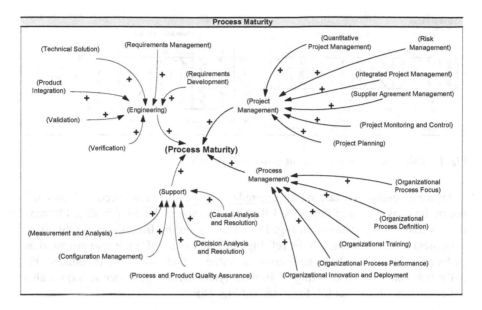

**Fig. 16** Secondary Evaluation Pattern: Process Maturity (Continuous Representation)

**End User Fears.** This EP (not shown here, but discussed in [35] and [42]) is based on experiences which allow to conclude that the introduction of an PAIS may cause end user fears, e.g., due to work profile change (i.e., job redesign) or changed social clues. Such fears often lead, for example, to emotional resistance of end users. This, in turn, can make it difficult to get the needed support from end users, e.g., during process analysis.

All primary and secondary EPs discussed can be considered as suggestions making similarities in BPM projects explicit. Thus they serve as a baseline and starting point for building more complex evaluation and simulation models.

## 4  Model Design Rules

Overall benefit of EcoPOST evaluation models depends on their quality. The latter, in turn, is determined by the syntactical as well as the semantical correctness of the evaluation model. Maintaining correctness of an evaluation model, however, can be a difficult task to accomplish.

### 4.1  Modeling Constraints for Evaluation Models

Rules for the correct use of flows and links are shown in Fig. 17A and Fig. 17B. By contrast, Fig. 18A – Fig. 18F show examples of *incorrect* models.

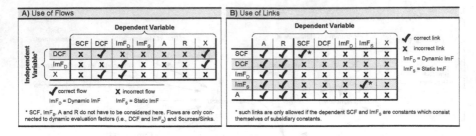

**Fig. 17** Using Flows and Links in our Evaluation Models

Dynamic evaluation factors, for example, may be only influenced by flows and not by links as shown in Fig. 18A. Likewise, flows must be not connected to auxiliary variables or constants (cf. Fig. 18B). Links pointing from DCFs (or auxiliary variables) to SCFs or static ImFs (cf. Fig. 18C and Fig. 18D) are also not valid as SCFs as well as static ImFs have constant values which cannot be influenced. Finally, flows and links connecting DCFs with dynamic ImFs (and vice versa) are also not considered as correct (cf. Fig. 18E and Fig. 18F).

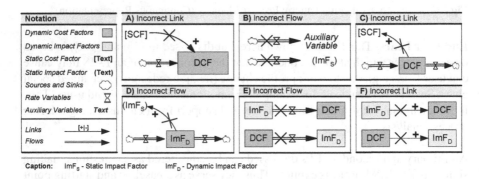

**Fig. 18** Examples of Incorrect Modeling

Several other constraints have to be taken into account as well when designing evaluation models. In the following let $EM = (V, F, L)$ be an evaluation model (cf. Definition 2.1). Then:

***Design Rule 1*** **(Binary Relations)** Every model variable must be used in at least one binary relation. Otherwise, it is not part of the analyzed evaluation context and can be omitted:

$$\forall v \in (S \,\dot\cup\, X) : \exists q \in (S \,\dot\cup\, X) \wedge ((v,q) \in F \vee (v,q) \in F) \tag{1}$$

$$\forall v \in (A \,\dot\cup\, C) : \exists q \in (A \,\dot\cup\, R) \wedge \exists \, (q,v,[+|-]) \in L \tag{2}$$

*Design Rule 2* (**Sources and Sinks**) Every state variable must be connected to at least one source, sink or other state variable. Otherwise it cannot change its value and therefore would be useless:

$$\forall v \in S : \exists q \in (S \cup X) \wedge ((v,q) \in F \vee (q,v) \in F) \qquad (3)$$

*Design Rule 3* (**Rate Variables**) Every rate variable is influenced by at least one link; otherwise the variable cannot change and therefore is useless:

$$\forall v \in R : \exists q \in (S \cup A \cup C) \wedge \exists (q,v,[+|-]) \in L \qquad (4)$$

*Design Rule 4* (**Feedback Loops**) There are no cycles consisting only of auxiliary variables, i.e., cyclic feedback loops must at least contain one state variable (cycles of auxiliary variables cannot be evaluated if an evaluation model is simulated):

$$\neg \exists < q_0, q_1, ..., q_r > \in A^{r+1} \; with \; (q_i, q_{i+1}, [+|-]) \in L \; for$$
$$i = 0, ..., r-1 \wedge q_0 = q_r \wedge \; q_k \neq q_l \; for \; k,l = 1, ..., r; k \neq l \qquad (5)$$

*Design Rule 5* (**Auxiliary Variables**) An auxiliary variable has to be influenced by at least two other static or dynamic evaluation factors or auxiliary variables (except for auxiliary variables used to represent table functions [35]):

$$\forall v \in A : \exists p,q \in (A \cup S \cup C) \wedge ((q,v,[+|-]) \in L \wedge (p,v,[+|-]) \in L) \qquad (6)$$

These modeling constraints provide basic rules for EcoPOST users to construct syntactically correct evaluation models.

## 4.2 Semantical Correctness of Evaluation Models

While syntactical model correctness can be ensured, this is not always possible for the semantical correctness of evaluation models. Yet, we can provide additional model design rules increasing the meaningfulness of our evaluation models.

*Design Rule 6* (**Transitive Dependencies**) Transitive link dependencies (i.e., indirect effects described by chains of links) are restricted. As example consider Fig. 19. Fig. 19A reflects the assumption that increasing end user fears result in increasing emotional resistance. This, in turn, leads to increasing business process costs. Consequently, the modeled transitive dependency between "End User Fears" and "Business Process Redesign Costs" is not correct, as increasing end user fears do not result in decreasing business process (re)design costs. The correct transitive dependency is shown in Fig. 19B. Fig. 19C illustrates the assumption that increasing process knowledge results in an increasing ability to (re)design business processes. An increasing ability to (re)design business processes, in turn, leads to decreasing process definition costs. The modeled transitive dependency between

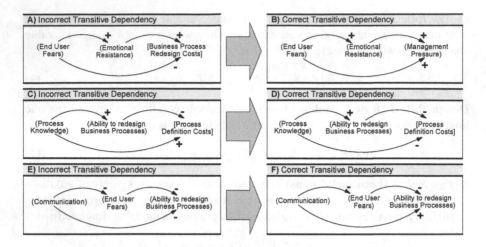

**Fig. 19** Transitive Dependencies (Simplified Evaluation Models)

"Process Knowledge" and "Process Definition Costs", however, is not correct, as increasing process knowledge does not result in increasing process definition costs (assuming that the first 2 links are correct). See Fig. 19D for the correct transitive dependency.

Finally, Fig. 19E deals with the impact of communication (e.g., the goals of an PAIS project) on the ability to redesign business processes. Yet, the transitive dependency shown in Fig. 19E is not correct. The correct one is shown in Fig. 19F.

Altogether, two causal relations ("+" and "-") are used in the context of our evaluation models. Correct transitive dependencies can be described based on a *multiplication operator*. More precisely, transitive dependencies have to comply with the following three *multiplication laws* for transitive dependencies (for any $x, y \in \{+, -\}$):

$$+ * y = y \tag{7}$$

$$- * - = + \tag{8}$$

$$x * y = y * x \tag{9}$$

The evaluation models shown in Fig. 19A and Fig. 19C violate Law 1, whereas the model shown in Fig. 19E violates the second one. Law 3 states that the "*" is commutative.

*Design Rule 7* (**Dual Links I**) A constant cannot be connected to the same auxiliary variable with both a positive and negative link:

$$\forall v \in C, \forall q \in A : \neg \exists\, l_1, l_2 \in L\ with l_1 = (v, q, -) \wedge l_2 = (v, q, +) \tag{10}$$

*Design Rule 8* (**Dual Links II**) A state variable cannot be connected to the same auxiliary variable with both a positive and negative link:

$$\forall v \in S, \forall q \in A : \neg \exists \, l_1, l_2 \in L \; with l_1 = (v, q, -) \wedge l_2 = (v, q, +) \qquad (11)$$

*Design Rule 9* (**Dual Links III**) An auxiliary variable cannot be connected to another auxiliary variable with both a positive and negative link:

$$\forall v \in A, \forall q \in A : \neg \exists \, l_1, l_2 \in L \; with l_1 = (v, q, -) \wedge l_2 = (v, q, +) \qquad (12)$$

Finally, there exist two additional simple constraints:

*Design Rule 10* (**Representing Cost Factors**) A cost factor cannot be represented both as SCF and DCF in one evaluation model.

*Design Rule 11* (**Representing Impact Factors**) An impact factor cannot be represented both as static and dynamic ImF in one evaluation model.

Without providing model design rules, incorrect evaluation models can be quickly modeled. This, in turn, does not only aggravate the derivation of plausible evaluations, but also hampers the use of the modeling and simulation tools [42] which have been developed as part of the EcoPOST framework.

## 5 Modeling Guidelines

To facilitate the use of our methodology, governing guidelines and best practices are provided. This section summarizes two categories of EcoPOST governing guidelines: (1) guidelines for evaluation models and (2) guidelines for simulation models.

In general, EcoPOST evaluation models can become large, e.g., due to a potentially high number of evaluation factors to be considered or due to the large number of causal dependencies existing between them. To cope with this complexity, we introduce guidelines for designing evaluation models (cf. Table 5). Their derivation is based on experiences we gathered during the development of our approach, its initial use in practice, and our study of general System Dynamics (SD) guidelines [63]. As example consider guideline EM-1 from Table 5. The distinction between SCFs and DCFs is a fundamental principle in the EcoPOST framework. Yet, it can be difficult for the user to decide whether a cost factor shall be considered as static or dynamic. As example take an evaluation scenario which deals with the introduction of a new PAIS "CreditLoan" to support the granting of loans at a bank. Based on the new PAIS, the entire loan offer process shall be supported. For this purpose, the PAIS has to leverage internal (i.e., within the bank) and external (e.g., a dealer) trading partners as well as other legacy applications for customer information and credit ratings. Among other things, this necessitates the integration of existing legacy applications. In case this integration is done by external suppliers, resulting costs can be represented as SCFs as they can be clearly quantified based on a contract or service agreement. If integration is done in-house, however, integration costs should be represented as DCFs as costs might be influenced by additional ImFs in this case. Other guidelines are depicted in Table 5.

**Table 5** Guidelines for Designing Evaluation Models

| GL | Description |
|---|---|
| EM-1 | Carefully distinguish between SCFs and DCFs. |
| EM-2 | If it is unclear how to represent a given cost factor represent it as SCF. |
| EM-3 | Name feedback loops. |
| EM-4 | Use meaningful names (in a consistent notation) for cost and impact factors. |
| EM-5 | Ensure that all causal links in an evaluation model have unambiguous polarities. |
| EM-6 | Choose an appropriate level of detail when designing evaluation models. |
| EM-7 | Do not put all feedback loops into one large evaluation model. |
| EM-8 | Focus on interaction rather than on isolated events when designing evaluation models. |
| EM-9 | An evaluation model does not contain feedback loops comprising only auxiliary variables. |
| EM-10 | Perform empirical and experimental research to generate needed data. |

To simulate EcoPOST evaluation models constitutes another complex task. The guidelines from Table 6 are useful to deal with it. Guideline SM-7, for example, claims to assess the usefulness of an evaluation model and related simulation results always in comparison with mental or descriptive models needed or used otherwise. In our experience, there often exists controversy on the question whether an evaluation model meets reality. However, such controversies miss the first purpose of a model, namely to provide insights that can be easily communicated.

**Table 6** Guidelines for Developing Simulation Models

| GL | Description |
|---|---|
| SM-1 | Ensure that all equations of a simulation model are dimensionally consistent. |
| SM-2 | Do not use embedded constants in equations. |
| SM-3 | Choose appropriately small time steps for simulation. |
| SM-4 | All dynamic evaluation factors in a simulation model must have initial values. |
| SM-5 | Use appropriate initial values for model variables. |
| SM-6 | Initial values for rate variables need not be given. |
| SM-7 | The validity of evaluation models and simulation outcomes is a relative matter. |

The governing guidelines and best practices represent a basic set of clues and recommendations for users of the EcoPOST framework. They support the modeler in designing evaluation models, in building related simulation models, and in handling dynamic evaluation factors. Yet, it is important to mention that the consideration of these guidelines does not automatically result in better evaluation and simulation models or in the derivation of more meaningful evaluation results. Notwithstanding, taking the guidelines increases the probability of developing meaningful models.

# 6   Practical Impact

We applied the EcoPOST framework to a complex BPM project from the automotive domain. We investigate cost overruns observed during the introduction of a large information system for supporting the development of *electrical and electronic (E/E) systems* (e.g., a multimedia unit in the car). Based on real project data, interviews with project members (e.g., requirements engineers, software architects, software developers), online surveys among end users, and practical experiences gathered in the respective BPM project, we developed a set of EcoPOST evaluation models and analyzed these models using simulation. Due to space limitations we cannot describe the complete case study in detail (for details see [35]).

## 6.1   The Case

An initial business case for the considered BPM project was developed prior to project start in order to convince senior management to fund the project. This business case was based on data about similar projects provided by competitors (evaluation by analogy) as well as on rough estimates on planned costs and assumed benefits of the project. The business case comprised six main cost categories: (1) *project management,* (2) *process management,* (3) *IT system realization,* (4) *specification and test,* (5) *roll-out and migration,* and (6) *implementation of interfaces.*

**Table 7** Analyzed Cost Categories

| | Description |
|---|---|
| 1 | *Process Management Costs*: This category deals with costs related to the (re)design of the business processes to be supported. This includes both the definition of new and the redesign of existing processes. As example of a process to be newly designed consider an E/E data provision process to obtain needed product data. As example of an existing process to be redesigned consider the basic E/E release management process. Among other things, process management costs include costs for performing interview-based process analysis and costs for developing process models. |
| 2 | *IT System Realization Costs*: This category deals with costs for implementing the new PAIS on top of process management technology. In our case study, we focus on the analysis of costs related to the use of the process management system, e.g., costs for specifying and implementing the business functions and workflows to be supported as well as costs for identifying potential user roles and implementing respective access control mechanisms. |
| 3 | *III. Online Surveys among End Users*: We conduct two online surveys among two user groups of the new PAIS (altogether 80 survey participants). The questionnaires are distributed via a web-based delivery platform. They slightly vary in order to cope with the different work profiles of both user groups. Goal of the survey is to confirm the significance of selected ImFs like "End User Fears" and "Emotional Resistance of End Users". |
| IV | *Specification and Test Costs*: This cost category sums up costs for specifying the functionality of the PAIS as well as costs for testing the coverage of requirements. This includes costs for eliciting and documenting requirements as well as costs for performing tests on whether requirements are met by the PAIS. |

In a first project review (i.e., measurement of results), it turned out that originally planned project costs are not realistic, i.e., cost overruns were observed – particularly concerning cost categories (2) and (3). In our case study, we analyzed cost overruns in three cost categories using the EcoPOST methodology (cf. Table 7).

To be able to build evaluation and simulation models for the three analyzed cost categories, we collected data. This data is based on four information sources (cf. Table 8), which allowed us to identify relevant cost and impact factors, i.e., evaluation factors that need to be included in the evaluation models to be developed. Likewise, the information sources also enables us to spot important causal dependencies between cost and impact factors and to derive evaluation models.

**Table 8** Data Collection

| | Description |
|---|---|
| I | *Project Data*: A first data source is available project data; e.g., estimates about planned costs from the initial business case. Note that we did not participate in the generation of this business case. |
| II | *Interviews*: We interview 10 project members (2 software architects, 4 software developers, 2 usability engineers, and 2 consultants participating in the project). Our interviews are based on a predefined, semi-structured protocol. Each interview lasts about 1 hour and is accomplished on a one-to-one basis. Goal of the interviews is to collect data about causal dependencies between cost and impact factors in each analyzed cost category. |
| III | *Online Surveys among End Users*: We conduct two online surveys among two user groups of the new PAIS system (altogether 80 survey participants). The questionnaires are distributed via a web-based delivery platform. They slightly vary in order to cope with the different work profiles of both user groups. Goal of the survey is to confirm the significance of selected ImFs like "End User Fears" and "Emotional Resistance of End Users". |
| IV | *Practical Experiences*: Finally, our evaluation and simulation models also build upon practical experiences we gathered when participating in the investigated BPM project. We have worked in this project as requirement engineers for more than one year and have gained deep insights during this time. Besides the conducted interviews, these experiences are the major source of information when designing our evaluation models. |

## 6.2  Lessons Learned

Based on the derived evaluation models and simulation outcomes, we were able to show that costs as estimated in the initial business case were not realistic. The simulated costs for each analyzed cost category exceeded the originally estimated ones. Moreover, our evaluation models provided valuable insights into the reasons for the occurred cost overruns, particularly into causal dependencies and resulting effects on the costs of the analyzed BPM project.

**Table 9** Lessons Learned

| LL | Description |
|---|---|
| LL-1 | Our case study confirms that the EcoPOST framework enables PAIS engineers to gain valuable insights into causal dependencies and resulting cost effects in BPM projects. |
| LL-2 | EcoPOST evaluation models are useful for domain experts and can support IT managers and policy makers in understanding an BPM project and decision-making. |
| LL-3 | BPM projects are complex socio-technical feedback systems which are characterized by a strong nexus of organizational, technological, and project-specific parts. Hence, all evaluation models include feedback loops. |
| LL-4 | Our case study confirms that evaluation models can become complex due to the large number of potential SCFs, DCFs and ImFs as well as the many causal dependencies existing between them. Governing guidelines (cf. Section 5) help to avoid too complex evaluation models. |
| LL-5 | Though our simulation models have been build upon data derived from four different data sources, it has turned out that it is inevitable to rely on hypotheses to build simulation models. |

Regarding the overall goal of the case study, i.e., the investigation of the practical applicability of the EcoPOST framework and its underlying evaluation concepts, our experiences confirm the expected benefits. More specifically, we can summarize our experiences by means of five lessons learned (cf. Table 9).

## 6.3  Critical Success Factors

We applied the EcoPOST framework in several case studies in the automotive domain. This has made us aware of a number of *critical success factors* which foster the transfer of the EcoPOST framework into practice.

*First*, it is important that EcoPOST users get enough time to become familiar with the provided evaluation concepts. Note that EcoPOST exhibits a comparatively large number of different concepts and tools, such that it will need some time to effectively apply them. In practice, this can be a barrier for potential users. However, this complexity quickly decreases through gathered experiences.

*Second*, it is crucial that results of EcoPOST evaluations are carefully documented. This not only enables their later reuse, it also allows to reflect on past evaluations and lessons learned as well as to reuse evaluation data. For that purpose, the *EcoPOST Cost Benefit Analyzer* can be used, which is a tool we developed to support the use of EcoPOST [35]. For example, it enables storage of complete evaluation scenarios, i.e., evaluation models and their related simulation models.

*Third*, evaluation models should be validated in an open forum where stakeholders such as policy makers, project managers, PAIS architects, software developers, and consultants have the opportunity to contribute to the model evolution process.

*Finally*, the use of EcoPOST has shown that designing evaluation models can be a complicated and time-consuming task. Evaluation models can become complex due to the high number of potential cost and impact factors as well as the many causal dependencies that exist between them. Evaluation models we developed to analyze a large BPM project in the automotive domain, for example, comprise more than ten DCFs and ImFs and more than 25 causal dependencies [35]. Besides additional efforts, this results in an exclusion of existing modeling experience, and prevents the reuse of both evaluation and simulation models. In response to this problem, we introduced reusable *evaluation patterns*.

# 7  Discussion

Evaluating IT investments regarding their costs, benefits, and risks is a complex task to accomplish. Even more complex are respective evaluations of BPM projects. In the previous chapters, we have shown that this difficulty stems from the interplay of the numerous technology-, organization-, and project-specific factors which arise in the context of BPM projects. In order to deal with this challenge, we have introduced the EcoPOST framework, a practically approved, model-based approach which enables PAIS engineers to better understand and investigate causal dependencies and resulting cost effects in BPM projects.

But are the evaluation concepts underlying the EcoPOST framework really suitable to capture causal dependencies and related effects? Is the use of simulation applicable in our context? What has been done and what still needs to be addressed in order to increase the expressiveness of conclusions derived using the EcoPOST framework? This section picks up these and other issues and discusses them.

**Complexity of IT Evaluation.** Today, IT evaluation is typically based on simple and static models as well as on intuition and experiences rather than on a profound analysis. Besides, rules of thumb such as *"invest to keep pace with technology"* or *"invest if competitors have been successful"* are often used as an evaluation baseline as well. Moreover, many financial evaluations exhibit an asymmetric consideration of short-term costs and long-term benefits, e.g., in order to increase "sympathy" for a potential IT investment by "proving" an extremely positive ROI.

Generally, existing IT evaluation approaches and software cost estimation techniques will lack satisfactory outcomes, in particular, if they are used at early planning stages of IT investments. Consequently, many IT projects (especially those dealing with innovative IT like BPM projects) often have – despite their potential strategic importance – a negative economic valuation result at an early stage (even if an asymmetric consideration of short-term costs and long-term benefits is intentionally enforced; see above). In particular, this situation may result in a high risk of false rejection of IT investments, i.e., decision makers need to avoid the problem of not routinely rejecting important IT investments based on results of too simple or inadequate evaluation techniques.

Likewise, these problems are valid in the context of PAIS and BPM projects as well. In particular, existing IT evaluation techniques are unable to take into account the numerous technology-, organization-, and project-specific evaluation factors arising in BPM projects. They are also unable to cope with the causal dependencies and interactions that exist between these factors and the resulting effects. Even process-oriented approaches such as the TSTS approach or the hedonic wage model (see [45] for details) do not allow to address these issues, i.e., they can only be applied to evaluate the impact of a PAIS and BPM technology on organizational business process performance and work performance.

**Special Case: Process-aware Information Systems.** The introduction of a PAIS – like the introduction of any large IT system – is typically associated with high costs. These costs need to be systematically analyzed and monitored during a BPM project. Yet, there exist no approaches to do so, for existing IT evaluation approaches and software cost estimation techniques are unable to deal with the complex interplay of the many cost and impact factors which arise in the context of such projects. Thus, decision makers thus often elude to less meaningful evaluation criteria (e.g., mere technical feasibility) and often rely on assumptions (e.g., regarding benefits such as improved business process performance) when justifying the costs of BPM projects. From the decision maker's viewpoint, this is rather insufficient.

What is needed, by contrast, is a comprehensive approach which enables PAIS engineers to investigate the complex interplay between cost and impact factors in BPM projects. Our EcoPOST framework picks up this challenge and particularly focuses on analyzing the dynamic interplay and causal dependencies of those factors that determine the complex costs of PAIS. Note that this has also been one major requirement guiding the development of the EcoPOST framework.

**Requirements Discussion.** More generally, we have identified various requirements for the design of an economic-driven evaluation approach for BPM projects: both requirements for performing economic-driven IT evaluation in general and more specific requirements for evaluating BPM projects. As a first requirement we have identified the performance of *cost-oriented evaluations* (R-1). The EcoPOST framework fully supports this requirement based on its evaluation models comprising both static and dynamic cost factors (SCFs and DCFs). We have also considered assistance for *decision support* as relevant requirement (R-2). However, this requirement is only partly fulfilled. Focus of the EcoPOST framework is on analyzing evolving dynamic cost and impact factors along the course of time. Decision support is only implicitly given by raising awareness about the complex causal dependencies and resulting cost effects emerging in BPM projects. Explicit criteria enhancing decision making, by contrast, are not provided. As a further requirement, we have identified the *derivation of plausible conclusions* (R-4). Deriving plausible conclusions, however, strongly depends on the availability of adequate[3] evaluation data underlying the developed evaluation and simulation models. Generally, the development of plausible evaluation and simulation models is a difficult task to accomplish. Notwithstanding, it is possible to scrutinize the overall suitability of evaluation models (though *"validation and verification of models is impossible"* [63]). As examples for respective actions consider the compliance of evaluation models with defined model design rules and the careful consideration of governing guidelines (cf. Chapter 7). Most important, however, is the availability of quantitative data. The experimental and empirical research activities we have described in Part III of this thesis have been important examples in this respect. Notwithstanding, there remain many unclarities about causal dependencies in BPM projects and we face the (common) problem of missing quantitative data. As final general requirements for economic-driven IT evaluation, we have identified the support of both *quantitative* (R-3) and *qualitative* (R-5) *conclusions*. Both requirements are supported by the EcoPOST framework. While qualitative conclusions can be derived based on our evaluation models and the causal dependencies and feedback loops specified by them, quantitative conclusions are only possible to some degree based on the simulation of evaluation models and the interpretation of respective outcomes.

Besides, we have identified six specific requirements for evaluating BPM projects. First, we have recognized *control of evaluation complexity* as a relevant requirement (R-6). In this context, the most important step – besides the provision of governing guidelines and the availability of adequate tool support – has been the introduction of *evaluation patterns* (EP). These predefined evaluation models can significantly reduce the complexity of building evaluation models as it is not always necessary to develop an evaluation model from scratch. Besides, *standardization of evaluation* has been identified as relevant requirement (R-7). Considering our clearly specified evaluation methodology (comprising seven consecutive steps), the provided

---

[3] We will denote evaluation data as "adequate", if the date clearly support the applicability of an evaluation approach. While some approaches require the availability of real project data, others additionally settle for data derived from on interviews, surveys, or focus-group sessions.

governing guidelines, and the availability of EPs, we consider this requirement as being fulfilled. Further, *reusing historical evaluation data* has been an important requirement as well (R-8). This requirement can be also considered as fulfilled considering both the availability of EPs and the possibility to store evaluation models and EPs in the model repository of the *EcoPOST Cost Benefit Analyzer*. Also, the *modeling of causal dependencies* has been an important requirement (R-9). We consider this requirement as being fulfilled as modeling causal dependencies is one fundamental notion underlying our evaluation models. As another requirement, we have demanded for a sufficient degree of *formalization* (R-10). In this context, it is important to mention that our evaluation and simulation models – like conventional System Dynamics models – have a sound theoretical foundation. Finally, we considered *tool-support* as crucial (R-11). In response to this, we described how the combination of a System Dynamics modeling and simulation tool and the *EcoPOST Cost Benefit Analyzer* supports the enforcement of EcoPOST evaluations. Hence, we consider this requirement as fulfilled.

**Using System Dynamics.** As discussed, we use *System Dynamics* (SD) for specifying our evaluation models. SD is a formalism for studying and modeling complex feedback systems, e.g. biological, environmental, industrial, business, and social systems [54, 47]. Its underlying assumption is that the human mind is excellent in observing the elementary forces and actions of which a system is composed (e.g., pressures, fears, delays, resistance to change), but unable to understand the dynamic implications caused by the interaction of a system's parts[4].

In BPM projects we have the same situation. Such projects are characterized by a strong nexus of organizational, technological, and project-driven factors. Thereby, the identification of these factors constitutes one main problem. Far more difficult is to understand causal dependencies between factors and resulting effects. Only by considering BPM projects as feedback systems we can really unfold the dynamic effects caused by these dependencies (i.e., by the interacting organizational, technological, and project-driven system parts).

**Benefits.** Based on the use of SD, the EcoPOST framework can unfold its benefits. In particular, the EcoPOST framework is the first available approach to systematically structure knowledge about BPM projects, to interrelate both hard facts and soft observations on respective projects, and to investigate and better understand causal dependencies and resulting effects emerging in them. Focus is on analyzing the complex interplay of organizational, project-specific, and organizational cost and impact factors. Not using the framework would imply that existing knowledge and experiences remain a mere collection of observations, practices, and conflicting incidents,

---

[4] System Dynamics can be easily confounded with *Systems Thinking* [71]. Generally, Systems Thinking utilizes the same kind of models to describe causal dependencies, but does not take the additional step of constructing and testing a computer simulation model, and does also not test alternative policies in a model (i.e., sensitivity analysis). A good overview on both approaches in the context of the evolution of the systems sciences is given in [60].

making it very difficult to derive conclusions regarding the costs of BPM projects. More specifically, the benefits of our approach can be summarized as follows:

- **Feedback**: We have denoted the modeling of feedback structures and causal dependencies between cost and impact factors as one major requirement. Our evaluation models pick up this requirement and enable not only the modeling of causal dependencies, but also the investigation of cyclic feedback structures.
- **Visualization**: Our evaluation models offer a simple way of visualizing both the structure and the behavior of interacting cost and impact factors in BPM projects. Thus, evaluation models can be easily communicated to decision makers[5].
- **Intangible Impacts**: Our evaluation models enable the PAIS engineer to investigate the effects of intangible impact factors (such as end user fears and process knowledge).
- **Delays**: BPM projects are typically faced with many delays (e.g., related to the evolution of dynamic cost and impact factors). These delays often develop over time due to internal or external influences. End user fears, for example, may be low at the beginning of a BPM project, but may quickly increase later. Our evaluation models allow to deal with such delays based on the notion of using state variables to represent both DCFs and dynamic ImFs.
- **Sensitivity Analysis**: Every modification of a variable in an evaluation model (respectively simulation model) results in various consequences. Some of these consequences can be anticipated and intended. Many others, however, are typically unanticipated and unintended. The opportunity to perform sensitivity analysis is of significant help in this context and can provide valuable insights into the consequences of changing variables.

Yet, our approach has also some limitations. The design of our evaluation models, for example, can constitute a time-consuming task – despite the availability of a library of predefined evaluation patterns. Also, evaluation models can become complex due to the large number of potential evaluation factors and the causal dependencies that exist between them. Further, the quantification of ImFs and their either linear or nonlinear effects (respectively their specification in simulation models) is difficult. When building simulation models, it often cannot be avoided to rely on assumptions. Generally, there remain many unclarities about causal relationships and feedback loops in BPM projects.

## 8 Summary

Though existing economic-driven IT evaluation and software cost estimation approaches have received considerable attention during the last decades, it is difficult to apply existing approaches to BPM projects. Reason is that existing approaches are unable to take into account the dynamic evolution of BPM projects caused by the numerous technological, organizational and project-specific facets influencing BPM projects. It is this evolution which often leads to complex and unexpected cost

---

[5] In [2] and [64], this benefit is confirmed for the management of software projects.

effects in BPM projects making even rough cost estimations a challenge. What is therefore needed is a comprehensive approach enabling BPM professionals to systematically investigate the costs of BPM projects. In response, this chapter takes a look at both well-known and often unknown cost factors in BPM projects, shortly discusses existing IT evaluation and software cost estimation approaches with respect to their suitability to be used in BPM projects, and finally introduces the Eco-POST framework. This new framework utilizes evaluation models to describe the interplay of technological, organizational, and project-specific BPM cost factors, and simulation concepts to unfold the dynamic behavior and costs of BPM projects.

Finally, there is a number of *critical success factors* (CSF) which foster the transfer of the EcoPOST framework into practice. *First*, it is important that users of the EcoPOST framework get enough time to get familiar with the underlying evaluation concepts and the provided tools. Using the EcoPOST framework will be difficult, particularly if it is initially used. It exhibits a comparatively large number of different evaluation concepts and tools and it will need some time for users to effectively apply them. In practice, this can be a problematic barrier for potential users. However, this complexity quickly decreases through gathered experiences. *Second*, it is crucial that results of EcoPOST evaluations are carefully documented. This allows not only to fall back on these results at a later day. It also allows to reflect past evaluations and problems respectively lessons learned. *Third*, evaluation models should be validated in an open forum where stakeholders such as policy makers, project managers, BPM architects, software developers, and consultants have the opportunity to contribute to the model development process.

# References

1. Alexander, C., Ishikawa, S., Silverstein, M.: A Pattern Language. Oxford University Press (1979)
2. Barros, M.D.O., Werner, C.M.L., Travassos, G.H.: Evaluating the Use of System Dynamics Models in Software Project Management. In: Proc. 20th Int'l. System Dynamics Conference (2002)
3. Bestfleisch, U., Herbst, J., Reichert, M.: Requirements for the Workflow-based Support of Release Management Processes in the Automotive Sector. In: Proc. 12th European Concurrent Engineering Conference (ECEC 2005), pp. 130–134 (2005)
4. Boehm, B., Abts, C., Brown, A.W., Chulani, S., Clark, B.K., Horowitz, E., Madachy, R., Reifer, D., Steece, B.: Software Cost Estimation with Cocomo 2. Prentice Hall (2000)
5. Boyd, D.F., Krasnow, H.S.: Economic Evaluations of Management Information Systems. IBM Systems Journal 2(1), 2–23 (1963)
6. Brassel, K.-H., Möhring, M., Schumacher, E., Troitzsch, K.G.: Can Agents Cover All the World? In: Simulating Social Phenomena. LNEMS, vol. 456, pp. 55–72 (1997)
7. Brynjolfsson, E.: The Productivity Paradox of Information Technology. Comm. of the ACM 36(12), 66–77 (1993)
8. Cardoso, J.: Control-flow Complexity Measurement of Processes and Weyuker's Properties. In: Proc. Int'l. Enformatika Conference, vol. 8, pp. 213–218 (2005)

9. Cardoso, J., Mendling, J., Neumann, G., Reijers, H.: A Discourse on Complexity of Process Models. In: Proc. Int'l. Workshop on Business Process Design (BPI 2006), pp. 115–126 (2006)
10. Carr, N.G.: IT doesn't Matter. Harvard Business Review (HBR), vol. 5 (2003)
11. Carr, N.G.: Does IT matter? Harvard Business Press (2004)
12. CMMi. Capability Maturity Model Integration. Software Engineering Institute (2006)
13. Dadam, P., Reichert, M.: The ADEPT project: A decade of research and development for robust and flexible process support - challenges and achievements. Computer Science - Research and Development 23(2), 81–97 (2009)
14. Dumas, M., van der Aalst, W.M.P., ter Hofstede, A.H.: Process-aware Information Systems: Bridging People and Software through Process Technology. Wiley (2005)
15. Eberlein, R.J., Hines, J.H.: Molecules for Modelers. In: Proc. 14th SD Conference (1996)
16. Forrester, J.W.: Industrial Dynamics. Productivity Press (1961)
17. Glykas, M.: Fuzzy Cognitive Maps: Advances in Theory, Methodologies, Tools and Applications. STUDFUZZ. Springer (2010)
18. Goh, K.H., Kauffman, R.J.: Towards a Theory of Value Latency for IT Investments. In: Proc. 38th Hawaii Int'l. Conf. on System Sciences (HICSS 2005), Big Island, Hawaii (2005)
19. Gruhn, V., Laue, R.: Complexity Metrics for Business Process Models. In: Proc. 9th Int'l. Conf. on Business Information Systems, BIS 2006 (2006)
20. Hackman, R.J., Oldham, G.R.: Development of the Job Diagnostic Survey. Journal of Applied Psychology 60(2), 159–170 (1975)
21. Hackman, R.J., Oldham, G.R.: Motivation through the Design of Work: Test of a Theory. Organizational Behavior & Human Performance 16(2), 250–279 (1976)
22. Hallerbach, A., Bauer, T., Reichert, M.: Capturing variability in business process models: The Provop approach. Journal of Software Maintenance and Evolution: Research and Practice 22(6-7), 519–546 (2010)
23. Jensen, F.V.: Bayesian Networks and Decision Graphs. Springer (2002)
24. Kartseva, V., Hulstijn, J., Tan, Y.-H., Gordijn, J.: Towards Value-based Design Patterns for Inter-Organizational Control. In: Proc. 19th Bled E-Commerce Conference (2006)
25. Kindler, E.: On the Semantics of EPCs: Resolving the Vicious Circle. Data Knowledge Engineering 56(1), 23–40 (2006)
26. Lanz, A., Weber, B., Reichert, M.: Workflow Time Patterns for Process-Aware Information Systems. In: Bider, I., Halpin, T., Krogstie, J., Nurcan, S., Proper, E., Schmidt, R., Ukor, R. (eds.) BPMDS 2010 and EMMSAD 2010. LNBIP, vol. 50, pp. 94–107. Springer, Heidelberg (2010)
27. Latva-Koivisto, A.: Finding a Complexity Measure for Business Process Models. Research Report, Helsinki University of Technology (2001)
28. Lenz, R., Reichert, M.: IT support for healthcare processes - premises, challenges, perspectives. Data and Knowledge Engineering 61(1), 39–58 (2007)
29. Liehr, M.: A Platform for System Dynamics Modeling - Methodologies for the Use of Predefined Model Components. In: Proc. 20th Int'l. System Dynamics Conference (2002)
30. Lim, J.H., Richardson, V.J., Roberts, T.L.: Information Technology Investment and Firm Performance: A Meta-Analysis. In: Proc. 37th Hawaii Int'l. Conf. on System Sciences (HICSS 2004), Big Island, Hawaii (2004)
31. Meineke, L.: Wirtschaftlichkeitsanalysen - Basis zur Durchfuehrung von IT-Projekten. IT Management 12, 12–17 (2003)

32. Mendling, J., Moser, M., Neumann, G., Verbeek, H.M.W(E.), van Dongen, B.F., van der Aalst, W.M.P.: Faulty EPCs in the SAP Reference Model. In: Dustdar, S., Fiadeiro, J.L., Sheth, A.P. (eds.) BPM 2006. LNCS, vol. 4102, pp. 451–457. Springer, Heidelberg (2006)
33. Mendling, J., Moser, M., Neumann, G., Verbeek, H.M.W., van Dongen, B.F., van der Aalst, W.M.P.: A Quantitative Analysis of Faulty EPCs in the SAP Reference Model. BPM Center Report, BPM-06-08, BPMcenter.org (2006)
34. Müller, D., Herbst, J., Hammori, M., Reichert, M.: IT Support for Release Management Processes in the Automotive Industry. In: Dustdar, S., Fiadeiro, J.L., Sheth, A.P. (eds.) BPM 2006. LNCS, vol. 4102, pp. 368–377. Springer, Heidelberg (2006)
35. Mutschler, B.: Analyzing Causal Dependencies on Process-aware Information Systems from a Cost Perspective. PhD Thesis, University of Twente (2008)
36. Mutschler, B., Reichert, M.: A Survey on Evaluation Factors for Business Process Management Technology. Technical Report TR-CTIT-06-63, Centre for Telematics and Information (CTIT), University of Twente (2006)
37. Mutschler, B., Reichert, M.: On Modeling and Analyzing Cost Factors in Information Systems Engineering. In: Bellahsène, Z., Léonard, M. (eds.) CAiSE 2008. LNCS, vol. 5074, pp. 510–524. Springer, Heidelberg (2008)
38. Mutschler, B., Reichert, M.: Evaluation Patterns for Analyzing the Costs of Enterprise Information Systems. In: van Eck, P., Gordijn, J., Wieringa, R. (eds.) CAiSE 2009. LNCS, vol. 5565, pp. 379–394. Springer, Heidelberg (2009)
39. Mutschler, B., Reichert, M., Bumiller, J.: Towards an Evaluation Framework for Business Process Integration and Management. In: 2nd Int'l. Workshop on Interoperability Research for Networked Enterprises Applications and Software (INTEROP), Enschede, The Netherlands (2005)
40. Mutschler, B., Reichert, M., Bumiller, J.: Designing an economic-driven evaluation framework for process-oriented software technologies. In: Proc. 28th Int'l Conf. on Software Engineering (ICSE 2006), pp. 885–888 (2006)
41. Mutschler, B., Reichert, M., Bumiller, J.: Unleashing the effectiveness of process-oriented information systems: Problem analysis, critical success factors and implications. IEEE Transactions on Systems, Man, and Cybernetics 38(3), 280–291 (2008)
42. Mutschler, B., Reichert, M., Rinderle, S.: Analyzing the Dynamic Cost Factors of Process-Aware Information Systems: A Model-Based Approach. In: Krogstie, J., Opdahl, A.L., Sindre, G. (eds.) CAiSE 2007 and WES 2007. LNCS, vol. 4495, pp. 589–603. Springer, Heidelberg (2007)
43. Mutschler, B., Reichert, M., Rinderle, S.: Analyzing the Dynamic Cost Factors of Process-Aware Information Systems: A Model-Based Approach. In: Krogstie, J., Opdahl, A.L., Sindre, G. (eds.) CAiSE 2007 and WES 2007. LNCS, vol. 4495, pp. 589–603. Springer, Heidelberg (2007)
44. Mutschler, B., Weber, B., Reichert, M.: Workflow management versus case handling: Results from a controlled software experiment. In: Proc. ACM SAC 2008, pp. 82–89 (2008)
45. Mutschler, B., Zarvic, N., Reichert, M.: A Survey on Economic-driven Evaluations of Information Technology. Technical Report, TR-CTIT-07, University of Twente (2007)
46. Myrtveit, M.: Object-oriented Extensions to System Dynamics. In: Proc. 18th Int'l. System Dynamics Conference (2000)
47. Ogata, K.: System Dynamics. Prentice Hall (2003)
48. Pisello, T.: IT Value Chain Management - Maximizing the ROI from IT Investments: Performance Metrics and Management Methodologies Every IT Stakeholder Should Know, Alinean, LLC (2003), www.alinean.com

49. Reichert, M., Rinderle, S., Kreher, U., Dadam, P.: Adaptive process management with ADEPT2. In: Proc. Int'l Conf. on Data Engineering, ICDE 2005 (2005)
50. Reichert, M., Rinderle-Ma, S., Dadam, P.: Flexibility in Process-Aware Information Systems. In: Jensen, K., van der Aalst, W.M.P. (eds.) ToPNoC 2009. LNCS, vol. 5460, pp. 115–135. Springer, Heidelberg (2009)
51. Reifer, D.J.: Making the Software Business Case - Improvement by the Numbers. Addison-Wesley (2001)
52. Reijers, H.A., van Wijk, S., Mutschler, B., Leurs, M.: BPM in Practice: Who Is Doing What? In: Hull, R., Mendling, J., Tai, S. (eds.) BPM 2010. LNCS, vol. 6336, pp. 45–60. Springer, Heidelberg (2010)
53. Reijers, H.A., van der Aalst, W.M.P.: The Effectiveness of Workflow Management Systems - Predictions and Lessons Learned. Int'l. J. of Inf. Mgmt. 25(5), 457–471 (2005)
54. Richardson, G.P., Pugh, A.L.: System Dynamics - Modeling with DYNAMO. Productivity Press (1981)
55. Rinderle, S., Weber, B., Reichert, M., Wild, W.: Integrating Process Learning and Process Evolution – A Semantics Based Approach. In: van der Aalst, W.M.P., Benatallah, B., Casati, F., Curbera, F. (eds.) BPM 2005. LNCS, vol. 3649, pp. 252–267. Springer, Heidelberg (2005)
56. Rinderle-Ma, S., Reichert, M., Weber, B.: On the Formal Semantics of Change Patterns in Process-Aware Information Systems. In: Li, Q., Spaccapietra, S., Yu, E., Olivé, A. (eds.) ER 2008. LNCS, vol. 5231, pp. 279–293. Springer, Heidelberg (2008)
57. Sassone, P.G.: Cost-Benefit Methodology for Office Systems. ACM Transactions on Office Information Systems 5(3), 273–289 (1987)
58. Sassone, P.G.: Cost Benefit Analysis of Information Systems: A Survey of Methodologies. In: Proc. Int'l. Conf. on Supporting Group Work (GROUP 1988), Palo Alto, pp. 73–83 (1988)
59. Scholl, H.J.: Agent-based and System Dynamics Modeling: A Call for Cross Study and Joint Research. In: Proc. 34th Hawaii Int'l. Conf. on System Sciences, HICSS 2001 (2001)
60. Schwaninger, M.: System Dynamics and the Evolution of Systems Movement - A Historical Perspective. Discussion Paper, Nr. 52, University of St. Gallen, Switzerland (2005)
61. Senge, P.M.: The 5th Discipline - The Art and Practice of the Learning Organization, 1st edn. Currency Publications (1990)
62. Smith, H., Fingar, P.: IT doesn't matter - Business Processes Do. Meghan Kiffer Press (2003)
63. Sterman, J.D.: Business Dynamics - Systems Thinking and Modeling for a Complex World. McGraw-Hill (2000)
64. Sycamore, D., Collofello, J.S.: Using System Dynamics Modeling to Manage Projects. In: Proc. 23rd Int'l. Conf. on Computer Software and Applications Conference, vol. 15(4), pp. 213–217 (1999)
65. Ventana Systems. Vensim (2006), http://www.vensim.com/
66. van der Aalst, W.M.P., ter Hofstede, A.H.M., Kiepuszewski, B., Barros, A.P.: Advanced Workflow Patterns. In: Scheuermann, P., Etzion, O. (eds.) CoopIS 2000. LNCS, vol. 1901, pp. 18–29. Springer, Heidelberg (2000)
67. Weber, B., Mutschler, B., Reichert, M.: Investigating the effort of using business process management technology: Results from a controlled experiment. Science of Computer Programming 75(5), 292–310 (2010)
68. Weber, B., Reichert, M., Rinderle-Ma, S.: Change patterns and change support features - enhancing flexibility in process-aware information systems. Data and Knowledge Engineering 66(3), 438–466 (2008)

69. Weber, B., Reichert, M., Wild, W., Rinderle-Ma, S.: Providing integrated life cycle support in process-aware information systems. Int'l Journal of Cooperative Information Systems 18(1), 115–165 (2009)
70. Weber, B., Sadiq, S., Reichert, M.: Beyond rigidity - dynamic process lifecycle support: A survey on dynamic changes in process-aware information systems. Computer Science - Research and Development 23(2), 47–65 (2009)
71. Weinberg, G.M.: An Introduction to General Systems Thinking. Dorset House Publishing (2001)
72. Winch, G., Arthur, D.J.W.: User-Parameterised Generic Models: A Solution to the Conundrum of Modelling Access for SMEs? System Dynamics Review 18(3), 339–357 (2003)

# Managing Organizational Intellectual Capital

Effimia Pappa[1], Michail Giakoumis[2], Viktoria Voxaki[3], and Michael Glykas[3,4]

[1] ATP, Innsbruck Planungs GmbH
[2] ProBank S.A, Chios Branch, Greece
[3] Department of Financial and Management Engineering, University of the Aegean, Chios, Greece
[4] Faculty of Business and Management, University of Wallangong, Dubai

## 1 Introduction

Today we live in a transitional period, where the main features are: international competition, radical technological changes, faster flow of information and communication, the increasing complexity of business and the expansion of "globalization."

The rapidly changing and intensely competitive environment, companies are required to meet the new demands presented. We are entering an era where the traditional pillars of economic power, namely capital, land, raw materials and technology, are not the determining factors of success in a business. The future is determined now by the ability to explore the most valuable resource, business knowledge (22).

In the business world, there has been a transition from "information age" to "knowledge era". Great contribution to this development is the proliferation of information technology and telecommunications.

It is necessary to create mechanisms of organizational learning in modern organizations and to lay emphasis on the development of intellectual capital in order to obtain the conditions of continuous innovation and to develop and maintain a competitive advantage, ensuring long-term survival (6).

The knowledge of management is an important qualification, perhaps the most important asset nowadays. The ability of an enterprise to generate wealth is based on knowledge and the capabilities of its people. Today, many companies see themselves as learning organisations pursuing the objective of continuous improvement in their knowledge assets (26). This means that knowledge assets are fundamental strategic levers in order to manage business performance and the continuous innovations of a company (19, 21, 24, 3).

If there is one feature of the new economy is the dominance of intellectual capital. The companies that thrive in this new environment identify themselves as learning organizations and seek to continuously improve the intellectual capital.

## 2 Definition of Intellectual Capital

Today, advances in technology and telecommunications have created "intangible benefits" that did not exist until now and now have become necessary for the operation of businesses. Each company consists of all physical and intangible assets.

M. Glykas (Ed.): Business Process Management, SCI 444, pp. 195–220.
springerlink.com

The intellectual capital consists of the Human Capital (Human Capital) , the Structural Chapter (Structural Capital) and the Relational Capita (35).

### 2.1.1 Human Capital

Human Capital is the generic term for the competences, skills and motivation of the employees. The Human Capital of the organisation comprise all qualities and professional skills the employee into the organisation. It is «owned» by the employee and leaves along with him the organisation. Human capital can be defined as a combination of employee's competence, attitude and creativity (fig. 1).

| Employees' competence | Strategic leadership of the management |
|---|---|
|  | Qualities of the employees<br>Learning ability of the employees<br>Efficiency of employee training<br>The employees' ability to participate in policy making and management<br>Training of key technical and managerial employees |
| Employees' attitude | Identification with corporate values |
|  | Satisfaction degree<br>Employees' turnover rate<br>Employees' average serviceable life |
| Employees' creativity | Employee's creative ability<br>Income on employees' original ideas |

**Source:** (10)

**Fig. 1** The indices of human capital

### 2.1.2 Structural Capital

Structural Capital is the generic term for all structures deployed by the employees to carry out the business processes. Structural capital is everything in an organization that supports employees (human capital) in their work. Structural capital is the supportive infrastructure that enables human capital to function.

The structural capital is owned by the enterprises and remains with the organisation to the largest extent when the employee leaves the company. In detail, structural capital can be classified into company culture, organizational structure, organizational learning, operational process, and information system (fig.2).

| Corporate culture | Construction of company's culture |
|---|---|
|  | Employee's identification with company's perspective |
| Organizational structure | Clarification of relationship among authority, responsibility and benefit |
|  | Validity of enterprise controlling system |
| Organizational learning | Construction and utilization of inner information net<br>Construction and utilization of company repository |
| Operation process | Business process period |
|  | Product quality level<br>Corporate operating efficiency |
| Information system | Mutual support and cooperation between employees |
|  | Availability of enterprise information<br>Share of knowledge |

Source: (10)

**Fig. 2** The indices of structural capital

### 2.1.3 Relational Capital

Relational Capital is the generic term for all relationships to external groups and persons established by the organisation, e.g. customer relationships, supplier relationships, and the relationships to other partners and the public. In this, Relational capital is classified into basic marketing capability, market intensity and customer's loyalty (Fig.3)

| Basic Marketing Capability | Construction and Utilization of the Customer Database |
|---|---|
|  | Customer service capability<br>Identifying ability of customer's needs |
| Market intensity | Market share |
|  | Market potential<br>Unit sales to customer<br>Brand and trademark reputation<br>Construction of sales channel |
| Customer loyalty indices | Customer satisfaction |
|  | Customer complaint<br>Customer outflow<br>Investment on customer relationship |

Source: (10)

**Fig. 3** The indices of Relational capital

## 3 The History of Intellectual Capital Development

Intellectual capital (IC) has been defined by Klein and Prusak (1994) as " a package of useful knowledge". It basically consists of knowledge, lore, ideas and innovations (32).

In 1997, Edvinsson and Malone use the terms of intangible capital (Intellectual Capital) and intangible assets (Intangible Assets).

The definition given is that «Intangible Assets , are those which have no physical existence, but contribute to the value of the company». In a first attempt to interpret it, it is proven that the main characteristic is the lack of physical - material existence. Despite this lack, they still contribute to create value for the company.

Edvinsson (1997) defined intangible capital as a combination of Human and Material and Structural Capital.

The proposal by Sveiby (1997), was the separation of intellectual capital in employees' skills, internal structure and external structure. According to Sveiby, the development of intellectual capital is based on the organization's staff.

Stewart in 1998, treats intellectual capital as spiritual material (Knowledge, information, intellectual property, experience), which can, together, collectively be used to create value. This is recognized as collective intellectual knowledge.

A year later, Professor Bontis and his associates in their research, use the term intangible resources, and recognize the category of intangible resources as intellectual capital. Thus, he defined intellectual capital as the sum of intangible resources and inputs generated, where intangible sources stand for each factor that contributes to the overall value of a business' process.

The attempt to define intellectual capital of a company becomes more intense in the mid 90's and peaked in the early new millennium. The biggest explosion of interest in the issue of intangible assets of enterprises, culminated in the last decades of the previous century, during which, most articles were written.

There have been many attempts to categorize intellectual capital by many academics and economists. Different efforts and different recommendations where made depending on the perspective of every writer about the concept of intangible assets and also depending on the objective on which each one focused on. But most scholars concluded to the separation of intellectual capital into 3 categories. There were simply different views regarding the designation of these categories. Thus, the first was creativity, knowledge and identification, the second was structural and human relations, the third was internal structure and external human capital (or in other capacities employees) and the last was human, organizational, operational and customer capital.

## 4 Organizational Intellectual Capital

Intellectual capital constitutes the most valuable organizational resource of a company. It represents a group of intangible resources of strategic value that does not appear in the financial organization of the company, in spite of contributing to the creation of organizational value. Intellectual capital is not only key to the creation of a competitive advantage but also for its long-term Maintenance.

Organizational Intellectual Capital is a strategic management instrument for assessing and developing the Intellectual Capital (IC) of an organisation. It shows how Intellectual Capital is linked to corporate goals, business processes and the business success of an organisation using indicators to measure these elements.

The organisation is embedded in the business environment (fig.4). Regularly, a vision of the founders and owners serves as general guiding principle for major decisions and strategic positioning. Depending on the business strategy, managerial decisions lead to operational measures. These measures serve to improve business processes and the utilisation of Intellectual Capital in those processes.

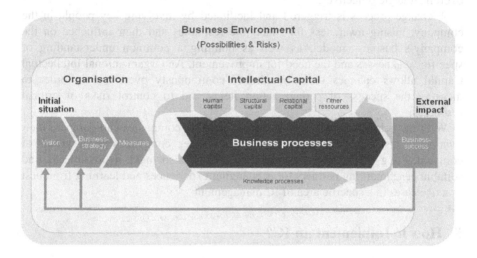

**Source:** (1)

**Fig. 4** Organizational Intellectual Capital Structural Model

Drawing up an Organizational Intellectual Capital in a company

- helps company determine strengths and weaknesses of strategic IC factors (diagnosis)
- prioritises improvement opportunities with the highest impact (decision support)
- supports the implementation of actions for organisational development (optimisation and innovation)
- enhances transparency and the involvement of employees (internal communication)

- diminishes strategic risks and controls the success of actions (monitoring)
- facilitates the communication of corporate value towards stakeholders (reporting)

In general, an Organizational Intellectual Capital helps owners and managers of organisations to facilitate the process of strategy development and strategy implementation. An Organizational Intellectual Capital assesses the internal capabilities, i.e. a firm's intangible resources, from the point of view of external strategic objectives, e.g. growth, market position, customer satisfaction etc. The core assets need strategic development and can be systematically identified by building on a consistent view of a representative team in the company. This way, the right action items for improving certain areas can be prioritised and linked to overall strategic objectives.

A change process is triggered and facilitated by involving key people in the company, raising awareness for intangible resources and their influence on the company's business model, as well as building a common understanding of specific weaknesses and the need for improvement. An Organizational Intellectual Capital allows changes to be monitored continuously over time in order to measure the success of certain action items and to control risks of critical resources.

With the help of this IC Benchmarking concept, companies can compare their own strengths and weaknesses in IC with other companies or a group of companies (e.g. within their industrial sector)and also make it possible to find suitable benchmarking partners for exchanging experiences and learning from best practice cases in a specific area of IC management.

## 5 How to Implement an IC?

The approach of conducting an IC is divided into five steps with each step building on the prior one (fig 5 ) The IC implementation is a workshop-based procedure involving a selected number of employees from the implementing organisation. The members of the IC project team are selected across units and hierarchies in order to ensure a comprehensive reflection of the company's Intellectual Capital. The people involved in the IC project team therefore range from representatives of the top management to staff from the operational level. The impartial position of an IC Moderator during IC workshop discussions provides a main benefit for the company's internal communication in general. It is an important basis for creating mutual understanding between strategic thinking of the management and the operational view of other Organizational Intellectual Capital project team members.

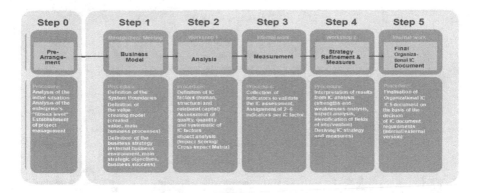

**Source:** (14)

**Fig. 5** Organizational Intellectual Capital Procedural Model

# 6 Support Material

An implementation, supporting material has been developed to simplify the Organizational Intellectual Capital implementation and guide the Organizational Intellectual Capital project team through the Organizational Intellectual Capital procedure.

The Organizational Intellectual Capital leaving the user to decide the extent to which the company's IC is to be analysed. Usually, a decision will be taken at the management meeting (Step 1) of representatives from the top management. The closer the look at the company's IC and its linkages with strategic objectives, the more questions will be raised and discussed during the Organizational Intellectual Capital workshops.

The supporting material offers basic information as well as advanced material for different levels of detail in the Organizational Intellectual Capital implementation process. While the basic supplementary material is provided to support a standard Organizational Intellectual Capital Organizational Intellectual Capital implementation (checklists/working sheets, Organizational Intellectual Capital Toolbox), the Organizational Intellectual Capital Extra Modules address advanced users who already have experience of implementing Organizational Intellectual Capital and/or want to go deeper into the Organizational Intellectual Capital methodology.

The following diagram provides a short overview on the Organizational Intellectual Capital working material directly related to the basic and advanced Organizational Intellectual Capital procedure:

A business that does not create value in a systematic way is not sustainable. Facing such a challenge entails that the business model the organisation has

| Organizational IC Support Material | | | | |
| --- | --- | --- | --- | --- |
| ✓ ✓ ✓ | Checklists/ Working sheets (Basic) | Organizational IC Toolbox (Basic) | | Extra Modules (Advanced) |
| Step 0: Pre-Arrangement | 0.1 Fitness Check 0.2 Project Planning | Organizational IC Toolbox Sheet "Participants List" | | M6 Follow-up Organizational IC |
| Step 1: Business Model | 1.1 Business Model | | | M1 Enhanced Business Model M2 Vision M3 Business Processes M4 |
| Step 2: IC Analysis | 2.1 Workshop 1 Procedure 2.2 Exploring Intellectual Capital 2.3 Common IC Factors | Organizational IC Toolbox Sheet "Definitions" Organizational IC Toolbox Sheet "QQS Assessment" | | M6 Follow-up Organizational IC |
| Step 3: Measurement | 3.1 Common IC Indicators | Organizational IC Toolbox Sheet "Indicators" | | M6 Follow-up Organizational IC |
| Step 4: Strategy Refinement & Measures | 4.1 Workshop 2 Procedure | Organizational IC Toolbox Sheet "QQS Overview" Organizational IC Toolbox Sheet "QQS-Bar-Charts" Organizational IC Toolbox Sheet "QQS Period Overview" Organizational IC Toolbox Sheet "QQS Period Chart" Organizational IC Toolbox Sheet "Weighting | | M1 Enhanced Business Model M4 External Environment M5 Learning Cycle M6 Follow-up Organizational IC |
| Step 5: Final Organizational IC Document | 5.1 Organizational IC Template | | | M6 Follow-up Organizational IC |

**Source:** (14)

**Fig. 6** Overview of Organizational Intellectual Capital support material

shaped is not only unique but also robust and adaptive at the same time. Whereas uniqueness is about creating distinctiveness, to be robust and adaptive a business model needs to be built around the organisation's core competencies and to be flexible enough to quickly respond to external influences. It assumes the organisation has adopted a knowledge-based approach to managing its business and is able to sense, anticipate and respond rapidly and effectively to customers' needs. To keep momentum an organisation needs to systematically assess its core competencies against other elements of the business model to ensure fitness and to be able to identify and capitalise on market opportunities. In this respect, the systematic dimension accounts for a big stake of a company's sustainability.

Hence, advancing in the understanding of the business model concept and how the organisation creates value holds tremendous promise for driving organisations to new levels of competitive fitness and higher levels of innovation.

# 7 Intellectual Capital Management Implementation Process

## 7.1 Pre - arrangement

Some fundamental principles must be followed when drafting the Organizational IC in order to ensure that the project runs smoothly. Especially for first-time adoption of Organizational IC is important that the Organizational IC project manager (person responsible for the Organizational IC in the enterprise) and the Organizational IC Moderator deal with these principles in detail. They coordinate and moderate the entire Organizational IC implementation process. Hence, the overall approach needs to be understood (in reasonable detail) and the Organizational IC project manager should be able to introduce the Organizational IC project team and other people involved in the project to its method and aims. In order to start the process and gather basic information, the following prerequisites have to be met:

- **Analysis of the initial situation**

In order to ensure that the management meeting is effective and productive, some information on the enterprise should be collected before the first meeting. Information on the background, history and if available specific cultural issues of the company are relevant. Ask the organisation for additional information on strategy, market development, market trends, etc.

Almost all organisations have some documents regarding the strategy, developments and trends of the markets, and even - though sometimes indirectly - possibilities and risks. These documents must be collected during the first visit and studied prior to the first workshop.

- **Analysis of the enterprise's "fitness" level for Organizational IC**

The implementing enterprises differ in size, maturity level and life-cycle stage. In the course of the project, these criteria have been aggregated under the term "fitness" level.

An organisation's "fitness" level affects the entire Organizational IC process, because it influences the determination whether basic or advanced implementation procedures should be applied. There may be circumstances in the life of an organisation that will work against a successful Organizational IC implementation, such as serious financial or strategic crises, internal conflicts, etc. or characteristics relative to the leadership style, culture or governance system autocratic, strongly corporative, poor transparency of its operations, etc. The presence of more than one of these elements will put the whole Organizational IC process at risk, thus making its implementation inadvisable, or will at least require a very experienced Organizational IC Moderator to cope with a difficult environment.

- **Establishment of project management**

The formation and reasonable composition of the Organizational IC project team plays a substantial part in the project. The members of the Organizational IC project team should be chosen from across all relevant divisions and hierarchy levels of the company. The view of the organisation as perceived by the team members will be reflected later in the Organizational IC document and should therefore be representative. The Organizational IC project team should comprise managers as well as operative employees. This will ensure that the discussion is down-to-earth and not only reflects the top management team's self-perception. Depending on the size of the organisation, the work should be done in one or more teams. It is important for these teams to regularly exchange information on the status of their work. Furthermore, sufficient time should be allocated to merge the results and develop a shared view, since considerable potential for discussion will arise.

The involvement of at least one representative from the top management in the team has proved to contribute to the success of the project. However, the Organizational IC project manager himself does not necessarily have to come from top management. Coordinating a heterogeneous team spanning the different hierarchy levels is not an easy task. Allow sufficient time to find appointments and to coordinate employees and implementation steps. Professional project management makes a significant contribution to the success of the project.

Persons involved in the Organizational IC project in brief

- The Organizational IC Moderator supports the Organizational IC project manager and accompanies the implementation process. The Organizational IC Moderator guides and leads the workshop discussions and documents the results in the Organizational IC Toolbox.
- The organisation's project manager is a person from the implementing company responsible for the Organizational IC project internally. He is responsible for organising the Organizational IC implementation: setting up the organisation's project team, fixing dates and communicating results to other employees and the company's management. The responsible person is in contact with the Organizational IC Moderator and assists the latter in preparing the Organizational IC workshops. The company's project manager should therefore basically also be acquainted with the Organizational IC method.
- Person from the enterprise's top management representing its overall strategic view. This person backs and promotes the Organizational IC project during its implementation. At the same time, he/she must communicate and support the implementation of Organizational IC results within the company afterwards. Therefore, the role is crucial to ensure sustainability of the Organizational IC.
- Heterogeneous team of 5 to 10 members from all units and hierarchies of the company. Usually, representatives from the most important units – operational and strategic are asked to join the Organizational IC project team in order to ensure a representative picture of the enterprise.

## 7.2  Business Model

A business system is defined as the way a company defines and differentiates its offers, defines the activities that properly match its strategy, selects its processes, configures and allocates its resources, enters the market, creates utility for its actual and potential customers and obtains a positive return from those activities.

Traditionally, as business systems evolved they observed an innovation phase strongly focused on vertical integration. They evolved processes of innovation in their structures and processes. New activities emerged as the result of disaggregation and reaggregation processes in the traditional value chains. In general, these activities are developed by means of electronics contexts, in which networks drive the interlinked phenomena of increasing returns and network effects. In the new economy (Kelly, 1998), web economy (Schwartz, 1999) or network economy (Shapiro and Varian, 1999), Tapscott et al. (2000) identify a new system of doing business (the business web), which, on the economic plane, has brought new proposals for value, new competition rules and procedures, new resources, capabilities and competences, new strategies and new, more sophisticated market approaches.

By default, the Organizational IC will be developed for the whole organisation. Since it may have to be adapted in some specific cases, the system boundaries for the Organizational IC must be defined by the management team. Afterwards, the company's business model should be described. The business model should include the value ceating model that shows what and how value is generated. The company's rough business strategy should also be roughly defined.

The background, history and specific cultural issues of the company should be analysed during the initial visit by interviewing the management staff (and other staff) of the company, in order to understand the company's situation and needs on a more detailed level. This is particularly important in order to develop any further strategies, especially the IC strategy, which should be in line with the business strategy. Discussion of the current situation and future orientation of the organisation forms the basis for all further steps.

An Organizational IC can be developed for the whole company, a department, a business process or any other part of the organisation. Especially for first-time adoption of an Organizational IC, it is important to consider which part of the organisation will be analysed. For several reasons - availability of employees, risk considerations, etc. – it may make sense to start with a prototype and then transfer the newly acquired knowledge into a second phase. Due to the fact that the participating companies are small and medium-sized the Organizational IC will be developed for the whole company in the majority of cases. The system boundaries should be set and defined as clearly as possible. Whatever decision is taken, it should be carefully documented and clearly stated in order to avoid any misunderstandings.

To define a company's value creating model two questions have to be answered:

- **What does the company actually sell (created value)?**

The created value is what the company actually offers to its customers. This can be a product, a service or a combination of both.

The following questions help to describe the value creating model of the company, i.e. the value the company intends to provide to its customers and how this value is produced. The value creating model is the kernel for any strategic considerations.

> What product or service does the business offer?
> How can customers benefit from this product or service?
> Which market segments / groups of customers are targeted?
> To whom will the proposition be appealing?
> From whom will resources be received?
> How are the products or services created?

> How are they going to be delivered to the customers?
> How will the customer pay for the product or service?
> What is the price/margin for the product or service offered to the customer?

- **How is this value produced (main business processes)?**

After identifying the value generated it is necessary to figure out how this value is generated. This can be done by identifying the value creating business processes. These are the different steps by which the product/service/value is produced and provided to the customer. They are the central, most important processes of an organisation. All other processes gather around them and need to be specifically defined for each individual company.

## 7.3 The "Systemic" Nature of the Business Model

A business model is a conceptualisation of how an organisation creates value for its customers and other stakeholders. This value creation process includes various processes or an ecology of them combining competencies and other resources through business processes in the way determined by its strategy and strategic objectives, and in accordance with its vision. The degree of *novelty and uniqueness* of this configuration as well as its *overall consistency and adaptive capacity* – both internal and external – are a source of competitive advantage. In particular, the dynamic *synergies* that the firm creates between its business processes and its knowledge base represent an opportunity for new ventures and appropriability.

To create competitive advantages, however, the organisation should manage this "systemic" aspect of business models; in other words, it should go through an exercise of decoupling and coupling its components. Decoupling is necessary to identify and understand the nature of the IC or knowledge components, while coupling is necessary to understand the functionality of the system – i.e. the relationships between its components. From an IC perspective, the business model could also be acknowledged as a *roadmap for building competencies and capabilities*. As stated by Johannesen et al (2005), "It is only when the knowledge base is integrated to transform input into output for the purpose of increasing values that the company's capability of execution is increased." (Note: For a more detailed discussion of the business model and its value creation potential see Module M1, Enhanced Business Model)

### 7.3.1 Business Strategy

The definition of business strategy is a long term plan of action designed to achieve a particular goal or set of goals or objectives. Strategy is management's game plan for strengthening the performance of the enterprise. It states how business should be conduct to achieve the desired goals. Without a strategy management has no roadmap to guide them.

Creating a business strategy is a core management function. It must be said that having a good strategy and executing the strategy well, does not guarantee success. Organisations can face unforeseen circumstances and adverse conditions through no fault of their own (30).

In order to develop strategic objectives, the business environment has to be examined. Keeping in mind the value creating model defined above, major possibilities and risks in the business environment should be explored and their influence on the company's business activities considered. Common features of the external environment include, for instance, buyer/supplier bargaining power, threat of substitutes, political, social and economic factors.

The external business environment exerts a remarkable influence on the activity of the organisation. In order to simplify analysis of the business environment, it can be divided into the micro-environment and the macro-environment (23). The micro-environment encompasses the driving factors in the company's closer environment, the marketplace in which the company acts. The main micro-environmental forces consequently originate from competitors, customers, suppliers and other stakeholders affecting the company's ability to make a profit. A change in any of these forces normally requires a company to re-assess the marketplace.

Analysis of macro-environmental forces helps to understand the "big picture" of the environment in which a business operates, allowing it to take advantage of the opportunities and minimize the threats faced by its business activities. The main factors considered are usually political, economical, socio-cultural, technological, environmental and legal forces (usually referred to as STEEP or PESTEL analysis).

PEST analysis stands for "Political, Economic, Social, and Technological analysis" and describes a framework of macro-environmental factors used in the environmental scanning component of strategic management. Some analysts added Legal and rearranged the mnemonic to SLEPT, inserting Environmental factors expanded it to PESTEL or PESTLE, which is popular in the UK. The model has recently been further extended to STEEPLE and STEEPLED, adding education and demographic factors. It is a part of the external analysis when conducting a strategic analysis or doing market research, and gives an overview of the different macro environmental factors that the company has to take into consideration. It is a useful strategic tool for understanding market growth or decline, business position, potential and direction for operations.

The growing importance of environmental or ecological factors in the first decade of the 21st century have given rise to green business and encouraged widespread use of an updated version of the PEST framework. STEER analysis systematically considers Socio-cultural, Technological, Economic, Ecological, and Regulatory factors.

To gain or maintain a sustainable competitive advantage for a company, it must be vigilant, watching for changes in the business environment. Ideally the business environment should be scanned continuously or at least on a regular basis. It must also be agile enough to alter its strategies and plans when the need arises. (37)

### 7.3.2 Main Strategic Objectives

As a starting point, some basic strategic objectives have to be determined by the company. The term "strategic objectives" refers to an organisation's articulated aims or responses to address major change or improvement, competitiveness or social issues and business advantages. Strategic objectives are generally focused both externally and internally and relate to significant customer, market, product, service, or technological opportunities and challenges identified in the business environment scanning. Broadly stated, they are what an organisation must achieve to remain or become competitive and ensure the organisation's long-term sustainability. Strategic objectives set an organisation's longer-term directions and guide resource allocations and redistributions.

In order to operationalise the strategic objectives, the company's management should define the desired business results the company wants to achieve. Business success comprises tangible (e.g. growth, revenue) and intangible (e.g. image, customer loyalty) business results.

## 7.4   IC Analysis

### Intellectual Capital

The IC analysis is broken down into three major parts:

- IC definition
- QQS Assessment
- Impact analysis

### 7.4.1 IC Definition

In addition to the business model identified in "business model", there are a large number of further (intangible) influencing factors which affect the efficiency and effectiveness of performance and the success of the organisation on the market. They are part of the organisation's Intellectual Capital (See Chapter 1.2, Organizational IC Structural Model).

As extensive research has shown, the following definitions thoroughly grasp the concept of Intellectual Capital:

– **Intellectual Capital (IC)** is divided into three categories:
   Human Capital (HC), Structural Capital (SC), and Relational Capital (RC). It describes the intangible resources of an organisation.
– **Human Capital (HC)** is defined as "what the single employee brings into the value adding processes".
– **Structural Capital (SC)** is defined as "what happens between people, how people are connected within the company, and what remains when the employee leaves the company"
– **Relational Capital (RC)** is defined as "the relations of the company to external stakeholders".

### 7.4.2 QQS Assessment

In order to identify the strengths and weaknesses of the IC factors, they must be assessed by the project team in a structured discussion. The IC factors are evaluated by self assessment, i.e. each factor is evaluated with regard to its current existing quantity, quality and systematic management by the Organizational IC project team.

For some IC factors, e. g. "Corporate Culture" or "Motivation", it is not possible to distinguish between quality and quantity, as these factors are characterised mainly by qualitative features. In these cases, quality and quantity cannot be evaluated separately and may therefore be merged into a single evaluation dimension.

### 7.4.3 Impact Analysis

Intangible resources are characterized by complex interactions which depend on the context and are regularly hard to understand from external perspectives. Simple cause and effect chains, e.g. in simple machines (switch on, machine runs), are of little use in the area of Intellectual Capital.

The challenge to be met is to manage these intangible resources. This is a highly complex task due to the ambiguity of interactions between influencing factors and the associated challenges of allocating resources efficiently.

Sensitivity analysis is one method for tackling this complexity which supports the analysis of interactions within an organisation and visualizes interdependencies (36).

The Organizational IC (OIC) procedure offers two possibilities for analysing the impact of a company's IC. Depending on the size and maturity of a company, either a simple (Impact Scoring) or a full version (Cross Impact Matrix) can be applied to assess and analyse the Intellectual Capital and its interrelations.

Impact analysis is a simple way of assessing the IC factors' impact on a company's business success. It is appropriate for companies going through the OIC implementation for the first time, as it provides fast results within a short time, maximising the cost – benefit relation. Applying the Impact Scoring reduces the OIC implementation process to two workshop days with the OIC project team. The simple version can sensitise inexperienced companies for the OIC methodology by limiting the complexity of interrelating IC factors to core information.

Start-up companies and micro-organisations may also prefer the Impact Scoring, as their organisational complexity is usually lower and might therefore not require an extensive analysis.

On the other hand, larger or more experienced companies with a higher level of complexity are advised to go through the full version of the impact analysis (Cross Impact Matrix). As the pilot implementations during a project have revealed, users already familiar with the OIC or management instruments in general will appreciate the additional information provided by the Cross Impact Matrix. It offers deeper insights into the complex interrelations between their intangible resources and the linkages to business success and strategy. The full version requires one more workshop day, i.e. altogether three workshops with the OIC project team.

Impact Scoring makes it possible to prioritise the fields for intervention. The OIC project team ranks the IC factors according to their impact on the organisation, i.e. the factor exerting the most influence on business success is ranked highest. The question to be answered is: “How important is this particular IC factor for achieving our strategic objectives?”

The full version helps to analyse the interrelations between IC factors. The interdependencies between IC factors are examined and their degree of influence on each other is analysed. Full impact analysis makes it possible to identify the interactions between the organisation’s IC and Business Processes and Business Success.

In contrast to the Impact Scoring, the Cross Impact Matrix analyses each factor with regard to its influence on other factors. Each IC factor is then analysed to determine whether it has no influence (0), weak influence (1), strong influence (2) or even an exponential influence (3) on other IC factors.

The Cross Impact Matrix is relatively comprehensive, but increases accuracy. Furthermore, the project team needs to deal with the subject in more detail and will therefore become more aware of IC relevant aspects.

**Cross Impact Matrix**

| No. | IC type | ID | IC Factor | HC-1 | HC-2 | HC-3 | SC-1 | SC-2 | SC-3 | RC-1 | RC-2 | Active Sum | Passive Sum | Relative Influence | Controll-ability |
|---|---|---|---|---|---|---|---|---|---|---|---|---|---|---|---|
| 1 | Human Capital | HC-1 | Professional competence | x | 1 | 1 | 0 | 1 | 1 | 1 | 0 | 5 | 4 | 10 % | 1,25 |
| 2 | | HC-2 | Social competence | 0 | x | 2 | 1 | 3 | 0 | 2 | 0 | 8 | 4 | 16 % | 2,00 |
| 3 | | HC-3 | Employee motivation | 2 | 1 | x | 3 | 3 | 0 | 2 | 1 | 12 | 9 | 24 % | 1,33 |
| 4 | Structural Capital | SC-1 | Corporate culture | 1 | 1 | 1 | x | 2 | 0 | 1 | 0 | 6 | 7 | 12 % | 0,86 |
| 5 | | SC-2 | Internal Co-operation and Knowledge Transfer | 1 | 0 | 2 | 2 | x | 2 | 2 | 0 | 9 | 11 | 18 % | 0,82 |
| 6 | | SC-3 | Information Technology & Explicit Knowledge | 0 | 0 | 1 | 0 | 1 | x | 1 | 0 | 3 | 4 | 8 % | 0,75 |
| 7 | Relational Capital | RC-1 | Customer Relationships | 0 | 1 | 2 | 1 | 1 | 1 | x | 1 | 7 | 9 | 14 % | 0,78 |
| 8 | | RC-2 | Investor Relationships | 0 | 0 | 0 | 0 | 0 | 0 | 0 | x | | 2 | | |
| 8 Total | | | | 4 | 4 | 9 | 7 | 11 | 4 | 9 | 2 | 50 | 50 | 100 % | 1,00 |

weak influence strong influence exponential influence

Example: The factor HC-1 "professional competence" exerts a weak influence (1) on HC-2 "social competence", i.e. if you enhance the qualification of employees, their social competence is influenced as well.

**Source:** (14)

**Fig. 7** Example of a Cross Impact Matrix working sheet

The main task of measurement is to find useful and appropriate indicators for the respective IC factors. The IC indicators help to measure the IC factors and their development over time on a quantitative basis, thus adding validity to the self-assessment. For perfect preparation, you should read the guideline for step 3 prior to the workshop in order to be able to brief the project team for step 3 at the end of the workshop.

## 7.5 Measurement

Between the first and the second workshop the team should do some internal work. They are supposed to determine IC indicators for the most important IC factors. These are necessary in order to measure the IC factors and monitor their development over time. In this sense, they add validity to the self-assessment in IC Analysis.

OIC measurement provides all the necessary information for briefing the OIC project team at the end of the first workshop and supporting the company's project manager as the person mainly responsible for determining IC indicators.

The company's project team should be informed about the definition of IC indicators, the requirements for determining IC indicators and their function in the whole OIC process.

The organisation's project manager is responsible for determining IC indicators. He is free to organise internal meetings in order to specify IC indicators. Since IC indicators are sometimes treated confidentially within a company, the project manager may need the support of the top management to access relevant data.

After briefing the OIC project team at the end of IC Analysis, the moderator should support the organisation's project manager and be available to answer any questions by phone or email. Please be aware that the moderator needs the determined IC indicators as an input for the second workshop. It is therefore important to ask the organisation's project manager to send the list of IC indicators prior to the second workshop.

In order to measure the IC factors the OIC project team has to determine related indicators. The IC indicators help to measure the IC factors and their development over time on a quantitative basis. Furthermore determining IC indicators is beneficial for monitoring measures for particular IC factors.

The enterprise's team should determine the IC indicators according to their particular business situation and status quo of IC. For example, if "Leadership ability" is assessed low, the OIC project team may justify this evaluation with the fact that there are simply not enough executives. The IC indicator "Number of executives" could be used as evidence.

IC indicators should be calculated on the basis of a clear definition. Additionally the data source should be of sufficient quality.

Frequently, a lot of key figures or management ratios are available within the organisation's various departments (Marketing, HR, Accounting, etc.). The team should take care not to choose figures simply because they are available, but should also choose figures which are useful and appropriate for measuring a particular IC factor. Individual project team members could be assigned to deliver specific indicators related to their domain.

## 7.6 Strategy Refinement & Measures

After the previous steps, all the information needed for defining IC measures and refining strategy will have been generated. The OIC project team should have identified the main IC factors and evaluated them with regard to their Quantity/ Quality and Systematic Management. Furthermore, they should have assessed the IC factors in terms of their relative importance. The moderator should also have a first list of IC indicators prepared by the organisation.

A presentation summarising all results of the IC analysis should be prepared for the second workshop on the organisation's premises.

Based on these results, the moderator will lead the project team's discussion on the following basic questions:

- How should an OIC project team interpret the results of the IC analysis?
- Which IC factors have the highest potential for intervention?
- What does that imply for the organisation's IC strategy?
- Which measures should be implemented for the development of IC according to the strategy?

Strategy Refinement & Measures explains how to interpret the findings of the IC analysis and how to moderate the workshop. The conclusions drawn during this workshop will serve as the basis for any activities on the systematic management of the company's IC, as well as for the final ICS document and further communication of the results.

Strategy Refinement and Measures is broken down into two major parts:

- **Interpretation of results**
- Strengths-and-weaknesses analysis

The first step is a strengths-and-weaknesses analysis based on the QQS Assessment.

When discussing the QQS Overview with the project team, the moderator should ask the team if these results reflect the status quo of the company in their opinion. As it is based on the team's self-assessment, the focus should be on the relative differences between IC factors and not on the absolute level of evaluation.

By validating the assessment results in this way, the team responsible may analyse in more detail what has led to individual evaluations, in order to review specific evaluations that might be questioned. Then it is possible to refer to the reasoning given by the group for the specific evaluation in the first workshop.

Finally, the summary of the QQS Overview shows the overall evaluation of the three IC categories and of the three dimensions of evaluation. The moderator and the OIC project team can identify any strengths or weaknesses in the company's Human, Structural or Relational Capital and whether the main observed problems lie in Quantity, Quality or the Systematic Management of IC.

- Impact analysis

In general, the Impact Scoring reveals the relative importance of each IC factor compared to all other factors.

- Identify fields of intervention

The IC Management Portfolio displays the future potential of the different IC factors in a four quadrant matrix.

Develop ( signifies high relative importance and high potential for improvement ).

Stabilise ( shows a high relative importance but low potential improvement ).

No need for action ( displays factors with low relative importance and low potential for improvement ).

Analyse ( displays low relative importance but high potential for improvement ).

The IC factors' potential for intervention depends on the assessment of their status quo (QQS Assessment) and on their relative importance regarding the strategic objectives.

The basic assumption is that factors with a high improvement potential combined with a high relative importance will be the most effective fields for intervention. Measures for improvement and development will unfold the greatest impact and will have the best cost-benefit ratio or the highest return-on-investment. The essential question for the top management, namely "Where should we start to invest? Where can we get the maximum impact at minimum costs?" can be answered by systematically searching for the factors with the highest potential for intervention.

- **Deriving IC strategy and measures**

After the IC analysis, the company can extend its aims to a detailed description of its future business and IC strategy. In addition, measures can be derived for IC development.

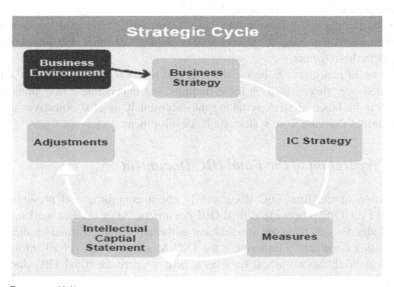

**Source:** (14)

**Fig. 8** Strategic Cycle

The IC strategy describes the organisation's position with regard to sub-areas of Intellectual Capital. It forms the basis of its care and further development.

The IC strategy is clearly derived from the business strategy and steers the measures to develop Intellectual Capital.

The success of the steps taken is measured and evaluated in the OIC and forms the starting point of a new cycle after adjustment to changes in the business environment.

The moderator should guide the team through this process by helping them to reflect on their basic strategic objectives in the light of the fields of intervention identified before.

The team should find a consensus on which intangible resources need development and which other IC resources might be needed to achieve the defined strategic objectives. Moreover, they may discover new strategic opportunities of which they were not aware before.

When the team has decided which IC factors need development as the first priority, they should discuss the measures to be taken in order to achieve an improvement. To support this discussion, the moderator should guide the team to reflect on the defined fields of intervention by taking another look at the QQS Assessment.

When a measure for developing a specific IC factor has been defined, the team should also think about how to measure the desired changes. Here, the IC indicators play an important role again. IC indicators make it possible to monitor the IC factors' development over time and answer the crucial question whether the implemented measures were successful.

Finally the team will have defined a set of measures aiming at the systematic development of particular IC factors as well as a set of indicators for measuring changes in these factors.

This set of measures can be viewed as a first rough IC strategy which could be elaborated over time. Based on these findings, the management may think about expanding its business strategy taking into account IC-related objectives and the opportunities deriving from systematic IC development.

## 7.7  Preparation of the Final OIC Document

Preparation of the final OIC document is about compiling and presenting the results of the OIC process in a final OIC document. After the last workshop, the OIC Moderator and the team should have gathered all the information needed to prepare this document. Furthermore the OIC Moderator will get all information needed to brief the workshop team as it will prepare the final OIC document internally.

The workshop team will prepare the OIC document internally. It therefore needs all the information for preparing the OIC document at the end of the second workshop. As this step is done internally, the process should be supported off-site, i.e. be available to answer any questions by phone or email. An internal version can be generated for management and internal communication purposes.

The OIC document has two major functions and its actual structure and content depend on the intended function. It can be used for internal purposes as a management tool and for external purposes as a communication tool. In the internal version all data should be disclosed whereas the external version might not show all data. The content therefore needs to be adjusted to the requirements of the two versions.

The OIC Moderator should talk with the team and especially with the management about which stakeholders the team wants to address with the OIC document. Depending on this, the participants can consider an appropriate structure and content. The exemplary OIC document template offered by the project should be of great use for the OIC project team. Additionally the moderator should inform the team about basic aspects which help to determine structure and content. The participants need to consider what the different stakeholders expect from such a document. Depending on the stakeholders' expectations and the company's own willingness to disclose information, they must decide which information is to be disclosed and how it is to be presented. Last but not least they must consider the kind of benefit the OIC document should generate for the organisation.

The OIC should be brought into the Corporate Identity layout and the insights need to be explained. The acquired data from "business model" to "strategy refinement & measures" make it clear that the findings can only be sensibly interpreted in the context of the organisation. Completely different conclusions may emerge, depending on the initial situation and the set strategic objectives. An OIC document which is used for communication must therefore provide a description of this context. Furthermore, an interpretation from the organisation's point of view should be given which helps to link all the facts and figures to the company's particular context. Based on these interpretations, the organisation should show which consequences are drawn and how the company will develop its Intellectual Capital to ensure future business success.

# 8 Preliminary Case Study

Regardless of the initial difficulties encountered by the organisations to envision and then describe their business models, they were able to recognise the different components (IC elements, key business processes, etc.) affecting its configuration – this goes for coupling abilities.

However, it was not until the organisations went through the cause-effect analysis (impact analysis) between business processes, business success factors and IC elements that they were able to see the whole picture and alter either the interrelationships or the components or both. The cause-effect analysis in particular focuses on the systemic nature of the business model enabling the project team to detect possible inconsistencies.

This was the case with the Engineering Business Unit of a production company when confronted with the almost total absence of synergies between the company's different business units , one of the EBU's strategic objectives for the period. A similar reaction was experienced by another project team when they discovered that the company was pouring resources into business processes that, as defined, were totally unrelated to the business success factors. This lack of connectivity within the business model, and particularly between the business processes and the success factors, meant that achievement of the company's strategic objectives were seriously at risk. Fortunately, these inconsistencies were already solved during the workshop in OIC "Strategy Refinement & Measures".

| No. | IC type | ID | IC Factor | 1 | 2 | 3 | 4 | 5 | 6 | 7 | 8 | 9 | 10 | Ranking Weighting Sum | Score |
|-----|---------|-----|-----------|---|---|---|---|---|---|---|---|---|----|----|----|
| 1 | Human Capital | HC-1 | Professional competence | 3 | 4 | 3 | 2 | 3 | 3 | 3 | 2 | 4 | 3 | 30 | 8 % |
| 2 | | HC-2 | Social competence | 4 | 3 | 5 | 4 | 4 | 5 | 5 | 4 | 3 | 5 | 42 | 12 % |
| 3 | | HC-3 | Employee motivation | 8 | 7 | 8 | 5 | 7 | 8 | 4 | 8 | 8 | 8 | 71 | 20 % |
| 4 | Structural Capital | SC-1 | Corporate culture | 5 | 5 | 4 | 7 | 5 | 7 | 6 | 7 | 5 | 4 | 55 | 15 % |
| 5 | | SC-2 | Internal Co-operation and Knowledge Transfer | 7 | 8 | 6 | 8 | 6 | 6 | 7 | 5 | 7 | 6 | 66 | 18 % |
| 6 | | SC-3 | Information Technology & Explicit Knowledge | 2 | 2 | 1 | 1 | 2 | 2 | 1 | 3 | 2 | 2 | 18 | 5 % |
| 7 | Relational Capital | RC-1 | Customer Relationships | 6 | 6 | 7 | 6 | 8 | 4 | 8 | 6 | 6 | 7 | 64 | 18 % |
| 8 | | RC-2 | Investor Relationships | 1 | 1 | 2 | 3 | 1 | 1 | 2 | 1 | 1 | 1 | 14 | 4 % |
| | | | | 36 | 36 | 36 | 36 | 36 | 36 | 36 | 36 | 36 | 36 | 360 | 100 % |

Impact Scoring — Ranking Team Member

8 Total          Highest Rank 8     Maximum possible total 36

**Source:** (14)

**Fig. 9** Impact Scoring

# 9  Conclusion

In the last decade management literature has paid significant attention to the role of knowledge for global competitiveness in the 21st century. It is recognized as a durable and more sustainable strategic resource to acquire and maintain competitive advantages (2, 7, 12). Today, intellectual capital (IC) is widely recognized as the critical source of true and sustainable competitive advantage (18). Knowledge is the basis of IC and is therefore at the heart of organizational capabilities. The need to continuously generate and grow this knowledge base has never been greater. Organizational capabilities are based on knowledge. Thus, knowledge is a resource that forms the foundation of the company's capabilities. Capabilities combine to become competencies and these are core competencies when they represent a domain in which the organization excels.

The organizations require different kinds and levels of knowledge to be successful. The organizations need to have knowledge to develop their goods. To determine the markets, they need to have knowledge about their customers and opponents; they also need to benefit of knowledge if they want to have a good coordination and power during their work; they need to use knowledge to improve the procedure of their intellectual capital and main capabilities Mutual effects among the knowledge process, intellectual capital, learning point and invention can be studied and considered in the organization and to create an effective knowledge-base strategy for this purpose. Knowledge-base strategy is a response that connects the special characteristics and measurements of a company with local advisability in which they work for it. In the sophisticated and searching world, the organizations must be so clever and strong to use some opportunities including the combination of new explorations and available knowledge use, sharing and supporting knowledge, intellectual capital flow management *(10)*.

Intellectual capital is important to both society and organisations. It can be a source of competitive advantage for businesses and stimulate innovation that leads to wealth generation.

Technological revolutions, the rise to pre-eminence of the knowledge-based economy and the networked society have all led to the realisation that successful companies excel at fostering creativity and perpetually creating new knowledge.

Companies depend on being able to measure, manage and develop this knowledge. Management efforts therefore have to focus on the knowledge resources and their use. Intangibles and how they contribute to value creation have to be appreciated so that the appropriate decisions can be made to protect and enhance them. There must also be a credible way of reporting those intangibles to the market to give the investment community comprehensive information to assist in valuing the company more accurately (28).

The aim of intellectual capital is to explain the difference between the book and market value of a company. In that way, a company's value can be determined more precisely, that could be important for investors. The measurement of intellectual capital has the another very significant dimension that enables better business managing. Only if we know a company's basic values, we will be able to manage them and to maximize their growth as the greatest values of the business activities are not al- ways visible in financial reports. Intellectual capital can be considered as a liabilities' balance sheet item showing the origin of some intangible (16).

# APPENDIX

## Example of an IC Definition List (Screenshot From ICS Toolbox)

| | | Definitions | |
|---|---|---|---|
| | | The information entered in column E will be transferred automatically to the subsequent working sheets! | |
| IC type | ID | IC Factor (english) | Definition (english) |
| **Human Capital** | HC-1 | Professional competence | The expertise gained within the organisation or in the employee's career: professional training, higher education, training courses and seminars, as well as practical work experiences gained on-the-job. |
| | HC-2 | Social competence | The ability to get on well with people, communicate and discuss in a constructive manner, nurturing trust-enhancing behaviour in order to enable a comfortable co-operation. Furthermore the learning ability, the self-conscious handling of critique and risks as well as the creativity and flexibility of individual employees are embraced in the term 'social competence' |
| | HC-3 | Employee motivation | The motivation to play a part within the organisation, to take on responsibility, committed to the fulfilment of tasks and the willingness for an open knowledge exchange. Typical sub areas are for example satisfaction with the labour situation, identification with the organisation, sense and participation of achievement. |
| | HC-4 | Leadership ability | The ability to administrate and motivate people. Develop and communicate strategies and visions and their empathic implementation. Negotiation skills, assertiveness, consequence and credibility as well as the ability to create a scope of self dependant development belong to this IC factor. |
| **Structural Capital** | SC-1 | Corporate culture | The business culture comprises all values and norms, influencing joint interaction, knowledge transfer and the working manner. Compliance to rules, good manners, "Do's and Don'ts" and the handling of failures are important aspects in the process. |
| | SC-2 | Internal Co-operation and Knowledge Transfer | The manner how employees, organisational units and different hierarchy levels exchange information and co-operate together (e.g. conjoint projects). The focused knowledge transfer among employees. Furthermore the focused knowledge transfer between generations is noticeable. |
| | SC-3 | Management Instruments | Tools and instruments supporting the efforts of the leadership and therefore have an impact on the way how decisions are made and what information paths are incorporated in the decision-making process. |

**Source:** (14)

# References

1. Alwert, K., Bornemann, M., Kivikas, M.: Intellectual Capital Statement – Made in Germany. Guideline. Published by the Federal Ministry for Economics and Technology, Berlin (2004),
   http://www.bmwi.de/BMWi/Redaktion/PDF/W/
   wissensbilanz-made-in-germany-leitfaden,
   property=pdf,bereich=bmwi,sprache=de,rwb=true.pdf
2. Barney, J.B.: Firm resources and sustained competitive advantage. Journal of Management 17(1), 99–120 (1991)
3. Boisot, M.H.: Knowledge Assets: Securing Competitive Advantage in the Information Economy. Oxford University Press, Oxford (1998)
4. Bontis, N., Dragonetti, N.C., Jacobsen, K., Roos, G.: The knowledge toolbox: A Review of the tools available to measure and Manage Intangible Resource. European Management Journal 17(4), 391–402 (1999)
5. Brennan, N., Connell, B.: Intellectual Capital: Current Issues and Policy Implications. Paper Presented at the 23rd Annual Congress of the European Accounting Association, Munich, March 29-31 (2000)

6. Chaston, I., Badger, B., Sadler-Smith, E.: Organizational learning: an empirical assessment of process in small U.K. manufacturing firms. Journal of small business Management 39(2), 139 (2001)
7. Drucker, P.F.: The coming of the new organization. Harvard Business Review (1988)
8. Edvinsson, L., Malone, M.S.: Intellectual Capital-the Proven way to establish your company's real value by measuring its hidden brainpower, p. 22. Judy Piatkus Ltd., London (1997)
9. European Commission, project "Intellectual Capital Statement – Made in Europe" (InCaS), DG Research under the EU 6th Framework Programme (2010),
   http://www.inthekzone.com/pdfs/Intellectual_Capital_
   Statement.pdf
10. European Journal of Social Sciences – Increasing the Intellectual Capital in Organization: Examining the Role of Organizational Learning, 14(1) (2010)
11. Granstrad, O.: The Economics and the Management of Intellectual Property. Edward Elgar Publishing, Cheltenham (1999)
12. Grant, R.M.: Contemporary Strategy Analysis. Blackwell, Oxford (1991)
13. Gunther, T.: Controlling Intangible Assets under the Framework of value-based management. Kostenrechnungspraxis No Sonderheft 1, 53–62 (2001)
14. InCas, InCaS Intellectual Capital Statement – Made in Europe (2010),
   http://www.inthekzone.com/pdfs/Intellectual_Capital_
   Statement.pdf
15. Johannessen, J.-A., Olsen, B., Olaisen, J.: In- tellectual capital as a holistic management philosophy: a theoretical perspective. International Journal of Information Management 25(2), 151–171 (2005)
16. Šaponja, L.D., Šijan, G., Milutinovíc, S.: Intellectual Capital: Part of a Modern Business Enterprise of the Future,
   http://www.fm-kp.si/zalozba/ISBN/961-6486-71-3/231-243.pdf
17. Marr, B., Chatzkel, J.: Intellectual capital at the crossroads: managing, measuring and reporting of IC. Journal of Intellectual Capital 5(2), 224–229 (2004)
18. Marr, B., Neely, IA.: Organizational performance measurement in the emerging digital age. Int. J. Business Performance Management 3(2-4), 191–215 (2001)
19. Marr, B., Schiuma, G.: Measuring and managing intellectual capital and knowledge assets in new economy organisations. In: Bourne, M. (ed.) Handbook of Performance Measurement, Gee, London (2001)
20. Marr, B., Schiuma, G., Neely, A.: The dynamics of value creation: mapping your intellectual performance drivers. Journal of Intellectual Capital 5(2), 312–325 (2004)
21. Mouritsen, et al.: Developing and Managing Knowledge through intellectual capital statements. Journal of Intellectual Capital 3(1), 10–29 (2002)
22. Nonaka, I.: A Dynamic Theory of Organizational Knowledge Creation. Organization Science 5(1), 14–37 (1994)
23. Porter, M.: Competitive Strategy. The Free Press, New York (1980)
24. Prahalad, C.K., Hamel, G.: The core competence of the corporation. Harvard Business Review 68(3), 79–91 (1990)
25. Quinn, J.B.: Intelligent Enterprise: A Knowledge and Service Based Paradigm for Industry. Free Press, New York (1992)
26. Schwartz, S.H., Lehmann, A., Roccas, S.: Multimethod probes of basic human values. In: Adamopoulos, J., Kashima, Y. (eds.) Social Psychology and Culture Context: Essays in Honor of Harry C. Triandis. Sage, Newbury Park (1999)

27. Senge, P.M.: The Fifth Discipline: The Art and Practice of the Learning Organization, Doubleday/Currency, New York (1990)
28. Shapiro, C., Varian, H.: A Strategy Guide to the Network Economy. Harvard Business School Press, USA (1999)
29. Starovic, D., Marr, B.: Understanding corporate value:managing and reporting lectual capital. CIMA. Chartered Institute of Management Accountants, Cranfield (2010)
30. Stewart, T.: Intellectual Capital- the new wealth of organization, 1st edn. Nicolas Brealey Publishing, London (1998)
31. Stewart, S.: A Definition of Business Strategy (2008), http://www.rapid-business-intelligence-success.com/definition-of-business-strategy.html
32. Sveiby, K.E.: The New Organizational Wealth. Berrett- Koehler, San Francisco (1997)
33. Sullivan, P.H.: Value-Driven Intellectual Capital: How to Convert Intangible Corporate Assets into Market Value. John Wiley, New York (2000)
34. Sveiby, K.E.: Intellectual Capital – The New Wealth of Organizations, 1st edn. Nicolas Brealey Publishing, London (1997)
35. Tapscott (ed.): Creating Value in the Network Economy. Harvard Business Review, Boston (1999)
36. VMI, Intellectual Capital Statement (2007), http://www.psych.lse.ac.uk/incas/page38/page39/files/VMI.pdf
37. Vester, F.:: Die Kunst vernetzt zu denken: Ideen und Werkzeuge für einen neuen Umgang mit Komplexität, Stuttgart (1999); Viedma, J.M.: In Search of an Intellectual Capital Comprehensive Theory, ICICKM 2006: 3rd International Conference on Intellectual Capital, Knowledge Management and Organisational Learning Pontificia Universidad Católica de Chile, October 19-20, (2006)
38. Wikipedia, PESTEL Analysis (2010), http://dictionary.sensagent.com/steep+analysis/en-en/

# Social Networking Sites as Business Tool: A Study of User Behavior

Efthymios Constantinides[1,*], Carlota Lorenzo-Romero[2],
and María-del-Carmen Alarcón-del-Amo[2]

[1] Faculty of Management and Governance, University of Twente,
Enschede, The Netherlands
P.O. Box 217
7500 AE Enschede, The Netherlands
e.constantinides@utwente.nl
[2] Business Administration, University of Castilla-La Mancha, Albacete, Spain
Faculty of Economics and Business, Marketing Department
Plaza de la Universidad, 1
02071 Albacete, Spain.
{Carlota.Lorenzo,MCarmen.Alarcon}@uclm.es

**Abstract.** Social Networking Sites (SNS) are second generation web applications allowing the creation of personal online networks; the social networking domain has become one of the fastest growing online environments connecting hundreds of millions of people worldwide. Businesses are increasingly interested in SNS as sources of customer voice and market information but are also increasingly interested in the SNS as the domain where promising marketing tactics can be applied; SNS can be also used as business process management (BPM) tools due to powerful synergies between BPM and SNS. Marketers have various options: SNS can be engaged as tools of customer engagement, social interaction, and relationship building but also as channels of information, collaboration and promotion. Understanding the adoption motives and adoption process of these applications is an essential step in engaging the SNS as part of the marketing toolbox.

In order to analyze the factors influencing the acceptance and use of SNS in The Netherlands a Technology Acceptance Model (TAM) was developed and tested. The findings indicate that there is a significant positive effect of the ease of use of SNS on perceived usefulness. Both variables have a direct effect on the intention to use the SNS and an indirect effect on the attitude towards the applications. Moreover the study has shown that intention to use SNS has a direct and positive effect on the degree of final use of SNS. Results demonstrate empirically that the TAM can explain and predict the acceptance of SNS by users as new communication system connecting them to peers and businesses.

This study presents an overview of the SNS user behavior and underlines the importance of using these applications as new communication technology.

---

* Corresponding author.

M. Glykas (Ed.): Business Process Management, SCI 444, pp. 221–240.
springerlink.com      © Springer-Verlag Berlin Heidelberg 2013

**Keywords:** Social Networking Sites, Technology Acceptance Model, user behavior in SNS, Structural Equation Modeling, Business Process Management.

# 1 Introduction

Social Networking Sites (SNS) belong to the second generation of Internet applications commonly known as Web 2.0 of Social Web. SNS are a relatively recent Internet phenomenon, the first examples appeared at the end of the 20th Century. Today hundreds of millions of web users are connected through SNS worldwide, many of them having fully integrated SNS into their everyday life (Boyd and Ellison 2007; Subrahmanyam et al. 2008). According to data from ComScore Media Metrix (2011) about digital trends in the European market, Europe shows the highest growth in SNS reach from December 2009 until December 2010 (up 10.9 percentage points), compared with North America (up 6.6 points), Latin America (up 5.5 points) and Middle East-Africa (up 2.7 points). North America has the highest number of Internet users who are SNS users (89.8%), followed by Latin America (87.7%) and Europe (84.4%). According to the same source The Netherlands is worldwide the country with the highest penetration in SNS like LinkedIn and Twitter.

SNS are considered of great importance for individuals and businesses, since they support the maintenance of existing social ties and the establishment of new connections between users (Donath and Boyd 2004; Ellison et al. 2006; Ellison et al. 2007; Boyd and Ellison 2007). These connections can be vital for facilitating various group tasks (Sproull and Kiesler 1991; Preece and Maloney-Krichmar 2003), controlling and decreasing bad online behavior or efforts to manipulate co-users (Reid 1999; Donath 1998), and allow the build-up of different types of social capital (Resnick 2001; Ramayah 2006). These are some of the potential benefits of social networking (Wellman 2001).

The objective of this study is to explain the essence of SNS and develop a Technology Acceptance Model (TAM) (Davis 1989) enabling the analysis the factors that influence the level of acceptance and use of SNS. In this research line, Chiu et al. (2006) integrate a model for investigating the motivations behind people's knowledge sharing in virtual communities. We drew from the basic perceptions of the TAM and applied it to the adoption of SNS as new communication system based on web technology. The study focuses on the SNS adoption in a mature Internet market specific market, The Netherlands. This country has a leading position in the use of the Internet in Europe. 83% of the population is regular Internet users –connecting to the Internet at least once a week– and 74% of the population has broadband connection. In both aspects The Netherlands is ranking n° 1 in Europe (European Commission 2010). Moreover, within Europe, 85.1% of Internet users from The Netherlands are SNS users (ComScore Media Metrix 2011).

The study is a step towards better understanding the mechanisms of adoption of SNS by users; this is a vital step for strategists willing to exploit the possibilities of using SNS as part of the corporate strategy.

# 2 Theoretical Review

## 2.1 Social Networking Sites for Business Process Management

SNS act as an extension of face-to-face interaction (Kujatch 2011). Boyd and Ellison (2007) define SNS as "services based on Internet that allow individuals to build a public or semi-public profile within a system, create a list of other users that share a connection, and see and navigate through their list of connections and of those created by others within the system". A more recent definition is proposed by Kwon and Wen who define SNS as "websites that allow building relationships online between persons by means of collecting useful information and sharing it with people. Also, they can create groups which allow interacting amongst users with similar interests" (Kwon and Wen 2010). SNS specifically offer the users a space where they can maintain and create new relationships, as well as share information (Kolbitsch and Maurer 2006). According to O'Dell (2010), SNS are becoming more popular than search engines in some countries. In fact, the role of SNS as marketing instruments is a subject attracting also substantial research attention (Constantinides and Fountain 2008; Waters et al. 2009; Hogg 2010; Spaulding 2010).

Next to the networking possibilities SNS can also empower their users as participants in the marketing process; this because online networks offer users the possibility to obtain more information about companies, brands and products (often in the form of user reviews) and make better buying decisions (Lorenzo et al. 2009). SNS can therefore become a useful input for the business process management (BPM).

BPM is defined as all efforts in an organization to analyze and continually improve fundamental activities such as manufacturing, marketing, communications and other major elements of company's operations (Zairi 1997; Trkman 2010). A business process is a complete, dynamically coordinated set of activities or logically related tasks that must be performed to deliver value to customers or to fulfill other strategic goals (Guha and Kettinger 1993; Strnadl 2006; Trkman 2010). BPM has received considerable attention recently by business administration because companies are interested in increasing customer satisfaction, reducing cost of doing business and launching new products and services at low cost (Trkman, 2010).

Distributed work and virtual teams are getting more important and most business processes involve several collaborating individuals (Ho et al. 2009). Social Media-based collaborative tools, such as SNS, blogs or forums, can be used to support the design, execution, and management of business processes (Richter et al. 2009, 2010; Vom Brocke et al. 2011).

According to Vom Brocke et al. (2011) an emergent application area of BPM can be identified at the intersection between Social Media and business process. Both traditional BPM and Social Media can address the management of work activities (Hammel and Breen 2007). According to Erol et al. (2010, pp. 453-454), social software provides a number of new tools and options, that should be considered when designing business processes:

- Self identification. Any individual who would like to contribute to an activity may do so and thus identifies themselves as competent to carry out such activity.
- Transparency. All work results are openly available to anyone.
- Signing. The performing individual signs all work activities upon completion.
- Logging. All activities are logged to provide a history of work activities.
- Open modification. Anyone can modify contributions by other individuals.
- Discussion. Comments on work results and suggestions for modifications can be discussed an even directly linked to content pieces.
- Banning. Actors exhibiting inappropriate behaviour may be banned.

Bruno et al. (2011) explain that the social software, the basis of the Social Media, offers four very valuable features: Weak ties, egalitarianism, social production and mutual service provisioning. Using these features it is possible to achieve integration and responsiveness as required by BPM. According to these authors, the Social Media:

- Facilitate the motivation to participate by consumers and workers and thus supports the organizational integration.
- Allow the social production which facilitates sharing knowledge and empowering human agents at the micro level and thus supports the organizational integration.
- The egalitarianism feature of Social Media facilitating knowledge sharing, participates also in the organizational integration.

According to Khoshafian (2008), there is a powerful synergy between BPM and SNS. BPM continuous improvement has different phases: model development, execution, and performance monitoring. SNS and collaboration can be organized along both time and place dimensions. A BPM user community spans across functional units, cross departmental value chains, trading partners and other BPM communities. Each community and each option of networking in the taxonomy of BPM undergoes continuous improvement phases and the BPM societies can leverage SNS solutions. SNS are often void of process context; BPM provides the process context of the collaboration and networking. So while business processes provide the context of the collaboration, SNS support and augments the various activities of the BPM continuous improvement lifecycle.

Due to the importance of SNS in the BPM, Bruno et al. (2011) affirm that companies should investigate how to run social processes on SNS by taking advantage of the dynamic teams and the informal interactions they provide. Traditional BPM life cycles are based on the assumption that it is possible to predefine the control and information flows needed during the BPM life cycle. Social Media breaks with this assumption and replaces the *a priori* control of quality by an *a posteriori* one. In Social Media in general, and SNS in particular, individuals are allowed to contribute; therefore, end users are included more in the BPM life cycle and play an active part of the BPM life cycle. Furthermore, there is no strict separation between contributor and consumer. Instead there is an exchange of contributions and each contribution to the BPM life cycle can be seen as a service provided by an end user or an involved stakeholder. Bruno et al. (2011) argue that SNS present some risk for companies. In an environment where everybody may provide information, some information or activities could damage the reputation of the company.

Regarding the potential of SNS as business tools, these can play different roles a part of marketing strategy (Constantinides and Fountain 2008; Waters et al. 2009; Tikkanen et al. 2009; Hogg 2010; Spaulding 2010); there are various ways that SNS can provide business value:

- SNS allow the creation of a corporate profile with information about the company, its products and/or services that could be used to create an interest group within these websites.
- SNS may be used as a tool for customer service and relationship marketing, thanks to the possibility of keeping customers informed about updates made by the company, and as a direct, quick and simple communication channel.
- They can also reinforce brand strength, as they enable companies to communicate with their target customers, enhancing brand loyalty and trust. They also create an innovative brand image by participating in these media.
- They can mean net savings in marketing expenses, as participating in a social network is relatively cheap, and certain marketing campaigns may no longer be needed. Furthermore, "worth-of-mouth" communication is much more effective.
- It is possible to advertise in SNS, using banners, buying keywords, creating events, sponsoring contents, etc.
- Professional social networks enable employee recruitment and management of business contacts.
- Customer information available in SNS voluntarily uploaded by the users allows companies to obtain a great amount of information about their customers, their personality and lifestyle as well as information on their trust in the Internet, perceived ease of use, perceived risk, attitudes to SNS, and so on.
- Companies can use SNS as source of customer voice for the development or testing of new products or services.
- Finally, based on user analysis and segmentation companies could selectively inform their customers even on personalized level, about their products or services, provide them with useful and interesting information or use the SNS as customer service channels.

## 2.3 Research Hypotheses

In this study the TAM framework was used to explain the adoption of SNS by Dutch users; this because of the efficacy of this model to predict the adoption of any technology (Mathieson 1991; Venkatesh and Davis 1996; Gefen et al. 2003a, b; Vijayasarathy 2004; Shih and Fang 2004; Ha and Stoel 2009). Stam and Stanton (2010), based on models related to acceptance of technology such as TAM and UTAUT, examine acceptance of workplace technology changes from the viewpoint of popular psychological theories. Pelling and White (2009) examine the theory of planned behavior (TPB) applied to analyze the young people's use of SNS.

Regarding the online technologies in particular, the TAM can explain certain models of e-collaboration between users (Dasgupta et al. 2002); e-collaboration is one of the central features of SNS. According to the reasoning behind the use of models based on the TAM for technology adoption, there is a direct and positive effect between attitude, intention to use, and final use of a technology that an individual chooses to adopt; the TAM has been studied with a variety of populations and technology, but there is not much research about the use of this model to explain and predict acceptance of social technologies (e.g. Willis (2008) found that the TAM explains social networking technology).

The relationship between attitude and intention to use of an online system is common and essential in many behavioral models. This relationship has been demonstrated by several researchers in different contexts: Adoption of information technology and information systems (Davis 1989; Mathieson 1991; Davis et al. 1992; Taylor and Todd 1994, 1995a, b; Bernadette 1996; Harrison et al. 1997; Karahanna et al. 1999; Bhattacherjee 2000; Chen et al. 2002; Van Der Heijden 2003), adoption of the Web (Fenech 1998; Lederer et al. 2000; Lin and Lu 2000; Moon and Kim 2001; Porter and Donthu 2006), embracing of e-commerce (Gefen and Straub 1997, 2000; Bhattacherjee 2000; Chen et al. 2002; Pavlou 2002; Pavlou and Fygenson, 2006), and the use of Web 2.0 in general (Mo and Coulson, 2008; Shin and Kim 2008), and some 2.0 tools such as virtual communities (Papadopoulou 2007; Shin 2008), and SNS (Willis 2008). In this line, Sicilia and Ruiz (2010) propose that the amount of information has differential effects on consumers' information processing and attitudes. As results they obtained that while information processing diminishes under high levels of information, attitudes remain favorable.

Therefore, it is evident that the attitude towards a given technology has a positive effect on the intention to use this technology, which leads us to propose the following hypothesis:

**H1. The attitude towards SNS has a positive and significant effect on the intention to use these websites.**

Some of the TAM-based studies have included the current use of technology (Davis et al. 1992; Henderson and Divett 2003; Shang et al. 2005) and the intention to use as response variables (Mathieson 1991; Lin and Lu 2000; Luarn and Li 2005). Other studies included both concepts and observed a causal relationship between them (Taylor and Todd 1994, 1995b; Igbaria et al. 1997; Horton et al. 2001; Wu and Wang 2005; Shang et al. 2005). In this line, we have introduced both final variables, as we believe that the variable "Intention to Use" acts as intermediary between the effect exerted by the perceptions (ease of use and perceived usefulness) and final use of the individual. Therefore, we propose the following hypothesis:

**H2. The intention to use SNS has a positive and significant effect on the final use of these websites.**

In the TAM the perceived usefulness directly affects the use through the intention to use. Davis et al. (1989) argue that although the direct effect of a belief (i.e. the perceived usefulness) on the intention to use is contrary to the premises of the Theory of Reasoned Action (TRA) proposed by Fishbein and Ajzen (Fishbein and Ajzen 1975), studies provide the theoretical justification, as well as empirical evidence of direct links between perceived usefulness and intention to use (Triandis 1977; Brinberg 1979; Bagozzi 1982; Davis et al. 1989; Mathieson 1991; Igbaria 1993; Taylor and Todd 1995a, b; Chuan-Chuand and Lu 2000; Liaw and Huang 2003; Wang et al. 2003; Bhattacherjee and Premkumar 2004). Furthermore, Lee et al. (2003) indicate that the relationship between the perceived usefulness and intention to use in the context of the TAM is statistically supported since there are 74 studies that show a significant relationship between both variables. Willis (2008) found a positive and significant relation between both constructs within SNS. Therefore, we propose the third hypothesis:

**H3. The perceived usefulness of SNS has a positive and significant effect on the intention to use them.**

Davis et al. (1992) suggests an indirect relationship between perceived ease of use and the intention to use, mediated by the perceived usefulness. In addition other studies confirm this indirect relationship (Davis et al. 1989; Karahanna and Straub 1999). More recent empirical studies have found that perceived ease of use has a positive and significant effect on the intention to use, defined as wish to use (Lee et al. 2003; Ramayah 2006; Wise et al. 2010). When the interaction with the technology is easier, the feeling of efficiency by the user should be greater and hence the intention to use it should be greater (Chung 2005). Willis (2008) obtained significant and positive effects between both variables after the empirical analysis applied to SNS. Based on the theoretical assumption, we propose the following hypothesis:

**H4. The perceived ease of use of SNS has a positive and significant effect on the intention to use them.**

According to Castañeda et al. (2007), the ease of use has a double impact on the attitude, because of self-efficacy and instrumentality. The efficiency or effectiveness (self-efficacy) is one of the factors of intrinsic motivation for a person (Bandura 1982; Young et al. 2009), therefore this effect is directly related to the attitude. On the other hand the ease of use can also be instrumental (instrumentality), contributing to increase of performance. This increase means less effort, thanks to the ease of use, allowing getting more work done with the same effort (Taylor and Todd 1994). This instrumental effect on the attitude occurs via perceived usefulness as the original TAM postulates (Castañeda et al. 2008). Pelling and White (2009) reveal that high-level SNS use is influenced by attitudinal, normative, and self-identity factors. Furthermore, this effect has been amply demonstrated in empirical studies about adoption of new technologies

(Davis et al. 1989, 1992; Venkatesh and Davis 1996, 2000; Agarwal and Prasad 1999; Venkatesh 2000; O'Cass and Fenech 2003; Liaw and Huang 2003; Shih 2004; Shang et al. 2005; Barker 2009; Wilson et al. 2010). On the basis of the above we propose the following hypotheses:

**H5. The perceived ease of use of SNS has a positive and significant effect on the attitude toward these sites.**
**H6. The perceived ease of use of SNS has a positive and significant effect on the perceived usefulness of using them.**

In the TAM, the ease of use and the perceived usefulness are considered beliefs that are postulated a priori and are considered constructs which determine the attitude (Davis, 1993). This assertion is based on a pillar of the TRA arguing that attitudes toward a behavior are influenced by relevant beliefs (Fishbein and Ajzen 1975; Davis et al. 1989, 1992). Furthermore, there is empirical evidence of these relationships (Davis 1993; Venkatesh and Davis 2000). Therefore, we propose the following hypothesis:

**H7. The perceived usefulness of SNS has a positive and significant effect on the attitude toward these sites.**

Following the assumptions explained in the previous section, we obtain an initial model (Fig. 1) illustrating the adoption of SNS.

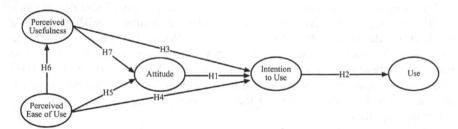

**Fig. 1** Proposal of a model to explain the SNS acceptance

## 3  Materials and Methods

For this study we used an online survey presented to a panel of SNS users in October 2010.

The final sample consisted of 400 Dutch SNS users from 16 to 74 years old; it was drawn using a non-probability method by quota sampling to ensure that various subgroups of the target population are proportionally represented in the sample with regard to gender, age and region of residence.

Table 1 shows the constructs used in our study adapted from previous studies and were all measured by multiple items five-points Likert-type scales, with the exception of the use construct.

**Table 1** Final items included in the model (content validity)

| Construct | Indicator | Description | Measurement | Source (Content validity) |
|---|---|---|---|---|
| Use (USE) | USE1 | How often do you participate in Social Networking Sites? | Five-points scale: Less than once a week - Several times a day | Moon and Kim (2001) |
| | USE2 | On average, how many hours do you access Social Networking Sites per week? | Five-points scale: Less than 1 hour - More than 25 hours | Legris et al. (2003) Shih and Fang (2004) |
| Perceived Usefulness (PU) | PU1 | I consider that the functions of SNS are useful for me | | Moon and Kim (2001) |
| | PU2 | Using the SNS contributes to interaction with others people | | Sánchez et al. (2007) |
| | PU3 | Using SNS enables me to access a lot of information | | Willis (2008) |
| | UP4 | Overall, the SNS are useful | | Rodríguez et al. (2009) |
| Perceived Ease of Use (PEU) | PEU2 | Learning to work with SNS is easy for me | | Venkatesh (2000) Moon and Kim (2001) Pikkarainen et al. (2004) Muñoz (2008) Shin (2008) Willis (2008) Rodríguez et al. (2009) |
| | PEU4 | I find it easy to get a SNS to do what I want it to do | | |
| | PEU6 | It is easy to remember how to use SNS | | |
| | PEU7 | My interaction with SNS is clear and understandable | | |
| | PEU9 | Everyone can easily use SNS | Five-points Likert-scale: Strongly disagree- Strongly agree | |
| | PEU11 | Overall, I think that SNS are easy to use | | |
| Attitude (A) | A1 | Using SNS is a good idea | | Moon and Kim (2001) Rodríguez et al. (2009) Sicilia and Ruiz (2010) |
| | A2 | It is fun to participate in SNS | | |
| | A3 | I agree with the existence of SNS | | |
| | A4 | It is nice to connect to SNS | | |
| | A5 | Using SNS seems to me a positive idea | | |
| Intention to Use (IU) | IU1 | It is probable that I will participate or continue participating in SNS | | Moon and Kim (2001) Mathwick (2002) Chan and Lu (2004) Castañeda (2005) Muñoz (2008) Willis (2008) Rodríguez et al. (2009) |
| | IU2 | It is true that I will share or continue sharing information on SNS | | |
| | IU3 | I intend to begin or continue using SNS | | |
| | IU4 | I will recommend others to use SNS | | |

# 4 Results

## 4.1 Reliability and Validity Assessment

A confirmatory factor analysis (CFA) was conducted jointly for all the constructs making up the model, with the aim of assessing the measurement reliability and validity. The structural equation modeling (SEM) techniques were applied using the statistics package EQS 6.1.b.

Reliability of the constructs is presented in Table 2 and demonstrates high-internal consistency of the constructs. With an exploratory analysis, we found that the item-total correlation, which measures the correlation of each item with the sum of the remaining items that constitute the scale, is above the minimum of 0.3 recommended by Nurosis (1993). In each case, Cronbach's Alpha exceeded Nunnally and Bernstein's (1994) recommendation of 0.70, except in the USE scale. Composite reliability (CR) represents the shared variance among a set of observed variables measuring an underlying construct (Fornell and Larcker 1981). Generally a CR of at least 0.60 is considered desirable (Bagozzi 1994). This requirement is met for each factor. Average variance extracted (AVE) was also calculated for each construct, resulting in AVEs greater than 0.50 (Fornell and Larcker 1981).

**Table 2** Internal consistency and convergent validity

| Variable | Indicator | Factor Loading | Robust t-value[a] | Cronbach's alpha | Composite reliability (CR) | Average Variance Extracted (AVE) |
|---|---|---|---|---|---|---|
| Use (USE) | USE1 | 0.891 | 11.834 | 0.580 | 0.71 | 0.56 |
| | USE2 | 0.574 | 6.946 | | | |
| Perceived Usefulness (PU) | PU1 | 0.775 | 14.706 | 0.846 | 0.90 | 0.60 |
| | PU2 | 0.738 | 14.478 | | | |
| | PU3 | 0.701 | 13.589 | | | |
| | UP4 | 0.844 | 18.474 | | | |
| Perceived Ease of Use (PEU) | PEU2 | 0.741 | 14.224 | 0.9 | 0.91 | 0.62 |
| | PEU4 | 0.783 | 18.279 | | | |
| | PEU6 | 0.918 | 20.608 | | | |
| | PEU7 | 0.870 | 18.829 | | | |
| | PEU9 | 0.589 | 10.896 | | | |
| | PEU11 | 0.794 | 14.724 | | | |
| Attitude (A) | A1 | 0.774 | 14.032 | 0.912 | 0.91 | 0.67 |
| | A2 | 0.845 | 15.088 | | | |
| | A3 | 0.812 | 14.244 | | | |
| | A4 | 0.831 | 17.098 | | | |
| | A5 | 0.842 | 16.820 | | | |
| Intention to Use (IU) | IU1 | 0.875 | 16.718 | 0,885 | 0.89 | 0.66 |
| | IU2 | 0.792 | 17.938 | | | |
| | IU3 | 0.850 | 17.101 | | | |
| | IU4 | 0.735 | 14.940 | | | |

[a]=p<0,01
Robust goodness of fit indices: $\chi^2$ (179 degree of freedom, df) = 311.59; $\chi^2/df$=1,74; NFI= 0,912; NNFI= 0,953; CFI=0,960; RMSEA=0,043.

Regarding content validity all the items included in the scale have been tested in the academic literature on the Internet (Cronbach 1971; Vila et al. 2000). Due to lack of valid scales adapted to SNS adoption, it was necessary to adapt the initial scales (Table 1).

Convergent validity is verified by analyzing the factor loadings and their significance. The t scores obtained for the coefficients in Table 2 indicate that all factor loadings were significant (p<.001). Also the size of all standardized loadings is higher than 0.50 (Sanzo et al. 2003; Steenkamp and Geyskens 2006), and the averages of the item-to-factor loadings are higher than 0.70 (Hair et al. 2006). This finding provides evidence supporting the convergent validity of the indicators (Anderson and Gerbing 1988).

Evidence for discriminant validity of the measures was provided in three ways (Table 3). First, none of the 95 per cent confidence intervals of the individual elements of the latent factor correlation matrix contained a value of 1.0 (Anderson and Gerbing 1988). Second, the shared variance between pairs of constructs was always less than the corresponding AVE (Fornell and Larcker 1981), except for scales related to attitude and intention to use, whose AVE is lower and equal, respectively, to the squared correlation (0.714). Third, Bagozzi argues that existing discriminant validity is acceptable if the correlations between the variables in the confirmatory model are not much higher than 0.8 points and in our study this argument is supported (Bagozzi 1994).

Therefore, construct validity was verified by assessing the convergent validity and discriminant validity of the scale (Vila et al. 2000).

**Table 3** Discriminant validity of the theoretical construct measures

|       | USE      | PU       | PEU             | A               | IU              |
|-------|----------|----------|-----------------|-----------------|-----------------|
| USE   | 0.56[a]  | [0.215 , 0.471] | [0.144 , 0.376] | [0.343 , 0.559] | [0.325 , 0.565] |
| PU    | 0.118    | 0.60[a]  | [0.161 , 0.397] | [0.527 , 0.739] | [0.496 , 0.716] |
| PEU   | 0.068    | 0.078    | 0.62[a]         | [0.336 , 0.560] | [0.461 , 0.641] |
| A     | 0.203    | 0.401    | 0.201           | 0.67[a]         | [0.795 , 0.895] |
| IU    | 0.198    | 0.367    | 0.304           | 0.714           | 0.66[a]         |

[a] The diagonal represents the AVE, while above the diagonal de 95 per cent confidence interval for the estimated factors correlations is provided, below the diagonal, the shared variance (squared correlations) is represented.

## 4.2 Results of Causal Model

The proposed conceptual model was tested using structural equation modeling. The overall fit of the model is acceptable because the goodness of statistics shows values greater than the commonly accepted. In the next paragraphs, each hypothesis will be justified according to results obtained; these can be observed in Table 4 and Fig. 2.

**Table 4** Structural model results

| Hypothesis | Path | Standardized path coefficients | Confidence level[a] | Robust t-value[b] |
|---|---|---|---|---|
| H1 | Attitude → Intention to Use | 0.676 | ** | 9.805 |
| H2 | Intention to Use → Use | 0.459 | ** | 7.353 |
| H3 | Perceived Usefulness → Intention to Use | 0.122 | * | 2.144 |
| H4 | Perceived Ease of Use → Intention to Use | 0.215 | ** | 5.355 |
| H5 | Perceived Ease of Use → Attitude | 0.295 | ** | 4.784 |
| H6 | Perceived Ease of Use → Perceived Usefulness | 0.278 | ** | 4.445 |
| H7 | Perceived Usefulness → Attitude | 0.551 | ** | 7.516 |

[a] Confidence level: **=p<0,01; *=p<0,05
[b] Robust goodness of fit indices: $\chi^2$ (182 df) = 317,11; $\chi^2$/df=1,74; NFI= 0,910; NNFI= 0,953; CFI=0,959; RMSEA=0,043

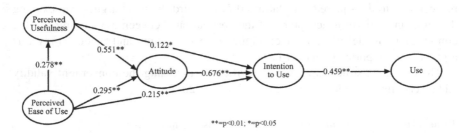

**Fig. 2** Estimation of research model

The results show that attitude towards social networks sites has a significant positive effect on intention to use ($\beta$=0.676, p<0.01) and therefore hypothesis 1 was supported.

At the same time, intention to use has a significant positive effect on final use of social networks sites ($\beta$=0.459, p<0.01), and thus, hypothesis 2 was supported also.

Perceived ease of use has a significant positive effect on intention to use ($\beta$=0.215, p<0.01), attitude ($\beta$=0.295, p<0.01) and perceived usefulness ($\beta$=0.278, p<0.01); thus, hypotheses 4, 5 and 6 were supported.

Perceived usefulness has a significant positive effect on attitude ($\beta$=0.551, p<0.01) and intention to use ($\beta$=0.122, p<0.05); thus, hypotheses 3 and 7 were also supported.

Summarizing we can confirm that all relationships between constructs analyzed in the TAM are translated and tested positively in the SNS context. In consequence, the typical adoption behavior of new technologies seems to be applicable to the adoption behavior of users of SNS.

## 5 Conclusions and Practical Implications for Business Process Management

The adoption and use of SNS by users and companies has dramatically increased in last years. SNS offer to people new ways to build and maintain social networks, share information, generate and edit content, participate in social activities online. They also allow the locating of individuals sharing the same backgrounds and interests based on the characteristics published in personal profiles. SNS have contributed to increasing the number of contacts maintained by the individuals, as well as facilitating the strengthening of links between them. Next to the social effects the SNS have allowed individuals to build up a strong position against businesses. This emerging customer empowerment is having its roots in increasing access to information sources but also in the increasing ability of customers to participate to commercial processes. As Riegner (2007) explains, user participation has far reaching commercial implications: Consumers exercise great and increasing influence on product offerings and on the strategies used to sell them.

There is a powerful synergy between BPM and SNS. While business processes provide the context of the collaboration, SNS supports and augments the various activities of the BPM applications' continuous improvement. SNS allow the integration of users into business process management. The benefits of combining BPM and SNS are facilitated by the completely new approach for putting together the inputs of different people (Erol et al. 2010). Therefore, introducing SNS provides several opportunities for the BPM, mainly when the process demands a high level of communication and collaboration among performing actors. The basic principles of openness and ease of usage of SNS are the pillars of the wide acceptance of SNS. However, these new ways of working may require considerable time for acceptance (Erol et al. 2010).

All these possibilities have led to an increasing professional and academic interest in the study of the Social Web in general and the SNS in particular. Researchers are interested in the potential of SNS as part of the marketing strategy and as new interaction environments between consumers and business. Several businesses are already successfully testing ways of utilizing these environments by tapping customer generated information relevant to their brands and products or create their own SNS where customers can participate and interact with the company and their peers. Many businesses already invest substantial marketing resources in SNS where customers are expressing and communicating    their ideas, tastes, preferences, worries, etc. (Constantinides and Fountain 2008; Peres et al. 2010).

This research has confirmed that the new technology adoption process as stipulated by the TAM approach applies to the adoption process of SNS; the perceived ease of use of SNS in The Netherlands has a positive effect on the perceived usefulness of these websites and both variables have a positive and direct influence on the intention to use the SNS and an indirect effect on the intention to use through the attitude towards the website. Moreover, it has shown that intention to use SNS has a direct and positive effect on the degree of adoption and use of SNS.

The findings of this study reveal that the attitudes of SNS users are important in predicting intention to use. To attract the participation in the SNS, online businesses and SNS providers need to develop strategies to cultivate positive attitudes toward the use of SNS. The results indicate that providing user-friendly and useful websites are very important preconditions for generating positive attitudes.

The changes in attitudes occur rapidly and in less time than the sense of usefulness or perceived ease of use (Thompson and Hunt 1996). As Yang and Yoo (2004) stated, many theories and programs have been developed for the change to positive attitudes, such as the direct influence of individuals (e.g. improving people's motivations, abilities, memories or mood states), the improvement of contextual clues (e.g. classical conditioning), or the consideration of persuasive messages (e.g. the credibility of the message, memory message, two face communication, etc.).

Although attitudes can change more quickly, continuous efforts should also be given to maintain the attitude, which is temporary, unstable, and malleable (Thompson and Hunt 1996). The motivations, skills, experience and education, all influence the development and maintenance of attitude. Therefore, maintenance and change of attitude should be considered as a complementary tool to traditional application techniques that can be used to improve user acceptance of new technologies (Yang and Yoo 2004).

This work complements the Willis study demonstrating empirically that the TAM could explain and predict the acceptance of SNS as new connection and communication technology when businesses integrate such tools into their marketing strategy (Willis 2008). In this line, SNS marketers must offer useful and simple social networking applications a positive attitude of users towards the applications, something that will improve their intention to use them. Customers experience the interactivity, the functionality and the access to information as factors enhancing their perception of usefulness. Simplicity and user friendliness of SNS seem to be important elements of the Perceived ease of Use as influencer of the customer attitudes and intention to use the application.

In sum, this study has supported all hypotheses and in this sense it could be useful for businesses eager to introduce these new interactive support systems in their websites and obtain better results in various areas: branding, customer advocacy and trust, customer engagement etc.

As a main limitation of this study we can mention the fact that the survey was answered exclusively by Dutch people. In order to be able to generalize the results of this research the study must be repeated using sample populations from different countries or geographic territories. A second limitation is that the model used is quite simple; however this study is an explorative one aiming at testing the suitability of original TAM in the adoption process of SNS in by demographically diverse populations.

As future research, it is possible that we could improve the results obtained in this study by using a more complex model including other variables like trust, satisfaction, and psychological variables (Wilson et al. 2010; Wise et al. 2010), in order to analyze the influence of the internal variables on the adoption level of the SNS by users. Other directions for further research are the execution of this study in several countries and conduct cross-cultural comparative analyses related to the adoption and use of SNS by users according to their cultural differences.

# References

Agarwal, R., Prasad, J.: Are Individual Differences Germane to the Acceptance of New Information Technologies? Decision Sciences 30(2), 361–391 (1999)

Anderson, E., Gerbing, D.W.: Structural equation modeling in practice: A review and recommended two-step approach. Psychological Bulletin 103, 411–423 (1988)

Baggozzi, R.P.: A field investigation of causal relations among cognition, affect, intentions and behaviour. Journal of Marketing Research 19, 562–584 (1982)

Bagozzi, R.P.: Structural Equation Model in Marketing Research. Basic Principles. Blackwell Publishers, MA (1994)

Bandura, A.: Self-Efficacy mechanism in Human Agency. American Psychologist 37(2), 122–147 (1982)

Barker, V.: Older adolescents' motivations for social network site use: The influence of gender, group identity, and collective self-esteem. CyberPsychology, Behavior, and Social Networking 12, 209–213 (2009)

Bernadette, S.: Empirical evaluation of the revised technology acceptance model. Management Science 42(1), 85–93 (1996)

Bhattacherjee, A.: Acceptance of e-commerce services: The case of electronic rokerages. IEEE Transactions on Systems, Man, and Cybernetics – Part A: Systems and Humans 30(4), 411–420 (2000)

Bhattacherjee, A., Premkumar, G.: Understanding changes in beliefs and attitude toward Information Technology usage: A theoretical model and longitudinal test. MIS Quarterly 28(2), 229–254 (2004)

Boyd, D.M., Ellison, N.B.: Social network sites: Definition, history, and scholarship. Journal of Computer-Mediated Communication 13(1), Article 11 (2007), http://jcmc.indiana.edu/vol13/issue1/boyd.ellison.html

Brinberg, D.: An examination of the determinants of intention and behavior: A comparison of two models. Journal of Applied Social Psychology 6, 560–575 (1979)

Bruno, G., Dengler, F., Jennings, B., Khalaf, R., Nurcan, S., Prilla, M., Sarini, M., Schmidt, R., Silva, R.: Key challenges for enabling agile BPM with social software. Journal of Software Process: Improvement and Practice 23, 297–326 (2011)

Castañeda, J.A.: El comportamiento del usuario de Internet: Análisis de los antecedentes y consecuencias de la fidelidad [The Internet user behavior: An analysis of the antecedents and consequences of fidelity]. Doctoral Thesis. Marketing Department. University of Granada (2005)

Castañeda, J.A., Muñoz-Leiva, F., Luque, T.: Web Acceptance Model (WAM): Moderating effects of user experience. Information & Management 44, 384–396 (2007)

Chan, S., Lu, M.: Understanding internet banking adoption and use behavior: A hong kong perspective. Journal of Global Information Management 12(3), 21–43 (2004)

Chen, L., Gillenson, M., Sherrel, D.: Enticing online consumers: an extended technology acceptance perspective. Information and Management 39, 705–719 (2002)

Chiu, C.M., Hsu, M.H., Wang, E.T.G.: Understanding knowledge sharing in virtual communities: An integration of social capital and social cognitive theories. Decision Support Systems 42, 1872–1888 (2006)

Chuan-Chuan, J., Lu, H.: Towards an understanding of the behavioural intention to use a web site. International Journal of Information Management 20(3), 197–208 (2000)

Chung, D.: Something for nothing: understanding purchasing behaviors in social virtual environments. CyberPsychology & Behavior 8(6), 538–554 (2005)

ComScore Media Metrix (2011), The 2010 Europe digital year in review, http://www.comscore.com/esl/Press_Events/Presentations_Whit epapers/2011/2010_Europe_Digital_Year_in_Review (accessed March 20, 2011)

Constantinides, E., Fountain, S.: Web 2.0: Conceptual foundations and Marketing Issues. Journal of Direct, Data and Digital Marketing Practice 9(3), 231–244 (2008)

Cronbach, L.J.: Test validation, educational measurement. In: Thorndike, R.L. (ed.) American Council on Education, Washington, pp. 443–507 (1971)

Dasgupta, S., Granger, M., McGarry, N.: User acceptance of ecollaboration technology: An extension of the technology acceptance model. Group Decision Negotiation 11, 87–100 (2002)

Davis, F.D.: Perceived usefulness, perceived ease of use and user acceptance of Information Technology. MIS Quarterly 13(3), 319–340 (1989)

Davis, F.D.: User Acceptance of Information Technology: System Characteristics, User Perceptions and Behavioral Impacts. International Journal of Man-Machine Studies 38, 475–487 (1993)

Davis, F.D., Bagozzi, R.P., Warsaw, P.R.: Extrinsic and Intrinsic Motivation to use Computers in The Workplace. Journal of Applied Social Psychology 22(14), 1111–1132 (1992)

Davis, F.D., Bagozzi, R.P., Warshaw, P.R.: User acceptance of computer technology: A comparison of two theoretical models. Management Science 35(8), 982–1003 (1989)

Donath, J., Boyd, D.: Public displays of connection. Technology Journal 22(4), 71–82 (2004)

Donath, J.S.: Identity and deception in the virtual community. In: Kollock, P., Smith, M. (eds.) Communities in Cyberspace. Routledge, London (1998)

Ellison, N.B., Steinfield, C., Lampe, C.: The benefits of Facebook "friends": Social capital and college students' use of online social network sites. Journal of Computer-Mediated Communication 12, 1143–1168 (2007)

Ellison, N.B., Heino, R., Gibbs, J.: Managing impressions online: Self-presentation processes in the online dating environment. Journal of Computer-Mediated Communication 11 (2006), http://jcmc.indiana.edu/vol11/issue2/ ellison.html (accessed March 20, 2011)

Erol, S., Granitzer, M., Happ, S., Jantunen, S., Jennings, B., Johannesson, P., Koschmider, A., Nurcan, S., Rossi, D., Schmidt, R.: Comgining BPM and social software: Contradiction or chance? Journal of Software Maintenance and Evolution: Research and Practice 22, 449–476 (2010)

Fenech, T.: Using perceived ease of use and perceived usefulness to predict acceptance of the World Wide Web. Computer Networks & ISDN Systems 30, 629–630 (1998)

Fishbein, M., Ajzen, I.: Belief, Attitude, Intention and Behavior: An Introduction to Theory and Research. Addison-Wesley, Reading (1975)

Fornell, C., Larcker, D.F.: Evaluating structural equations models with unobservable variables and measurement error. Journal of Marketing Research 18, 39–50 (1981)

Gefen, D., Straub, D.W.: Gender differences in the perception and use of E-mail: An extension to the technology acceptance model. MIS Quarterly 21(4), 389–400 (1997)

Gefen, D., Straub, D.W.: The relative importance of perceived ease of use in IS adoption: A study of e-commerce adoption. Journal of Association for Information Systems 1(8), 1–28 (2000)

Gefen, D., Karahanna, E., Straub, D.W.: Inexperience and experience with online Stores: The importance of TAM and Trust. IEE Transactions on Engineering Management 50(3), 307–321 (2003a)

Gefen, D., Karahanna, E., Straub, D.W.: Trust and TAM in online shopping: An integrated Model. MIS Quarterly 27(1), 51–90 (2003b)

Guha, S., Kettinger, W.J.: Business process reengineering. Information Systems Management 10(3), 13–22 (1993)

Ha, S., Stoel, L.: Consumer e-shopping acceptance: Antecedents in a technology acceptance model. Journal of Business Research 62, 565–571 (2009)

Hair, J.F., Black, W.C., Babin, B.J., Anderson, R.E., Tatham, R.L.: Multivariate data analysis. Prentice-Hall, New York (2006)

Hamel, G., Breen, B.: The Future of Management. Harvard Business School Press (2007)

Harrison, D.A., Mykytyn, P.P., Riemenschneider, C.K.: Executive decisions about adoption of information technology in small business: Theory and empirical tests. Information Systems Research 8(2), 171–195 (1997)

Henderson, R., Divett, M.: Perceived uselfulness, ease of use and electronic supermarket use. International Journal of Computer Studies 59, 383–395 (2003)

Ho, D.T.Y., Jin, Y., Dwivedi, R.: Business process management: A research overview and analysis. In: Proceedings of the 15th Americas Conference on Information Systems, San Francisco, CA (2009)

Hogg, T.: Inferring preference correlations from social networks. Electronic Commerce Research and Applications 9, 29–37 (2010)

Horton, R., Buck, T., Waterson, P.E., Clegg, C.: Explaining intranet use with the technology acceptance model. Journal of Information Technology 16, 237–249 (2001)

Igbaria, M.: User acceptance of microcomputer technology. An empirical test. International Journal of Management Science 21(1), 73–90 (1993)

Igbaria, M., Zinatelli, N., Cragg, P., Cavaye, A.L.M.: Personal computing acceptance factors in small firms: a structural equation model. MIS Quarterly 21(3), 279–302 (1997)

Karahanna, E., Straub, D.W.: The psychological origins of perceived usefulness and ease of-use. Information & Management 35(4), 237–250 (1999)

Karahanna, E., Straub, D.W., Chervany, N.L.: Information Technology adoption across time: A cross-sectional comparison of pre-adoption and post-adoption beliefs. MIS Quarterly 23(2), 183–213 (1999)

Khoshafian, S.: MyBPM: Social Networking for Business Process Management. In: Fischer, L. (ed.) BPM and Workflow Handbook Spotlight on Human-Centric BPM, pp. 187–196. Future Strategies, Inc., Florida (2008)

Kolbitsch, J., Maurer, H.: The transformation of the Web: How emerging communities shape the information we consume. Journal of Universal Computer Science 2(12), 187–213 (2006)

Kwon, O., Wen, Y.: An empirical study of the factors affecting social network service use. Computers in Human Behavior 26(2), 254–263 (2010)

Lederer, A.L., Maupin, D.J., Sens, M.P., Zhuang, Y.: The technology acceptance model and the World Wide Web. Decision Support Systems 29, 269–282 (2000)

Lee, Y., Kozar, K.A., Larsen, K.R.T.: The Technology Acceptance Model: Past, present, and Future. Communications of the Association for Information Systems 12, 752–780 (2003)

Legris, P., Ingham, J., Collerette, P.: Why do people use information technology? A critical review of the technology acceptance model. Information and Management 40(3), 191–204 (2003)

Liaw, S.S., Huang, H.-M.: An investigation of user attitudes toward search engines as an information retrieval tool. Computers in Human Behavior 19(6), 751–765 (2003)

Lin, C.C., Lu, H.: Towards an Understanding of the Behavioral Intention to Use a Web site. International Journal of Information Management 20, 197–208 (2000)

Lorenzo, C., Constantinides, E., Gómez-Borja, M.A.: Effects of Web Experience on virtual retail purchase preferences. International Retail and Marketing Review 5(1), 1–15 (2009)

Luarn, P., Li, H.-H.: Toward an understanding of the behavioural intention to use mobile banking. Computers in Human Behavior 21(6), 873–891 (2005)

Mathieson, K.: Predicting user intentions: Comparing the Technology Acceptance Model with the Theory of Planned Behavior. Information Systems Research 2(3), 173–191 (1991)

Mathwick, C.: Understanding the online consumer: A typology of online relational norms and behaviour. Journal of Interactive Marketing 16(1), 40–55 (2002)

Mo, P.K.H., Coulson, N.S.: Exploring the communication of social support within virtual communities: a content analysis of messages posted to an online HIV/AIDS support group. CyberPsychology, Behavior, and Social Networking 11, 371–374 (2008)

Moon, J., Kim, Y.: Extending the TAM for a World-Wide-Web context. Information and Management 38, 217–230 (2001)

Muñoz, F.: La adopción de una innovación basada en la Web. Análisis y modelización de los mecanismos generadores de confianza [The adoption of a Web-based innovation. Analysis and modeling of the mechanisms that generate trust]. Doctoral Thesis, Marketing Department, University of Granada (2008)

Nunnally, J., Bernstein, I.H.: Psychometric Theory. McGraw-Hill, New York (1994)

Nurosis, M.J.: SPSS. Statistical Data Analysis. Spss Inc. (1993)

O'Cass, A., Fenech, T.: Web retailing adoption: exploring the nature of Internet users web retailing behaviour. Journal of Retailing and Consumer Services 10, 81–94 (2003)

O'Dell: Is Facebook getting bigger than Google? (2010), http://mashable.com/2010/06/08/social-network-stats/ (accesed March, 2011)

Papadopoulou, P.: Applying virtual reality for trust-building ecommerce environments. Virtual Reality 11(2), 107–127 (2007)

Pavlou, P.: A theory of Planned Behavior Perspective to the Consumer Adoption of Electronic Commerce. MIS Quarterly, 1–51 (2002)

Pavlou, P.A., Fygenson, M.: Understanding and predicting electronic commerce adoption: An extension of the Theory of Planned Behavior. MIS Quarterly 30(1), 115–144 (2006)

Pelling, E.L., White, K.M.: The theory of planned behavior applied to young people's use of social networking web sites. CyberPsychology, Behavior, and Social Networking 12, 755–759 (2009)

Peres, R., Muller, E., Mahajan, V.: Innovation diffusion and new product growth models: A critical review and research directions. International Journal of Research in Marketing 27, 91–106 (2010)

Pikkarainen, T., Pikkarainen, K., Karjaluoto, H., Pahnila, S.: Consumer acceptance of online banking: An extension of the Technology Acceptance Model. Internet Research: Electronic Networking Applications and Policy 14(3), 224–235 (2004)

Porter, C.E., Donthu, N.: Using the technology acceptance model to explain how attitudes determine Internet usage: The role of perceived access barriers and demographics. Journal of Business Research 59, 999–1007 (2006)

Preece, J., Maloney-Krichmar, D.: Online Communities. In: Jacko, J., Sears, A. (eds.) Handbook of Human-Computer Interaction, pp. 596–620. Lawrence Erlbaum Associates Inc., Mahwah (2003)

Ramayah, T.: Interface characteristics, perceived ease of use and intention to use an online library in Malaysia. Information Development 22(2), 123–133 (2006)

Reid, E.: Hierarchy and power: Social control in cyberspace. In: Smith, M.A., Kollock, P. (eds.) Communities in Cyberspace, pp. 107–133. Routledge, London (1999)

Resnick, P.: Beyond bowling together: Sociotechnical capital. In: Carroll, J. (ed.) HCI in the New Millennium, pp. 247–272. Addison-Wesley, Boston (2001)

Richter, D., Riemer, K., vom Brocke, J.: Internet social networking – Distinguishing the phenomenon from its manifestations. In: Proceedings of the 17th European Conference on Information Systems, Verona, Italy (2009)

Richter, D., Riemer, K., vom Brocke, J.: Social transactions on social network sites: Can transaction cost theory contribute to a better understanding of Internet social networking. In: Proceedings of th 23rd Bled eConference, Bled, Slovenia (2010)

Riegner, C.: Word of mouth on the Web: The impact of Web 2.0 on consumer purchase decisions. Journal of Advertising Research 47, 436–447 (2007)

Rodríquez, N., Liñares, S., De la Llana, M.: The Main Determinants of Web 2.0 Acceptance: The Case of Youtube. In: 8th International Marketing Trends Congress, Paris (2009)

Sánchez, M.J., Rondán, F.J., Villarejo, A.F.: Un modelo empírico de adaptación y uso de la Web. Utilidad, facilidad de uso y flujo percibidos [An empirical model of adaptation and use of the Web. Usefulness, ease of use and perceived flow]. Cuadernos de Economía y Dirección de Empresa 30, 153–179 (2007)

Sanzo, M., Santos, M., Vázquez, R., Álvarez, L.: The Effect of Market Orientation on Buyer-Seller Relationship Satisfaction. Industrial Marketing Management 32(4), 327–345 (2003)

Shang, R., Chen, Y., Shen, L.: Extrinsic versus intrinsic motivations for consumers to shop on-line. Information and Management 42(3), 401–413 (2005)

Shih, H.P.: Extended Technology Acceptance Model of Internet utilization behaviour. Information & Management 41(6), 719–729 (2004)

Shih, Y., Fang, K.: The use of a decomposed Theory of Planned Behavior to study Internet banking in Taiwan. Internet Research: Electronic Networking Applications and Policy 14(3), 213–223 (2004)

Shin, D.H.: Understanding purchasing behaviors in a virtual economy: Consumer behavior involving virtual currency in Web 2.0 communities. Interacting with Computers 20, 433–446 (2008)

Shin, D.H., Kim, W.Y.: Applying the technology acceptance model and flow theory to cyworld user behavior: implication of the web 2.0 user acceptance. CyberPsychology, Behavior, and Social Networking 11, 378–382 (2008)

Sicilia, M., Ruiz, S.: The Effect of Web-Based information availability on consumers' processing and attitudes. Journal of Interactive Marketing 24, 31–41 (2010)

Spaulding, T.J.: How can virtual communities create value for business. Electronic Commerce Research and Applications 9, 38–49 (2010)

Sproull, L., Kiesler, S.: Connections: New ways of working in the networked organization. MIT Press, Cambridge (1991)

Stam, K.R., Stanton, J.M.: Events, emotions, and technology: examining acceptance of workplace technology changes. Information Technology & People 23, 23–53 (2010)

Steenkamp, J.E.B.M., Geyskens, I.: What drives the perceived value of web sites? A cross-national investigation. Journal of Marketing 70(3), 136–150 (2006)

Strnadl, C.F.: Aligning business and it: The process-driven architecture model. Information Systems Management 23(4), 67–77 (2006)

Subrahmanyam, K., Reich, S., Waechter, N., Espinoza, G.: Online and offline social networks: Use of social networking sites by emerging adults. Journal of Applied Development Psychology 29, 420–433 (2008)

Taylor, S., Todd, P.A.: Descomposition and crossover effects in the theory of planned behavior: A study of consumer adoption intentions. International Journal of Research in Marketing 12, 137–155 (1994)

Taylor, S., Todd, P.A.: Understanding Information Technology usage: A test of competing models. Information Systems Research 6(2), 144–176 (1995a)

Taylor, S., Todd, P.A.: Assessing IT usage: The role of prior experience. MIS Quarterly 19(4), 561–570 (1995b)

Thompson, R.C., Hunt, J.G.J.: In the black box of alpha, beta, and gamma change: using a cognitive-processing model to assess attitude structure. Academy of Management Review 21, 3 (1996)

Tikkanen, H., Hietanen, J., Henttonen, T., Rokka, J.: Exploring virtual worlds: success factors in virtual world marketing. Management Decision 47(8), 1357–1381 (2009)

Triandis, H.C.: Interpersonal behavior Monterey. Brooks/Cole, CA (1977)

Trkman, P.: The critical success factors of business process management. International Journal of Information Management 30(2), 125–134 (2010)

Van Der Heijden, H.: Factors Influencing the Usage of Websites: The Case of a Generic Portal in the Netherlands. Information & Management 40(6), 541–549 (2003)

Venkatesh, V.: Determinants of perceived ease of use: integrating control, intrinsic motivation, and emotion into the technology acceptance model. Information Systems Research 11(4), 342–365 (2000)

Venkatesh, V., Davis, F.D.: A Model of the Antecedents of Perceived Ease of Use: Development and Test. Decision Sciences 27(3), 451–481 (1996)

Venkatesh, V., Davis, F.D.: A theoretical extension of the Technology Acceptance Model: Four Longitudinal Field Studies. Management Science 46(2), 186–204 (2000)

Vijayasarathy, L.R.: Predicting consumer intentions to use online shopping: The case for an augmented technology acceptance model. Information & Management 41(6), 747–762 (2004)

Vila, N., Küster, I., Aldás, J.: Desarrollo y validación de escalas de medida en marketing [Development and validation of measurement scales in marketing]. In: Quaderns de Treball, vol. 104. University of Valencia, Spain (2000)

Vom Brocke, J., Becker, J., Maria Braccini, A., Butleris, R., Hofreiter, B., Kapocius, K., De Marco, M., Schmidt, G., Seidel, S., Simons, A., Skpal, T., Stein, A., Stieglitz, S., Suomi, R., Vossen, G., Winter, R., Wrycza, S.: Current and future issues in BPM research: A European perspective from the ERCIS Meeting 2010. Communications of the Association for Information System 28(25), 393–414 (2011)

Wang, Y.S., Wang, Y.M., Lin, H.H., Tang, T.I.: Determinants of user acceptance of internet banking: An empirical study. International Journal of Service Industry Management 14(5), 501–519 (2003)

Waters, R., Burnett, E., Lamm, A., Lucas, J.: Engaging stakeholders through social networking: How nonprofit organizations are using Facebook. Public Relations Review 35, 102–106 (2009)

Wellman, B.: Physical place and cyberplace: The rise of personalized networking. International Journal of Urban and Regional Research 25(2), 227–252 (2001)

Willis, T.: An evaluation of the technology acceptance model as a means of understanding online social networking behavior. Dissertation Abstracts International, 69 (2008)

Wilson, K., Fornasier, S., White, K.: Psychological predictors of young adults' use of social networking sites. Cyberpsychology, Behavior, and Social Networking 13, 173–177 (2010)

Wise, K., Alhabash, S., Park, H.: Emotional responses during social information seeking on Facebook. CyberPsychology, Behavior, and Social Networking 13, 555–562 (2010)

Wu, J.C., Wang, S.C.: What drives mobile commerce?: An empirical evaluation of the revised technology acceptance model. Information and Management 42(5), 719–729 (2005)

Yang, H.-D., Yoo, Y.: It's all about attitude: revisiting the technology acceptance model. Decision Support Systems 38, 19–31 (2004)

Young, S., Dutta, D., Dommety, G.: Extrapolating psychological insights from facebook profiles: A study of religion and relationship status. CyberPsychology, Behavior, and Social Networking 12, 347–350 (2009)

Zairi, M.: Business process management: A boundaryless approach to modern competitiveness. Business Process Management Journal 3(1), 64–80 (1997)

# Fuzzy Cognitive Maps in Social and Business Network Analysis

George Stakias[1,2], Markos Psoras[1,2], and Michael Glykas[1,2,3]

[1] Aegean Technopolis and
[2] Financial and Management Engineering, University of the Aegean, Greece
[3] Faculty of Business and Management, University of Wallangong, Dubai

**Abstract.** Social networks have gain a vast momentum and a widespread audience in recent years and have by far the biggest audience of users ever in the world targeted by marketeers and companies worldwide. The objective of this chapter is to propose a new complementary approach to Social Network Analysis (SNA) based on Fuzzy Cognitive Maps (FCMs).

Initially we provide an analysis of all types of Social Networks. The objective is to clarify basic concepts and make evident to the reader that social networks are part of our daily routine and that even when we don't realize it, we participate in them almost daily. We then present Fuzzy Cognitive Maps and their suitability for use in Social Network and Business Network analysis

In the final part of the chapter we present an innovative tool about the tool used for the construction of the FCMs called Visual FCM Modeler (VFCM). All concepts and the usage the tool is validate through a real world case study.

## 1 Introduction

The study of Social Network Analysis is of high importance. At present there is a wide use of social network services (e.g. facebook, twitter, myspace). Social networks are not only encountered online. There are social networks in our social life and in most cases we participate in them without knowing it. People participate in a number of activities through which influence or get influenced by others, thus shaping relationships of higher or lower importance with each other.

Social Networks are not composed by people exclusively. They might contain groups of people with same interests, enterprises or even whole countries.

The vast majority of people have had some experience of an online social network. Considering Facebook, the most well-known social network, we can see that each member has a wide variety of options: groups to join, pages to follow, other people to be friends with, applications to participate etc.

Through all these personal activities, others can shape an overview about a person, its behavior, its preferences and many other parts of its personality. With social network analysis, we are able to discover its main preferences and predict

M. Glykas (Ed.): Business Process Management, SCI 444, pp. 241–279.
springerlink.com          © Springer-Verlag Berlin Heidelberg 2013

things, such as the individuals with the highest probabilities to be friends with a particular person, the most probable groups to join, etc.

Apart from people social networks there also exist social networks of enterprises (also called business networks (Bns)). BNs can offer us information about business activities enterprises like the number of interacting enterprises, reliability and quality of enterprises participating in the network, their financial situation, even their market positioning and portions etc. In this chapter we will present the use of Fuzzy Cognitive Maps, as a mean of designing, monitoring and simulating simulate Social and Business Networks we are able to create simulations of social networks.

## 2 State of the Art in Social Networks

As social networking we may describe the compilation of individuals into certain groups, such as small rural communities or neighborhood subdivisions. Social networking is most popular online, but it is also possible in real life, especially in the workplace, universities, and high schools. The usual combination of social networks to the internet is because people search more to find others online.

A social network is a social structure made of nodes (which are generally individuals or organizations), often very complex, that are tied by one or more types of interdependency.

Social network analysis, analyze social relationships in terms of nodes and ties. Nodes are the individual actors within networks, and ties are the relationships between actors. A research in academic field shows that social networks operate on many levels, from families and small enterprises up to nations, and play a critical role in defining the way problems are solved, organizations function, and the degree individuals achieve their goals.

Simply, a social network is a map which contains all the ties between the nodes being studied. Social networks are often displayed in a social network diagram, where nodes are points and ties are lines.

### 2.1 Social Network Analysis

The interaction patterns describing social structure can be viewed as a network of relations (Radclife-Brown 1940) [7], therefore social network analysis is a relevant and highly useful tool for describing organizations and measuring the effects of organization systems.

Social network analysis views social relationships in terms of network theory consisting of nodes and ties (also called edges, links, or connections). Nodes are individual actors within the networks, and ties are the relationships between actors. There are various types of ties between nodes and also the resulting graph-based structures are often very complex. Research in a number of academic fields has shown that social networks operate on many levels, like the ones mentioned before and the degree which individuals succeed in achieving their goals. [4],[7]

A simple form of a social network could be a map of specified ties, such as friendship, between the nodes being studied. The nodes to which an individual is connected are the social contacts of that individual. The network can also be used to measure social capital – the value that an individual gets from the social network.

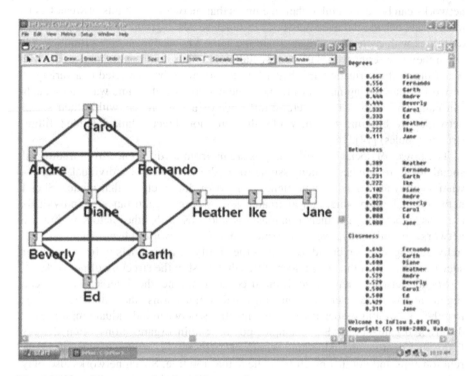

**Fig. 1** Image with an example of social network

Social network analysis has now evaluated from being a suggestive metaphor to an analytic approach to a paradigm, with its own theoretical statements, methods, social network analysis software, and researchers. Analysts specify from whole to part; from behavior to attitude; from structure to relation to individual. They either study whole networks, all ties containing specified relations in a defined population, or personal networks, ties that specified people have, mostly known as "personal communities".[9],[10],[11],[16]

Several analytic tendencies categorize social network analysis:

Groups are substructures of society: the approach is open to studying less-bounded social systems, from nonlocal communities to links among Web sites.

Rather than treating individuals (persons, organizations, states) as discrete units of analysis, it focuses on the effect caused by the structure of ties on individuals and their relationships.

Network analysis examines the extent to which structure and composition of ties can affect norms. The shape of a social network helps determine a network's usefulness to its individuals. More open networks, with many weak ties and social connections, are more likely to introduce new ideas and opportunities to their members than closed networks with many redundant ties. In general the tighter networks can be less useful to their members than networks with lots of weak ties to individuals outside the main network. In other words, a group of friends who only do things with each other already share the same knowledge and opportunities, but when they have connections to other **social worlds** it is common to have access to a wider range of information. It is better for individuals to be connected to a variety of networks, than having many ties to the same network. In the same way, individuals can gain more information, exercise influence or act as brokers within their social networks by bridging two networks that are not directly linked (called filling structural holes).[14],[16]

The power of social network analysis stems from its difference from traditional social scientific studies, which assume to be the attributes of individual actors -- whether they are friendly or unfriendly, smart or dumb, etc. -- that matter. Social network analysis produces an alternate view, where the attributes of individuals are less important than their relationships and ties with other actors within the network. In this way, we can explain many real-world phenomena, but the individual agency is reduced, as well as the ability of individuals to influence their success, because of the big number of ties that exist in the structure of network.

Social networks have also been used to examine the interaction between organizations, characterizing the informal connections that link executives together, as well as associations and connections between individual employees at different organizations. For example, power within organizations often derives from the degree an individual within a network is at the center of many relationships and not that much from the actual job title. Social networks also play a key role in hiring, in business success, and in job performance. Networks provide ways for companies to gather information, to gain a good place in the competitive environment and also to get adapted to its changes.

## 2.2  History of Social Network Analysis

A summary of the progress of social networks and social network analysis has been written by Linton Freeman.[1]

Precursors of social networks in the late 1800s include Émile Durkheim and Ferdinand Tönnies. Tönnies argued that social groups can exist as personal and direct social ties that either link individuals who share values and belief or impersonal, formal, and instrumental social links. Durkheim gave a non-individualistic explanation of social facts describing that social phenomena arise when interacting individuals constitute a reality that can no longer be accounted for in terms of the properties of individual actors. He categorized them between a traditional society – "mechanical solidarity" – which prevails if individual differences are minimized, and modern society – "organic solidarity" – that develops out of cooperation between differentiated individuals with independent roles.[123]

Georg Simmel, writing at the beginning of the twentieth century, was the first scholar to think directly in social network terms. He described in detail the nature of network size on interaction and to the likelihood of interaction in ramified, loosely-knit networks rather than groups (Simmel, 1908/1971).[67],[143]

Three main traditions in social networks appeared after a little bit. In the 1930s, J.L. Moreno[139] was the first to explore the systematic recording and analysis of social interaction in small groups, especially classrooms and work groups (sociometry), while a Harvard group led by W. Lloyd Warner and Elton Mayo explored interpersonal relations at work. In 1940, A.R. Radcliffe-Brown's[7] presidential address to British anthropologists urged the systematic study of networks. However, it took about 15 years before this call was followed-up systematically.

Social network analysis developed even more with the kinship studies of Elizabeth Bott in England in the 1950s and the 1950s-1960s urbanization studies of the University of Manchester group of anthropologists (centered around Max Gluckman and later J. Clyde Mitchell) investigating community networks in southern Africa, India and the United Kingdom. Furthermore, British anthropologist S.F. Nadel[8] codified a theory of social structure that was influential in later network analysis.

In the 1960s-1970s, a growing number of scholars worked to evolve all this theory, combining different tracks and traditions. One large group was centered around Harrison White[49] and his students at Harvard University: Ivan Chase, Bonnie Erickson, Harriet Friedmann, Mark Granovetter, Nancy Howell, Joel Levine, Nicholas Mullins, John Padgett, Michael Schwartz and Barry Wellman[9]. White's group had as object to rebel against the reigning structural-functionalist orthodoxy of then-dominant Harvard sociologist Talcott Parsons, leading them to devalue concerns with symbols, values, norms and culture. They also were opposed to the methodological individualism espoused by another Harvard sociologist, George Homans[26], which was endemic among the dominant survey researchers and positivists of the time. Mark Granovetter[9],[10] and Barry Wellman[9] are among the former students of White who have elaborated and popularized social network analysis.

Apart from the White's group, significant independent work was done by scholars elsewhere: quantitative analysts at the University of Chicago, including Joseph Galaskiewicz, Wendy Griswold, Edward Laumann, Peter Marsden, Martina Morris, and John Padgett; University of California Irvine social scientists interested in mathematical applications[14], centered around Linton Freeman, including John Boyd, Susan Freeman, Kathryn Faust, A. Kimball Romney and Douglas White); and communication scholars at Michigan State University, including Nan Lin and Everett Rogers. A substantively-oriented University of

Toronto sociology group developed in the 1970s, centered on former students of Harrison White: S.D. Berkowitz, Harriet Friedmann, Nancy Leslie Howard, Nancy Howell, Lorne Tepperman and Barry Wellman, and also including noted modeler and game theorist Anatol Rapoport.[49],[166]

The idea of network analysis is said to be written even in Ancient Greek's documents, while the main development of this field occurred in the 1930's by several groups in different traditional fields working independently[139].

They are also used in:

- Psychology
- Anthropology
- Mathematics

## 2.3  Applications

The social networking has various applications in many domains, such as in government agencies so as to know the opinion of the public, in dating, in business, in education and even in medicine [5].

One of the most recent examples in the domain of medicine was the use of SNA and network modeling approaches in epidemiology to help understand how patterns of human contact aid or inhibit the spread of diseases such as HIV in a population. The evolution of social networks can sometimes be modeled by using agent based models, providing insight into the interplay between communication rules, rumor spreading and social structure. Here is an interactive model of rumour spreading, based on rumour spreading from model on Cmol[5],[14],[29].

The diffusion of innovations theory explores social networks and their crucial role in influencing the spread of new ideas and practices. Change agents and opinion leaders often play major roles in spurring the adoption of innovations, although factors inherent to innovations also play a role.

According to Mark Granovetter[9], besides from the strong ties, more numerous weak ties could be important in seeking information and innovation.

A good example could be the one with the cliques, where every member knows more or less what the other members know. For this reason, in order to gain more information and insights, the members of the clique will have to look beyond the clique to its other friends and acquaintances. This is what Granovetter[10] called "the strength of weak ties".

Guanxi is a central concept in Chinese society (and other East Asian cultures) that can be summarized as the use of personal influence. Guanxi can be studied from a social network approach [89],[90],[92].

The small world phenomenon is the hypothesis that the chain of social acquaintances required to connect one arbitrary person to another arbitrary person anywhere in the world is generally short. The concept gave rise to the famous phrase six degrees of separation after a 1967 small world experiment by psychologist Stanley Milgram.

In his experiment, Milgram chose a sample of US individuals, which were asked to reach a particular target person by passing a message along a chain of acquaintances. The majority of chains in that study failed to complete, but it was shown that the average length of successful chains turned out to be about five intermediaries or six separation steps. These methods (and ethics) were questioned by an American scholar. Some further research showed that the degrees of connection could be higher. Academic researchers continue to explore this phenomenon as Internet-based communication technology has supplemented the phone and postal systems available during the times of Milgram. [32],[41],[68] A recent electronic small world experiment at Columbia University found that about five to seven degrees of separation are sufficient for connecting any two people through e-mail.

The study of socio-technical systems is loosely linked to social network analysis, and looks at relations among individuals, institutions, objects and technologies.

## 2.4  Principles in Social Network Analysis

### 2.4.1  Betweenness

The degree an individual lies between other individuals in the network; the extent a node is directly connected only to nodes that are not directly connected to each other; an intermediary; liaisons; bridges. Therefore, it's the number of individuals a person is connected to, indirectly through their direct links.

### 2.4.2  Closeness

The degree an individual is near all other individuals in a network (directly or indirectly). It reflects the ability to access information through the "grapevine" of network members. Thus, closeness is the inverse of the sum of the shortest distances between each individual and every other person in the network.

### 2.4.3  (Degree) Centrality

The count of the number of ties to other actors in the network.

### 2.4.4  Flow betweenness Centrality

The degree that a node contributes to a sum of maximum flow between all pairs of nodes (not that node).

### 2.4.5  Eigenvector Centrality

A measure of the importance of a node in a network. It assigns relative scores to all nodes in the network based on the principle that, connections to nodes having a high score contribute more to the score of the node in question.

### 2.4.6 Centralization

The difference between the n of links for each node divided by maximum possible sum of differences. A centralized network will have many of its links dispersed around one or a few nodes, while a decentralized network is one in which there is little variation between the n of links each node possesses.

### 2.4.7 Clustering Coefficient

A measure of the likelihood that two associates of a node are associates themselves. A higher clustering coefficient indicates a greater 'cliquishness'.

### 2.4.8 Cohesion

The degree that actors are connected directly to each other by cohesive bonds. Groups are identified as 'cliques' if every actor is directly tied to every other actor, 'social circles' if there is less stringency of direct contact, which is imprecise, or as structurally cohesive blocks if precision is wanted.

### 2.4.9 (Individual-Level) Density

The degree a respondent's ties know one another/ proportion of ties among an individual's nominees. Network or global-level density is the proportion of ties in a network relative to the total number possible (sparse versus dense networks).

### 2.4.10 Path Length

The distances between pairs of nodes in the network. Average path-length is the average of these distances between all pairs of nodes.

### 2.4.11 Radiality

Degree an individual's network reaches out into the network and provides novel information and influence

### 2.4.12 Reach

The degree any member of a network can reach other members of the network.

### 2.4.13 Structural Cohesion

The minimum number of members who, if removed from a group, would disconnect the group.

### 2.4.14 Structural Equivalence

Refers to the extent to which actors have a common set of linkages to other actors in the system. The actors don't need to have any ties to each other to be structurally equivalent.

### 2.4.15  Structural Hole

Static holes that can be strategically filled by connecting one or more links to link together other points. Linked to ideas of social capital: if you link to two people who are not linked you can control their communication.

## 2.5  *Network Analytic Software*

**Social network analysis software** is used to identify, represent, analyze, visualize or simulate nodes (e.g. agents, organizations, or knowledge) and edges (relationships) from various types of input data (relational and non-relational), including mathematical models of social networks. The output data can be saved in external files. Various input and output file formats exist.

The SNA software facilitates quantitative or qualitative analysis of social networks, by describing features of a network, either through numerical or visual representation.

There is a great variety of groups who can consist a social network: families, classrooms, project teams, soccer teams, nation-states, disease vectors, legislatures and of course membership on networking websites like Twitter or Facebook, or even Internet. [28],[160] The levels of network features may vary from individual nodes, dyads, triads, ties and/or edges, to the entire network. For example, node-level features can include network phenomena such as betweeness and centrality, or individual attributes such as age, sex, or income. SNA software generates these features from raw network data formatted in an edgelist, adjacency list, or adjacency matrix (also called sociomatrix), often combined with (individual/node-level) attribute data.

Network analysis tools allow researchers to investigate representations of networks of different size - from small (e.g. families, project teams) to very large (e.g. the Internet, disease transmission). The various tools provide mathematical and statistical routines that can be applied to the network model.

Visual representations of social networks are important to understand network data and convey the result of the analysis. Visualization is often used as an additional or standalone data analysis method. Network analysis tools are used to change the layout, colors, size and other properties of the network representation, with respect to visualization.

Some of the social network tools are:

- For scholarly research tools like *UCINet*, *Pajek*, *ORA*, the "statnet" suite of packages in "R"), and "GUESS" are popular.
- Examples of business oriented social network tools include *NetMiner*, *InFlow*.
- An open source package for Linux is Social Networks Visualizer or *SocNetV*
- Another generic open source package for Windows, Linux and OS X with interfaces to Python and R is "igraph"

- For Mac OS X a related package installer of *SocNetV* is available.
- For integrated egocentric data collection and visualization <u>SocioMetrica</u>

A systematic overview and comparison of a selection of software packages for social network analysis was provided by Huisman and Van Duijn.[160]

The most important software programs for the SNA are described in the Appendix.

## 2.6 Social Network Categories

Online communities can vary from informational, professional, educational, hobbies, academic, to news related. There are thousands of communities filled with active members who dedicate a fair portion of their day to participate in those social networks. There are also subcategories like the ones described:

### 2.6.1 Clique

A clique (IPA:/ˈklɪk/ in America, /ˈkliːk/ elsewhere) is an exclusive group of people who share common interests, views, purposes, patterns of behavior, or ethnicity. A clique is a smaller group of people belonging to a larger group. These people are more closely identified with one another than the remaining members of the group, and exchange something among themselves, such as friendship, affection, or information.

A clique has an informal structure, and it is composed from more than two people. All the members of the group have some type of relationship with one another, and thus the group is tightly knit together as a type of social network.[69]

### 2.6.2 Economic Network

Economic network or refereed network of independent individuals has the primary purpose of making a strong community in order to gain strength and perform as a significant player in relation to current market situation. Activities of economic network consist also of recruiting new members to join, reviewing, surveying, or providing a fresh perspective on existing community growth and strength.[27]

### 2.6.3 Social Network Service

Not to be confused with <u>social network analysis</u>, a type of social scientific model

A social network service uses <u>software</u> to build online <u>social networks</u> for communities of people who share interests and activities or who are interested in exploring the interests and activities of others. Most services are primarily web-based and provide a collection of various ways for users to interact, such as <u>chat</u>, <u>messaging</u>, <u>email</u>, <u>video</u>, <u>voice chat</u>, <u>file sharing</u>, <u>blogging</u>, <u>discussion groups</u>, and so on. The most common examples are Facebook, Twitter, MySpace, LinkedIn, Second Life etc.

These types of social network have completely changed the way we communicate and share information. Every day, thousands of people use them in order to get in touch with others, to talk to their friends or even to communicate for professional reasons. Of course, the information is shared much faster and easier. [68],[137],[142]

### 2.6.4 Business Network

Business networking is a socioeconomic activity by which groups of like-minded business people recognize, create, or act upon business opportunities. A business network is a type of social network whose reason for existing is business activity. There are several business networking organizations that create models of networking activity that, when followed, allow the business person to build new business relationships and generate business opportunities at the same time. Social network services are in fact types of business networks.

Business networking is used as a more cost-effective method in order to advertise or gain public relations. This is due to more personal commitment than company money demanded.

Taking an example, a business network may agree to meet weekly or monthly with the purpose of exchanging business leads and referrals with fellow members. To complement this activity, members often meet outside this circle, on their own time, and build their own one-to-one relationship with the fellow member.

Business networking may vary and can be conducted in a local business community, or on a larger scale via Internet. In the recent years, there has been a massive development in the Business networks, due to the wide use of Internet. In this way, Internet companies often set up business leads for sale to bigger corporations and companies looking for data sources.

Business networking can have a meaning also in the ICT domain, i.e. the provision of operating support to companies and organizations, and related value chains and value networks.

Most of today's systems are characterized as complex, with high dimension and a huge variety of factors. Until now, the conventional methods have offered us a lot of help concerning the research and the solution of many traditional problems. In the complex problems, however, and especially in those combined with complex dynamical systems, their effectiveness has proved to be limited. New methods were proposed; between them there is the failure detection and the identification qualities. In the next chapter we propose one of the most promising methods based on Fuzzy Cognitive Maps. [68],[127],[129]

## 3   Fuzzy Cognitive Maps (FCMs)?

Fuzzy Cognitive Maps are symbolic representations for the description and modeling of complex systems. They consist of concepts. Each one shows different aspects of system behavior. Concepts are related and they interact, showing the dynamics of the system. The human factor possesses a crucial role in the procedure. Moreover, human experience and knowledge can develop FCMs, if there are experts who know the operation and the "behavior" of the system under certain circumstances. FCMs show the cause and effect relationships amongst concepts, presenting system's behavior in a symbolic manner with accumulated knowledge about the system.

FCMs have been used in a variety of scientific areas, i.e. informatics, medicine, finance etc. Their use was mainly focused on the behavior of systems, but also for decision analysis and operations research. In the next sections we will present the underlying principles foron the construction and use of FCMs in complex systems. It will be also shown that the FCMs are really useful in exploiting the accumulated human knowledge concerning the operation of complex systems. In general, FCMs represents knowledge in a symbolic manner relating states, variables, events and inputs to an analogous and sometimes linear manner. [124]

## 3.1 Basic Theories

FCMs are consisted of nodes linked with weighted arcs depicted as a signed weighted graph with feedback. The weighted arcs linking the concept nodes, represent the causal relationship that exists among concepts. Concepts of an FCM represent key factors and characteristics of the modeled complex system and stand for events, goals, inputs, outputs states, variables and trends of the complex system being modeled. [124]

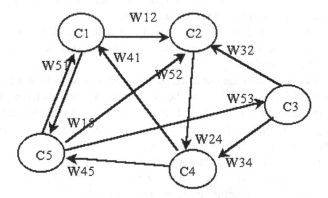

**Fig. 2** A graphic display of an example FCM

As we can see on the graph, causal variables are the concept variables at the origin of the arrows. On the other hand, effect variables are these on the terminal points of the arrows. For example $C_1$ impacts $C_2$, because $C_1$ is the causal variable, while $C_2$ is the effect variable ($C_1$-> $C_2$). Each concept is represented with a letter $A_i$, which represents its value and it results from the transformation of the real value of the system's variable, for which this concept stands, in the interval $[0,1]$. Causality between concepts allows degrees of causality and not the usual binary logic, so the weights of the interconnections can range in the interval $[-1,1]$.

In other words, a Fuzzy Cognitive Map shows the degree of causal relationship between different variables (concepts). The causal relationships are expressed by fuzzy weights.

Relationships between concepts can be of three types: a) positive causality ($W_{ij}>0$), b) negative causality ($W_{ij}<0$), c) no relationship ($W_{ij}=0$). The value of $W_{ij}$ indicates the strength of influence of $C_i$ to $C_j$. Furthermore, $W_{ij}$ indicates whether the relationship between $C_i$ and $C_j$ is direct or inverse. The direction of causality shows if $C_i$ influences $C_j$ or vice versa, these parameters must be taken into account when a value is assigned to $W_{ij}$.

The formula for calculating values of concepts at each time step of a fuzzy cognitive map is the following:

$$A_i^t = f\left(k_1 \sum_{\substack{j=1 \\ j \neq i}}^{n} A_j^{t-1} W_{ji} + k_2 A_i^{t-1}\right)$$

Where $A_i$ is the value of the concept $C_i$ at time t, $A^{t-1}_i$ is the value of the concept $C_i$ at time t-1 and $A^{t-1}_j$ is the value of the concept $C_j$ at time t-1. The $k_1$ expresses the influence of the interconnected concepts in configuration of the new value of concept $A_i$ and $k_2$ represents the contribution of the new value. This formulation assumes that a concept links to itself with a weight equal to $W_{ii}=k_2$.

In case we assume that k1=k2=1, the formulation turns into:

$$A_i^t = f\left(\sum_{\substack{j=1 \\ j \neq i}}^{n} A_j^{t-1} W_{ji} + A_i^{t-1}\right)$$

Just like before, $A_i$ is the value of the concept $C_i$ at time t, $A^{t-1}_i$ is the value of the concept $C_i$ at time t-1 and $A^{t-1}_j$ is the value of the concept $C_j$ at time t-1 and $W_{ij}$ is the weight of the interconnection between $C_i$ and $C_j$.

Function f is used to squash the result between $[0,1]$. The function f is nothing more than the sigmoid continuous function:

$$f(x) = \frac{1}{1 + e^{-\lambda x}}$$

As written before, the nature of concepts can also be negative. In this case, their relationship may belong to [-1,1]. In order to achieve this, we use the function

$$f(x) = \tanh(x)$$

As we can see, the FCM mathematical model is simple. But what happens if we have to deal not with two, but with n concepts? The solution comes with the creation of two matrices: the nxn weight matrix W, which includes all the weights if the interconnections and the 1xn state vector A which includes all the values of the concepts. The formulation which combines W and A is the following:

$$\mathbf{A}^t = f(\mathbf{A}^{t-1}\mathbf{W} + \mathbf{A}^{t-1})$$

In the extreme case where we want a concept to be connected with itself, we use the following formula:

$$\mathbf{A}^t = f(\mathbf{A}^{t-1}\mathbf{W}^{new})$$

Where W is a matrix with $W_{ii}=1$

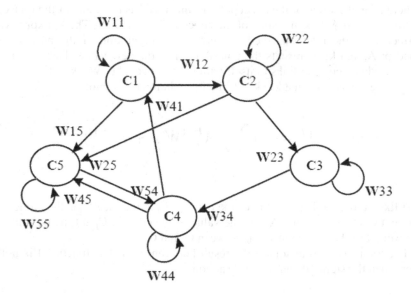

**Fig. 3** Picture with an example of FCM nodes and ties

## 3.2  Constructing Fuzzy Cognitive Maps

FCMs are used in order to solve complex systems as they can represent human knowledge on the operation of a complex system. They are developed by experts who use their own experience concerning a problem according to what they believe are the "variables" of the problem, they construct nodes and causal relationships between them. Nodes could be the main factors of a complex system, which means that they could stand for almost everything: events, goals, values, trends, actions etc. Experts also know which nodes influence other nodes and thus determine the degree of causal relationships between them. The degree of causal relationship is usually determined by applying experience or learning rules for choosing appropriate weights amongst nodes of FCMs.

In the next sections, we will describe the way FCMs can be constructed.

### 3.2.1  Assigning Numerical Weights

The most popular method of FCM construction is to give a complex system to experts. They are able, based on their experience, to determine the number of concepts and causal relationships between them. [124]

However, there is a significant problem, which has to do with the different kind of knowledge each expert has. In order to resolve this problem, we have to leave them work separately and then accumulate their knowledge in order to create a clear picture of the causal relationships. The results of all this procedure could be summarized in this formula below:

$$\mathbf{W} = f(\sum_{1}^{N} \mathbf{W}_k)$$

Where W is the overall weight matrix, $W_k$ is the weight matrix of each expert, N is the number of experts and f is the sigmoid function.

We must not forget, though, that each expert has different credibility on the knowledge of the system and this is a variable we can't ignore. For this reason, the credibility weight must be added in the previous formulation as:

$$\mathbf{W} = f(\sum_{1}^{N} b_k \mathbf{W}_k)$$

In order to determine the credibility, we should fix some predetermined rules which would "penalize" with low credibility (e.g. close to 0) the experts whose estimated causal relationships are far from the average causal relationships.

Professor Peter Groumpos (Fuzzy Cognitive Maps: Basic Theories and Their Application to Complex Systems, 2010) proposes an algorithm able to assign weights for each interconnection and credibility weights for experts, taking for granted that each expert works separately in order to assume his estimated causal relationships concerning a Fuzzy Cognitive Map.[124], [168]

In this algorithm, it is examined the sign of the causal weights. If there is a number of signs less than $\pi*N$, then the experts should re-examine their estimations concerning causal relationships between concepts. Otherwise, the procedure continues and the proposed weights are used in order to determine the final weight. If there are estimated weights far from the average weight, the experts are penalized with a lower credibility and their estimations will be partially taken into account. [168]

*Algorithm 1*

*Step 1: For all the* N *experts, set credibility weight* $b_k = 1$

*Step 2: For i,j=1 to n*

*Step 3: For each interconnection (* $C_i$ *to* $C_j$ *) examine the* N *weights* $W_{ij}^k$ *that each* $k_{th}$ *of the* N *experts has assigned.*

*Step 4: IF there are weights* $W_{ij}^k$ *with different sign and the number of weights with the same sign is less than* $\pi*N$

*THEN*

*ask experts to reassign weights for this particular interconnection and go to step3*

                    *ELSE*

*take into account the weights of the greater group with the same sign and consider that there are no other weights and penalize the experts who chose "wrong" sighed weight with a new credibility weight* $b_k = \mu_1 * b_k$

*Step 5: For the weights with the same sign, find their average value*

$$W_{ij}^{ave} = \frac{(\sum_{k=1}^{N} b_k W_k)}{N}$$

*Step 6: IF* $\left| W_{ij}^{ave} - W_{ij}^k \right| \geq \omega_1$ *THEN consider that there is no weight* $W_{ij}^k$, *penalize the* $k_{th}$ *expert* $b_k = \mu_2 * b_k$ *and go to step 5*

*Step 7: IF there have not examined all the* $n \times n$ *interconnection go to step 2*

                    *ELSE construct the new weight matrix* **W** *which has elements the weights* $W_{ij}^{ave}$

*END.*

### 3.2.2  Assigning Linguistic Variables for FCM Weights

Another methodology to construct a Fuzzy Cognitive is with the use of linguistic notions by experts. Every expert will determine the influence of one concept to the other as "negative" or "positive" and then he will describe the grade of influence with a linguistic variable such as "strong", "weak" and etc [168]

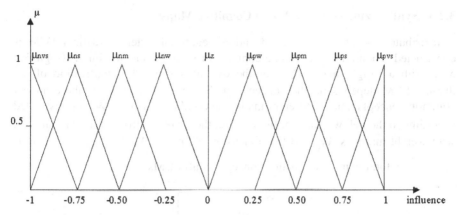

**Fig. 4** Terms of the linguistic variable influence

*Influence* of one concept on another, is interpreted as a linguistic variable taking values in the universe $U=[-1,1]$ and its term set $T(influence)$ could be:

$T(influence)$ = {negatively very strong, negatively strong, negatively medium, negatively weak, zero, positively weak, positively medium, positively strong, positively very strong}

Semantic rule $M$ is defined as follows and these terms are characterized by the fuzzy sets whose membership functions are shown in Figure 2:

$M$(negatively very strong)= the fuzzy set for "an influence below to -75%" with membership function $\mu_{nvs}$

$M$(negatively strong)= the fuzzy set for "an influence close to -75%" with membership function $\mu_{ns}$

$M$(negatively medium)= the fuzzy set for "an influence close to -50%" with membership function $\mu_{nm}$

$M$(negatively weak)= the fuzzy set for "an influence close to -25%" with membership function $\mu_{nw}$

$M$(zero)= the fuzzy set for "an influence close to 0" with membership function $\mu_z$

$M$(positively weak)= the fuzzy set for "an influence close to 25%" with membership function $\mu_{pw}$

$M$(positively medium)= the fuzzy set for "an influence close to 50%" with membership function $\mu_{pm}$

$M$(positively strong)= the fuzzy set for "an influence close to 75%" with membership function $\mu_{ps}$

$M$(positively very strong)= the fuzzy set for "an influence above to 75%" with membership function $\mu_{pvs}$

Linguistic variables that describe each interconnection are combined and the overall linguist variable will be transformed in the interval [-1,1]. A numerical weight for each interconnection will be the outcome of the defuzzifier, where the Center of Gravity method is used to produce this crisp weight. This method is less complex than the previous one, considering the fact that experts do not have to assign numerical causality weights but to describe the degree of causality among concepts. [124], [168]

### 3.2.3  Synthesizing Different Fuzzy Cognitive Maps

A distributed system is considered and for each subsystem a distinct FCM is constructed. Then all FCMs can be combined to one augmented Fuzzy Cognitive Map with a weight matrix $W$ for the overall system. The unification of the distinct FCM depends on the concepts of the segmental FCM, if there are no common concepts among different maps; the combined matrix $W$ is constructed according to the below equation. In this case, there are $K$ different FCM matrices, with weight matrices $W_i$ and the dimension of matrix $W$ is $n \times n$ where $n$ is equal to the total number of distinct concepts in all FCMs.

$$W = \begin{bmatrix} W_1 & & \\ & W_2 & 0 \\ & 0 & \ddots \\ & & & W_K \end{bmatrix}$$

(Fuzzy Cognitive Maps: Basic Theories And Their Application To Complex Systems, Prof. Peter.P.Groumpos, 2010)[168]

### 3.2.4  Neural Network Nature of Fuzzy Cognitive Maps

Fuzzy Cognitive Maps have been described as a hybrid methodology, because it utilizes characteristics of fuzzy logic and neural networks. The development and construction of FCMs have shown their fuzzy nature. Learning rules, used in Neural Networks theory, are used to train the Fuzzy Cognitive Map. Parameter learning of FCM concerns the updating of connection weights among concepts.

The construction of FCM is based on experts who determine concepts and weighted interconnections among concepts. This methodology may lead to a distorted model of the system because human factor is not always reliable. In order to refine the model of the system, learning rules are used to adjust weights of FCM interconnections. The Differential Hebbian learning rule has been proposed to be used in the training of a specific type of FCMs. The Differential Hebbian learning law adjusts the weights of the interconnection between concepts and grows a positive edge between two concepts if they both increase or decrease or it grows a negative edge if values of concepts move in opposite directions. Adjusting the idea of differential Hebbian learning rule in the framework of Fuzzy Cognitive Map, the following rule is proposed to calculate the derivative of the weight between two concepts.

$$w'_{ji} = -w_{ji} + s(A_j^{new})s(A_i^{old}) + s'(A_j^{new})s'(A_i^{old})$$

Where $S(x) = \dfrac{1}{1 + e^{-\lambda x}}$

Of course, there is more investigation needed in order to create appropriate learning rules for Fuzzy Cognitive Maps. These rules will give FCMs useful characteristics such as the ability to learn arbitrary non-linear mappings, capability to generalize to situations the adaptivity and the fault tolerance capability. [168]

## 3.3 Decision Analysis and Fuzzy Cognitive Maps

Decision analysis is based on a variety of quantitative methods that help people to choose between numbers of alternatives. The traditional decision analysis is used to indicate decisions with good outcomes, even though there is an uncertainty concerning the decision itself. Moreover, there may be a variety in the outcome of the decision, depending to factors like costs, utility, benefits etc.

Over the last years, several approaches have been investigated in the field of Decision Analysis, with the most popular to be used that of Decision Trees (DT). However, there has been little considering the combination of DT with FCMs. In this chapter the technique of combining a DT with a FCM model in Decision Analysis is presented.

The derived FCM model is subsequently trained using an unsupervised learning algorithm to achieve improved decision accuracy.

The DT-FCM's function is briefly outlined in Figure. If there is a large number of input data, then the quantitative data are used to induce a Decision Tree and qualitative data derived from experts' knowledge are used to construct the FCM model. The FCM's flexibility is enriched by the fuzzification of the derived fuzzy IF-THEN rules to assign weights direction and values. Finally, the FCM model (new weight setting and structure) is trained by the unsupervised NHL algorithm in order to reach a final decision. [124],[168]

This methodology can be used for three different circumstances. These circumstances are separated depending on the type of the initial input data: (1) when the initial data are quantitative, the DT generators are used and an inductive learning algorithm produce the fuzzy rules which then are used to update the FCM model construction; (2) when experts' knowledge is available, FCM model is constructed and through the unsupervised NHL algorithm is trained to calculate the target output concept responsible for the decision line; and (3) when both quantitative and qualitative data are available, the initial data are divided and each data type is used to construct DTs and FCMs separately. Then fuzzy rules induced from the inductive learning restructure the FCM model enhancing it. At the enhanced FCM model the training algorithm is applied in order to help FCM model to reach a proper decision.

The new technique has three major advantages. First, the association rules derived from the decision trees have a simple and direct interpretation and introduced in the initial FCM model to update its operation and structure. For example, a produced rule can be: If *variable 1* (input variable) has *feature A* Then *variable 2* (output variable) has *feature B*.

Second, the procedure that introduces the Decision Tree rules into an FCM also specifies the weight assignment through new cause-effect relationships among the FCM concepts. Third, this technique fares better than the best Decision Tree inductive learning technique and the FCM decision tool. [124],[168]

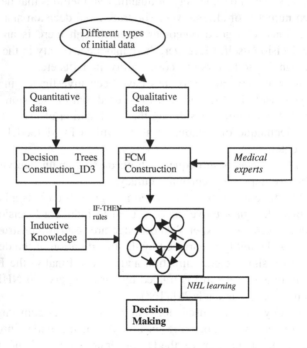

**Fig. 5** The decision making system constructed by Decision Trees and Fuzzy Cognitive Maps

# 4 Application of Fuzzy Cognitive Maps in Social Networks

In this section, we present the application of FCMs in Social and Business Networks. At the beginning we present the FCM tool we have developed used for the creation and simulation of FCMs and then the use of the tool in Social and Business Networks.

## 4.1   Visual FCM Modeler Tool (vfcm)

The vfcm tool is used to construct visual Fuzzy Cognitive Maps. It consists of several parts that make the construction of the FCMs easier for the user.

### 4.1.1   User Interface

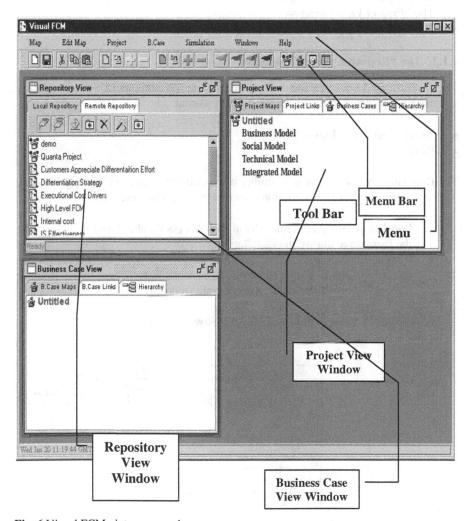

**Fig. 6** Visual FCM picture example

It can be seen in Figure below that the user interface has been implemented using Multiple Document Interface (MDI) technology.

The new User Interface incorporates all the basic components of a Windows based application. More specifically:

- A Menu Bar which includes a number of drop-down menus that include all the application's functionalities.
- A Tool Bar that includes the most important functionalities of the application.
- The "*Repository View*" window
- The "*Project View*" window
- The "Business Case View" window

### 4.1.2 The "Repository View" Window

The "*Repository View*" window is the place where the contents of the repositories are visually presented to the user. A repository contains :

- Fuzzy Cognitive Maps (FCMs)
- Projects

The Repository View can have a number of repositories where FCMs and Projects can be saved. The default installation contains two repositories:

- Local Repository contains a local MS Access database and it is stored in the local drive during the installation.
- Remote Repository contains the same database in MS SQL Server format and it is stored in a remote server. A user can have access to the Remote Repository only if he or she has the right to do so.

The Repository View Window is shown below:

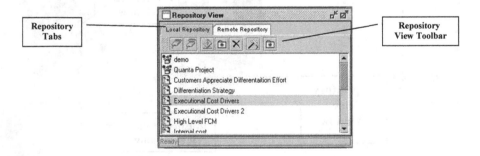

**Fig. 7** Repository view window

### 4.1.3 The "Project View" Window

In the "*Project View*" window a user can built a project (i.e. to apply node linking in a number of FCMs). The "*Project View*" window is shown below:

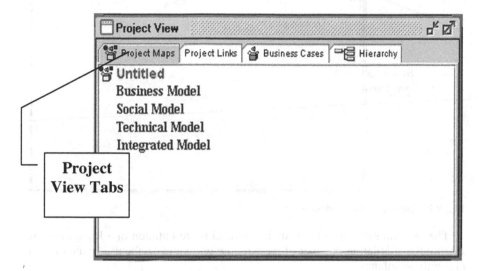

**Fig. 8** Project view window

The "*Project Maps*" tab contains four directories, which comprise the Project structure. In each directory a user can add FCMs taken from the repository. The FCMs that have been added into a Project can be opened from the "*Project View*" window. This is the place where node linking is applied and it is going to be described later. In general, two FCMs can be linked only if they have been added to a Project.

The "*Project Links*" tab will contain all the maps that are linked in a project as well as the concept node they are linked with.

The "*Business Case*" tab will contain all the business cases that have been created form a project. In general, a Business Case is a part of a Project or the whole Project that can be simulated.

The "*Hierarchy*" tab contains a graphical representation of the Project Hierarchy.

### 4.1.4 The "Business Case View" Window

The "Business Case View" window will contain a part of a Project that can be simulated. In other words a Business Case will contain a number of FCMs that have been taken from a Project and their corresponding node links. The "Business Case View" window is shown below:

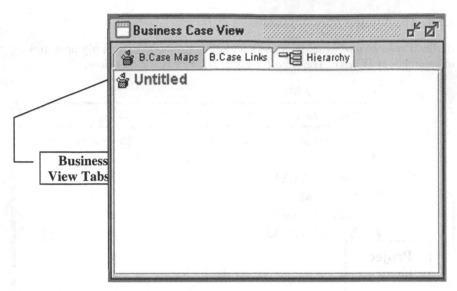

**Fig. 9** Business case view window

The "Hierarchy" tab will contain a graphical representation of a Business Case. During the simulation this part of the window will show to the user which map is currently simulated.

## 4.2   Creating a New FCM

In order to create a new FCM either the "*New Map*" menu item has to be selected or the New FCM toolbar item has to be selected. In both cases an untitled FCM window will open and a user can start creating the FCM.

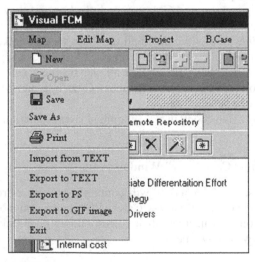

**Fig. 10** FCM new map option

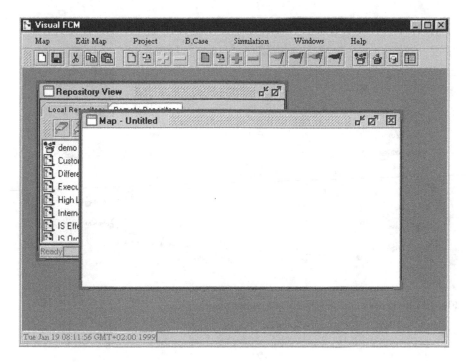

**Fig. 12** FCM new map

## 4.3 Case Study

In this section we show the use of the VFCM in Social Networking. We have taken the relationships (friends and groups) from the profile of a real person (Giorgos Stakias) as well as some relationships of his friends. Each person in a network is represented by a node and the relationships by an FCM relationship. Thus we were able to understand the relationship gravity between these people and the central node (Giorgos Stakias) as shown in the following Figure. The gravity changes according to the common interests that each person has with the central node. In this example, we took the relationships of friendship amongst people in a scientific group (University of the Aegean) and of a sports group (Olympiacos). We take for granted that the central node participates in both group and page.

We can see the relationship between individuals in the following diagrams:

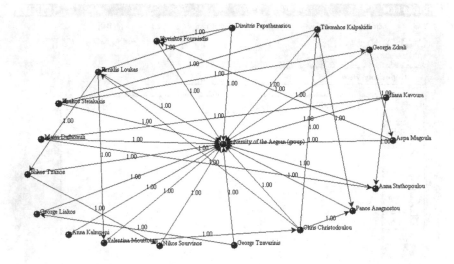

**Fig. 13** University of the Aegean Relationships diagram

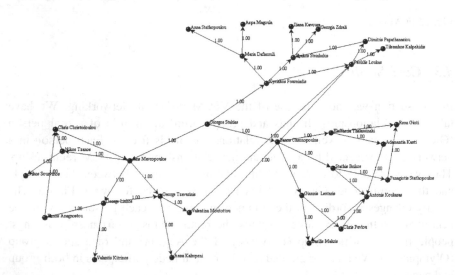

**Fig. 14** Olympiacos Group Relationships diagram

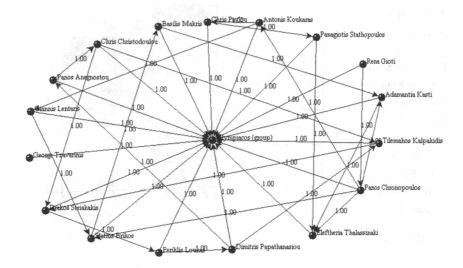

Supposing that friendship has a weight of 0,4 and participation in the group and the page has a weight of 0,3 each, the "relationship weight" between the central node and each person is described in the following table:

| People | Friends | Group | Page | Result (weight) |
|---|---|---|---|---|
| Dimitris Papathanasiou | | | | 1 |
| Chris Christodoulou | | | | 1 |
| Panos Anagnostou | | | | 1 |
| Drakos Steiakakis | | | | 1 |
| George Tzavarinis | | | | 1 |
| Maria Dafnomili | | | | 0,7 |
| Periklis Loukas | | | | 0,7 |
| Kyriakos Fourniadis | | | | 0,7 |
| Eleftheria Thalassinaki | | | | 0,7 |
| Stathis Brikos | | | | 0,7 |
| Giannis Lentaris | | | | 0,7 |
| Panos Chronopoulos | | | | 0,7 |
| George Liakos | | | | 0,7 |
| Nikos Tzanos | | | | 0,7 |
| Aris Mavropoulos | | | | 0,7 |
| Tilemahos Kalpakidis | | | | 0,7 |
| Georgia Zdrali | | | | 0,7 |
| Iliana Kavoura | | | | 0,7 |

**Fig. 15** Relationship weights table

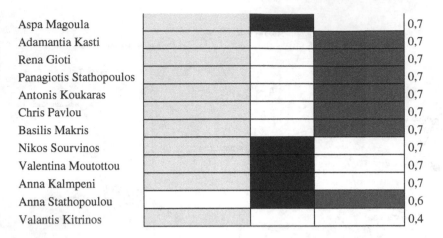

| | | | |
|---|---|---|---|
| Aspa Magoula | | | 0,7 |
| Adamantia Kasti | | | 0,7 |
| Rena Gioti | | | 0,7 |
| Panagiotis Stathopoulos | | | 0,7 |
| Antonis Koukaras | | | 0,7 |
| Chris Pavlou | | | 0,7 |
| Basilis Makris | | | 0,7 |
| Nikos Sourvinos | | | 0,7 |
| Valentina Moutottou | | | 0,7 |
| Anna Kalmpeni | | | 0,7 |
| Anna Stathopoulou | | | 0,6 |
| Valantis Kitrinos | | | 0,4 |

**Fig. 15** *(continued)*

From the table above, it is obvious that the first five persons have the most common interests with the central node, so they are influencing him more than the others. It can also be observed that people participating in the same group or page with the central node, are more likely to be proposed to each other for friendship from the social network. This can be applied to young entrepreneurs who want a network in order to start their job and get known to a small initial society in the beginning and to a larger social network as they become famous.

## 5  Conclusions

From what we have seen in this chapter, it is clear that Social Network Analysis is suitable for social networking. Social networks are the basic ingredient for the man as social being, joining a social network, it means that he is a part of the society where he lives, thinks and reacts. Using the VFCM Tool, we are able to define the relationship weights between the nodes and most importantly, all the participants in the Social Network, as well as their relationships.

Social Network Analysis can help into the analysis of a Business Network, this type of networks emphasize on business activities. Business networking is very important for enterprises as they are able to gain access to public and customers, built further their corporate image and reputation.

Fuzzy Cognitive Maps are a very useful tool of prevision in order to define the relationships between the nodes, so as to be able to watch the degree of importance of that relationship. In this chapter, we presented only a small part of what can FCMs be useful into. The results of our case study were presented in an SME conference dedicated to the subject concerning (CAE-PME, Athens, October 2010), where we attracted interest from professionals from different industrial sectors including medicine (cancer prediction), mathematics, economics, etc.

From the work being done in the chapter, we ascertained that Fuzzy Cognitive Maps combined with Social and Business Networks can offer a serious advantages both in academic research and applied case studies in the future.

## Appendix

### *Collection of Social Network Analysis Tools and Libraries*

| Product | Main Functionality | Input Format | Output Format | Platform | Cost | Notes |
|---|---|---|---|---|---|---|
| Igraph | Analysis and visualization of very large networks | .txt (edge list), .graphml, .gml, .ncol, .lgl, .net | .txt (edge list), .graphml, .dot, .gml, .ncol, .lgl, .net | Windows, Linux, Mac OS X | Open source (GNU GPL) | igraph is a C library for the analysis of large networks. It includes fast implementations for classic graph theory problems and recent network analysis methods like community structure search, cohesive blocking, structural holes, dyad and triad census and motif count estimation. |
| NetMiner | All-in-one Software for Network Analysis and Visualization | .xls(Excel), .csv(text), .dl(UCINET), .net(Pajek), .dat(StOCNET), .gml; NMF(proprietary) | .xls(Excel), .csv(text), .dl(UCINET), .net(Pajek), .dat(StOCNET), NMF(proprietary) | Microsoft Windows | Commercial ($16/user ~ $6,600/user) with free trial | NetMiner is an innovative software tool for exploratory analysis and visualization of network data. NetMiner allows you to explore your network data visually and interactively, and helps you to detect underlying patterns and structures of the network. Main features include analysis of large networks, comprehensive network measures and models, both exploratory & confirmatory analysis, interactive visual analytics, what-if network analysis, built-in statistical procedures and charts, full documentation (1,000+ pages of User's Manual), expressive network data model, facilities for data & workflow management, and user-friendliness. |

| Product | Main Functionality | Input Format | Output Format | Platform | Cost | Notes |
|---|---|---|---|---|---|---|
| ORA | Social Network Analysis | DyNetML, .csv | DyNetML, .csv | Windows | Freeware for non-commercial use | Risk assessment tool for locating individuals or groups that are potential risks given social, knowledge and task network information. Based on network theory, social psychology, operations research, and management theory a series of measures of "criticality" have been implemented into ORA. find those people, types of skills or knowledge and tasks that are critical from a performance and information security perspective. |
| Pajek | Analysis and Visualization of Large Scale Networks | .net, .paj, .dat(UCINET), .ged, .bs, .mac, .mol | .net, .paj, .dat(UCINET), .xml(graphML), .bs | Windows, Linux, Mac OS X | Freeware for non-commercial use | |
| SocioMetrica | EgoNet, LinkAlyzer, and VisuaLyzer applications | DyNetML, Excel, DL, text, UCINET | DyNetML, Excel, DL, text, UCINET, SPSS | Windows | Shareware | A set of applications for interview-based gathering of egocentric data (EgoNet), linking of data records through matching of node attributes (LinkAlyzer), and visualization (VisuaLyzer). VisuaLyzer also provides prototype functionality for analysis using a relational algebra model. A relational programming language, RAlog, derives and analyzes representations in this relation algebra. |

| Product | Main Functionality | Input Format | Output Format | Platform | Cost | | Notes |
|---------|-------------------|--------------|---------------|----------|------|--|-------|
| SocNetV | Social Networks Visualisation and Analysis Tool | .net (pajek), .dot (GrapViz), .sm/.net (Sociomatrix), .xml (GraphML) | .net (pajek), .dot (GrapViz), .sm/.net (Sociomatrix) | Linux, Windows, Mac (Qt toolkit needed) | Free (GPL3) | Software | SocNetV (Social Networks Visualiser) is an open-source GUI tool, developed on the cross-platform Qt toolkit. Its primary purpose is to enable the user to draw social networks by clicking on a canvas. It can also compute network properties and statistics (centralities, etc) as well as apply (limited) layout algorithms. |
| UCINET | Allrounder for Network Analysis | Excel, DL, text, Pajek format | proprietary (##.d & ##.h) | Windows | Shareware | | A comprehensive package for the analysis of social network data as well as other 1-mode and 2-mode data. Can handle a maximum of 32,767 nodes (with some exceptions) although practically speaking many procedures get too slow around 5,000 - 10,000 nodes. Social network analysis methods include centrality measures, subgroup identification, role analysis, elementary graph theory, and permutation-based statistical analysis. In addition, the package has strong matrix analysis routines, such as matrix algebra and multivariate statistics. |

# References

1. Freeman, L.: The Development of Social Network Analysis. Empirical Pres, Vancouver (2006)
2. Wellman, B., Berkowitz, S.D. (eds.): Social Structures: A Network Approach. Cambridge University Press, Cambridge (1988)
3. Freeman, L.: The Development of Social Network Analysis. Empirical Pres, Vancouver (2006); Wellman, B., Berkowitz, S.D., (eds.) Social Structures: A Network Approach. Cambridge University Press, Cambridge (1988)
4. Scott, J.: Social Network Analysis. Sage, London (1991)

5. Wasserman, S., Faust, K.: Social Network Analysis: Methods and Applications. Cambridge University Press, Cambridge (1994)
6. The Development of Social Network Analysis. Empirical Press, Vancouver
7. Radcliffe-Brown, A.R.: On Social Structure. Journal of the Royal Anthropological Institute 70, 1–12 (1940)
8. Nadel, S.F.: The Theory of Social Structure. Cohen and West, London (1957)
9. Granovetter, M.: Introduction for the French Reader. Sociologica 2, 1–8 (2007); Wellman, B.: Structural Analysis: From Method and Metaphor to Theory and Substance. In: Wellman, B., Berkowitz, S.D. (eds.) Social Structures: A Network Approach, pp.19–61. Cambridge University Press, Cambridge (1988)
10. Granovetter, M.: Introduction for the French Reader. Sociologica 2, 1–8 (2007); Wellman, B.: Structural Analysis: From Method and Metaphor to Theory and Substance. In: Wellman, B., Berkowitz, S.D. (eds.) Social Structures: A Network Approach, pp. 19-61. Cambridge University Press, Cambridge (1988); (see also Scott, 2000 and Freeman, 2004)
11. Wellman, B., Chen, W., Weizhen, D.: Networking Guanxi. In: Gold, T., Guthrie, D., Wank, D. (eds.) Social Connections in China: Institutions, Culture and the Changing Nature of Guanxi, pp. 221–41. Cambridge University Press (2002)
12. Could It Be A Big World After All?: Judith Kleinfeld article
13. Electronic Small World Experiment: Columbia.edu website; Six Degrees: The Science of a Connected Age, Duncan Watts
14. The most comprehensive reference is: Wasserman, Stanley, & Faust, Katherine: Social Networks Analysis: Methods and Applications. Cambridge University Press, Cambridge (1994); A short, clear basic summary is in Krebs, Valdis: The Social Life of Routers. Internet Protocol Journal 3, 14–25 (2000)
15. Moody, J., White, D.R.: Structural Cohesion and Embeddedness: A Hierarchical Concept of Social Groups. American Sociological Review 68(1), 103–127 (2003); online: (PDF file)
16. Social network map of social network scholars: Orgnet.comwebsite (retrieved on March 25, 2008)
17. Barnes, J.A.: Class and Committees in a Norwegian Island Parish. Human Relations 7, 39–58
18. Berkowitz, S.D.: An Introduction to Structural Analysis: The Network Approach to Social Research. Butterworth, Toronto (1982)
19. Brandes, U., Thomas, E. (eds.): Network Analysis: Methodological Foundations. Springer, Heidelberg (2005)
20. Breiger, R.L.: The Analysis of Social Networks. In: Hardy, M., Bryman, A. (eds.) Handbook of Data Analysis, pp. 505–526. Sage Publications, London (2004) Excerpts in pdf format
21. Burt, R.S.: Structural Holes: The Structure of Competition. Harvard University Press, Cambridge (1992)
22. Carrington, P.J., Scott, J., Wasserman, S. (eds.): Models and Methods in Social Network Analysis. Cambridge University Press, New York (2005)
23. Christakis, N., Fowler, J.H.: The Spread of Obesity in a Large Social Network Over 32 Years. New England Journal of Medicine 357(4), 370–379 (2007)
24. Doreian, P., Vladimir, B., Anuska, F.: Generalized Blockmodeling. Cambridge University Press, Cambridge (2005)
25. Freeman, L.C.: The Development of Social Network Analysis: A Study in the Sociology of Science. Empirical Press, Vancouver (2004)

26. Hill, R., Dunbar, R.: Social Network Size in Humans. Human Nature 14(1), 53–72 (2002); Google
27. Jackson, M.O.: A Strategic Model of Social and Economic Networks. Journal of Economic Theory 71, 44–74 (2003); pdf
28. Huisman, M., Van Duijn, M.A.J.: Software for Social Network Analysis. In: Carrington, P.J., Scott, J., Wasserman, S. (eds.) Models and Methods in Social Network Analysis, pp. 270–316. Cambridge University Press, New York (2005)
29. Krebs, V.: Social Network Analysis, A Brief Introduction (Includes a list of recent SNA applications Web Reference) (2006)
30. Lin, N., Burt, R.S., Cook, K. (eds.): Social Capital: Theory and Research. Aldine de Gruyter, New York (2001)
31. Mullins, N.: Theories and Theory Groups in Contemporary American Sociology. Harper and Row, New York (1973)
32. Müller-Prothmann, T.: Leveraging Knowledge Communication for Innovation. In: Framework, Methods and Applications of Social Network Analysis in Research and Development, Frankfurt a. M. et al., Peter Lang (2006) ISBN 0-8204-9889-0
33. Manski, C.F.: Economic Analysis of Social Interactions. Journal of Economic Perspectives 14, 115–136 (2000) viaJSTOR
34. Moody, J., White, D.R.: Structural Cohesion and Embeddedness: A Hierarchical Concept of Social Groups. American Sociological Review 68(1), 103–127 (2003)
35. Newman, M.: The Structure and Function of Complex Networks. SIAM Review 56, 167–256 (2003) pdf
36. Nohria, N., Eccles, R.: Networks in Organizations, 2nd edn. Harvard Business Press, Boston (1992)
37. Nooy, W.D., Mrvar, A., Batagelj, V.: Exploratory Social Network Analysis with Pajek. Cambridge University Press, Cambridge (2005)
38. Scott, J.: Social Network Analysis: A Handbook, 2nd edn. Sage, Newberry Park (2000)
39. Tilly, C.: Identities, Boundaries, and Social Ties. Paradigm Press, Boulder (2005)
40. Valente, T.: Network Models of the Diffusion of Innovation. Hampton Press, Cresskill (1995)
41. Wasserman, S., Katherine, F.: Social Networks Analysis: Methods and Applications. Cambridge University Press, Cambridge (1994)
42. Watkins, S.C.: Social Networks. In: Demeny, P., McNicoll, G. (eds.) Encyclopedia of Population, rev. ed., pp. 909–910. Macmillan Reference, New York (2003)
43. Watts, D.: Small Worlds: The Dynamics of Networks between Order and Randomness. Princeton University Press, Princeton (2003)
44. Watts, D.: Six Degrees: The Science of a Connected Age. W. W. Norton & Company (2004)
45. Wellman, B.: Networks in the Global Village. Westview Press, Boulder (1999)
46. Wellman, B.: Physical Place and Cyber-Place: Changing Portals and the Rise of Networked Individualism. International Journal for Urban and Regional Research 25(2), 227–252 (2001)
47. Wellman, B., Berkowitz, S.D.: Social Structures: A Network Approach. Cambridge University Press, Cambridge (1988)
48. Weng, M.: A Multimedia Social-Networking Community for Mobile Devices Interactive Telecommunications Program. Tisch School of the Arts/ New York University (2007)

49. White, H., Boorman, S., Breiger, R.: Social Structure from Multiple Networks: I Blockmodels of Roles and Positions. American Journal of Sociology 81, 730–780 (1976)
50. The International Network for Social Network Analysis (INSNA) - professional society of social network analysts, with more than 1,000 members
51. Organizational Network Mapping - SNA applied in business organizations
52. Virtual Center for Supernetworks
53. VisualComplexity.com - a visual exploration on mapping complicated and complex networks
54. Dynamic Centrality in Social Networks
55. Center for Computational Analysis of Social and Organizational Systems (CASOS) at Carnegie Mellon
56. NetLab at the University of Toronto, studies the intersection of social, communication, information and computing networks
57. Netwiki (wiki page devoted to social networks; maintained at University of North Carolina at Chapel Hill)
58. Network Science Center at the U.S. Military Academy at West Point, NY
59. Social Life of Routers - social network analysis applied to computer systems
60. FAS.research - network visualizations produced using social network analysis
61. Building networks for learning- A guide to on-line resources on strengthening social networking
62. Program on Networked Governance - Program on Networked Governance. Harvard University
63. http://www.worldlingo.com/ma/enwiki/en/Social_network
64. http://www.whatissocialnetworking.com/
65. http://en.wikipedia.org/wiki/Social_network
66. http://www.orgnet.com/sna.html
67. http://www.analytictech.com/networks/history.html
68. http://en.wikipedia.org/wiki/Social_networking_service#Business_applications
69. Tichy, N.: An Analysis of Clique Formation and Structure in Organizations. Administrative Science Quarterly 18(2), 194–208 (1973), doi:10.2307/2392063
70. Deutsch, M., Krauss, R.: Theories in Social Psychology. Basic Books, New York (1965)
71. Jones, E., Gerrard, H.: Foundations of Social Psychology. Wiley Books, New York (1965)
72. Kelley, H.H.: Two functions of reference groups. In: Swanson, G., Newcomb, T.M., Hartley, E. (eds.) Readings in Social Psychology, pp. 410–414. Henry Holt, New York (1952)
73. Hallinan, M.T.: Classroom Characteristics and Student Friendship Cliques. Social Forces 67(4), 898–919 (1989), doi:10.2307/2579707
74. Adler, P.A., Kless, S.J., Adler, P.: Socialization to Gender Roles: Popularity among Elementary School Boys and Girls. Sociology of Education 65(3), 169–187 (1992), doi:10.2307/2112807
75. Shin, Bickel: In: Kimble, C., Hildreth, P. (eds.) Communities of Practice: Creating Learning Environments for Educators. Information Age Publishing (2008) ISBN 1593118635
76. Lave, J., Wenger, E.: Situated Learning: Legitimate Peripheral Participation. Cambridge University Press, Cambridge (1991)

77. Wenger, E.: Communities of Practice: Learning, Meaning, and Identity, p. 318. Cambridge University Press, Cambridge (1998) ISBN 978-0-521-66363-2
78. Lesser, E.L., Fontaine, M.A., Slusher, J.A.: Knowledge and Communities. Butterworth-Heinemann (2000)
79. Wenger, E., McDermott, R., Snyder, W.M.: Cultivating Communities of Practice. HBS Press (2002)
80. Saint-Onge, H., Wallace, D.: Leveraging Communities of Practice. Butterworth Heinemann (2003)
81. Hildreth, P., Kimble, C.: Knowledge Networks: Innovation through Communities of Practice, pp. 1–59140. Idea Group Inc. Hidreth, London (2004) ISBN 1-59140-200-X
82. Smith, M.K.: Communities of practice. The Encyclopedia of Informal Education (2003)
83. Chua, A.: Book Review: Cultivating Communities of Practice. Journal of Knowledge Management Practice (October 2002)
84. van Winkelen, C.: Inter-Organizational Communities of Practice
85. Defense Acquisition Universitiy Community of Practice Implementation Guide, v3.0, Published by the Defense Acquisition University Press, Fort Belvior, Virginia 22060-5565 (October 2007)
86. Gannon-Leary, P.M., Fontainha, E.: Communities of Practice and virtual learning communities: benefits, barriers and success factors. ELearning Papers (September 26, 2007),
    http://www.elearningpapers.eu/index.php?page=doc&vol=5&do
    c_id=10219&doclng=6 (accessed November 2007)
87. National Library for Health Knowledge Management Specialist Library - collection of resources about communities of practice
88. Fan, Y.: Questioning Guanxi: Definition, Classification and Implications. International Business Review 11(5), 543–561 (2002)
89. Definition, meaning and application of guanxi in Chinese business life, scientific study on Guanxi in relation to business
90. China's modern power house, BBC article discussing the role of Guanxi in the modern governance of China
91. What is guanxi? Wiki discussion about definitions of guanxi, developed by the Publishers of Guanxi: The China Letter
92. Buderi, R., Huang, G.T.: Guanxi, The art of relationships, ISBN 0-7432-7322-2
93. China Characteristics - Regarding Guanxi
94. Cohen, J.: A just legal system, International Herald Tribune (December 11, 2007),
    http://www.iht.com/articles/2007/12/11/opinion/
    edcohen.php
95. Ansfield, J.: Where Guanxi Rules. Newsweek (December 17, 2007),
    http://www.newsweek.com/id/74369
96. MSNBC
97. Mashable
98. ABC News
99. Social Networking Goes Mobile (Business Week Online)
100. NY Times: Social Networking Moves to the Cellphone (itsmy.com, GyPSii, Facebook etc.)
101. "Social Nets Engage in Global Struggle" - 66% of MySpace and Facebook users come from North America: Adweekwebsite. name="adweek"/> (retrieved on January 15, 2008)

102. Bebo - most popular of its kind in UK: TechCrunch website (August 2007) (retrieved on January 15, 2008)
103. Hi5 popular in Europe: article from the PBS MediaShiftwebsite (retrieved on January 18, 2008)
104. "Why Users Love Orkut" - 55% of users are Brazilian: About.com website (retrieved on January 15, 2008)
105. Social Networking Goes Global, by comScore;Consumer Trends in Social Networking by Gian Fulgoni, comScore;Social Networks Size & Growth by Pipl
106. Roxanne Hiltz, S., Turoff, M.: The Network Nation. Addison-Wesley (1993, 1978)
107. Andrews, D.: The IRG Solution. Souvenir Press (1984)
108. Boyd, Ellison: p.3 (2007)
109. Boyd, Ellison: p. 3 (2007)
110. Weinreich, A.: cited by Boyd & Ellison, p. 3 (2007)
111. Rosen, C.: Virtual Friendship and the New Narcissism. New Atlantis (summer 2007) Available Online(PDF)
112. MySpace Page Views figures: BusinessWeekwebsite (2005)
113. "Social graph-iti": Facebook's social network graphing: article from The Economist's website (retrieved on January 19, 2008)
114. News Corporation buys MySpace: BBC.co.uk website
115. ITV buys Friends Reunited: BBC.co.uk website
116. Over 200 social networking sites: InfoJuice website (retrieved on January 19, 2008)
117. Vertical Social Networks;[2]: Nine Ways to Build Your Own Social Network, TechCrunch (July 24, 2007)
118. Boyd, D.: Friends, Friendsters, and MySpace Top 8: Writing Community Into Being on Social Network Sites. First Monday 11(12) (2006) (available online)
119. Gross, R., Acquisti, A.: Information Revelation and Privacy in Online Social Networks (The Facebook case). Pre-proceedings version. In: ACM Workshop on Privacy in the Electronic Society (WPES) (2005), Available Online, PDF
120. For example Mike Thelwall, MySpace, Facebook, Bebo: Social Networking Students, ALT: Online Newsletter (January 2008) available online; Also Mazer, J.P., Murphy, R.E., Simonds, C.J.: I'll See You On "Facebook": The Effects of Computer-Mediated Teacher Self-Disclosure on Student Motivation, Affective Learning, and Classroom Climate. Communication Education 56(1), 1–17 (2007) available online
121. Ellison, N.B., Steinfield, C., Lampe, C.: The Benefits of Network Sites. Journal of Computer Mediated Communication 12 (2007) available online (PDF)
122. Boyd, D.: Why Youth (Heart) Social Network Sites. In: Buckingham, D. (ed.) MacArthur Foundation Series on Digital Learning - Youth, Identity, and Digital Media Volume. MIT Press (2007) available online (PDF)
123. Boyd, D., Ellison, N.: Social Network Sites: Definition, History, and Scholarship. Journal of Computer-Mediated Communication 13(1) (October 2007) available online
124. Glykas, M.: Fuzzy Cognitive Maps: Advances in Theories, Methodologies, Applications and Tools. Springer (2010)
125. Boyd, D.: Research on Social Network Sites Known academic scholarship concerning social network sites, maintained
126. Social Networking: Now Professionally Ready, PrimaryPsychiatry.com website
127. Social Networks Impact the Drugs Physicians Prescribe According to Stanford Business School Research, Pharmalive.comwebsite
128. A New Generation Reinvents Philanthropy, Wall Street Journalwebsite

129. "Companies warned not to rush into social networking", implications of internal social networking in a business environment: News.com website (retrieved on January 22, 2008)
130. "Facebook, MySpace, and Co.: IHEs ponder whether or not to embrace social networking websites", implications of external social networking in education: TheFreeLibrary.comwebsite (retrieved on January 22, 2008)
131. The Value of Social Networking Tools Second Life Insider
132. Murdoch Will Earn a Payday from MySpace Forbes
133. Linked In Targeted Advertising LinkedIn
134. As Applications Blossom, Facebook Is Open for BusinessWired
135. Facebook Marketplace Guidelines Facebook
136. LinkedIn's Business Accounts LinkedIn
137. Flor, N.: Web Business Engineering. Addison-Wesley, Reading: Description of the autonomous business model used in social network services: article by Nick V. Flor at the InformIT website (2000)
138. Social network launches worldwide spam campaign E-consultancy.com (accessed September 10, 2007)
139. Moreno, M.A., Fost, N.C., Christakis, D.A.: Research ethics in the MySpace era. Pediatrics 121(1), 157–161 (2008), doi:10.1542/peds.2007-3015, PMID 18166570
140. "MySpace exposes sex predators", use of its content in the courtroom: Herald and Weekly Times (Australia) website (retrieved on January 19, 2008)
141. "Getting booked by Facebook", courtesy of campus police:Milwaukee Journal Sentinel website (retrieved on January 19, 2008)
142. Danah Boyd's list of articles about social network services
143. Boyd, D., Ellison, N.: Social Network Sites: Definition, History, and Scholarship. Journal of Computer-Mediated Communication 13(11) (October 2007)
144. Else, L., Turkle, S.: Living online: I'll have to ask my friends. New Scientist (2569). (September 20, 2006) (interview)
145. Glaser, M.: " Your Guide to Social Networking Online." PBS MediaShift
146. 9 Ways To Leverage Your Online Business Networking Activities (April 18, 2010)
147. Österle, H., Fleisch, E., Alt, R.: Business networking: shaping collaboration between enterprises (2, illustrated ed.). Springer (2001) ISBN 3540413510, 9783540413516
148. http://en.wikipedia.org/wiki/Business_networking
149. Barnes, J.A.: Class and Committees in a Norwegian Island Parish. Human Relations 7, 39–58
150. Berkowitz, S.D.: An Introduction to Structural Analysis: The Network Approach to Social Research. Butterworth, Toronto (1982)
151. Brandes, Ulrik, Erlebach, T. (eds.): Network Analysis: Methodological Foundations. Springer, Heidelberg (2005)
152. Breiger, R.L.: The Analysis of Social Networks. In: Hardy, M., Bryman, A. (eds.) Handbook of Data Analysis, pp. 505–526. Sage Publications, London (2004) Excerpts in pdf format
153. Burt, R.S.: Structural Holes: The Structure of Competition. Harvard University Press, Cambridge (1992)
154. Carrington, P.J., Scott, J., Wasserman, S.: Models and Methods in Social Network Analysis. Cambridge University Press, New York (2005)
155. Christakis, N., Fowler, J.H.: The Spread of Obesity in a Large Social Network Over 32 Years. New England Journal of Medicine 357(4), 370–379 (2007)

156. Doreian, P., Batagelj, V., Ferligoj, A.: Generalized Blockmodeling. Cambridge University Press, Cambridge (2005)
157. Freeman, L.C.: The Development of Social Network Analysis: A Study in the Sociology of Science. Empirical Press, Vancouver (2004)
158. Hill, R., Dunbar, R.: Social Network Size in Humans. Human Nature 14(1), 53–72 (2002), Google
159. Jackson, M.O.: A Strategic Model of Social and Economic Networks. Journal of Economic Theory 71, 44–74 (2003), pdf
160. Huisman, M., Van Duijn, M.A.J.: Software for Social Network Analysis. In: Carrington, P.J., Scott, J., Wasserman, S. (eds.) Models and Methods in Social Network Analysis, pp. 270–316. Cambridge University Press, New York (2005)
161. Krebs, V.: Social Network Analysis, A Brief Introduction (Includes a list of recent SNA applications Web Reference) (2006)
162. Lin, Nan, Burt, R.S., Cook, K. (eds.): Social Capital: Theory and Research. Aldine de Gruyter, New York (2001)
163. Mullins, N.: Theories and Theory Groups in Contemporary American Sociology. Harper and Row, New York (1973)
164. Müller-Prothmann, T.:: Leveraging Knowledge Communication for Innovation. In: Framework, Methods and Applications of Social Network Analysis in Research and Development, Frankfurt a. M. et al., Peter Lang (2006) ISBN 0-8204-9889-0
165. Manski, C.F.: Economic Analysis of Social Interactions. Journal of Economic Perspectives 14, 115–136 (2000) via JSTOR
166. Moody, J., White, D.R.: Structural Cohesion and Embeddedness: A Hierarchical Concept of Social Groups. American Sociological Review 68(1), 103–127 (2003)
167. Newman, M.: The Structure and Function of Complex Networks. SIAM Review 56, 167–256 (2003), pdf
168. Groumpos, P.: Basic Theories and Their Application to Complex Systems (2010)
169. Nohria, N., Eccles, R.: Networks in Organizations, 2nd edn. Harvard Business Press, Boston (1992)
170. Nooy, W.D., Mrvar, A., Batagelj, V.: Exploratory Social Network Analysis with Pajek. Cambridge University Press, Cambridge (2005)
171. Scott, J.: Social Network Analysis: A Handbook, 2nd edn. Sage, Newberry Park (2000)
172. Tilly, C.: Identities, Boundaries, and Social Ties. Paradigm press, Boulder (2005)
173. Valente, T.: Network Models of the Diffusion of Innovation. Hampton Press, Cresskill (1995)
174. Wasserman, S., Faust, K.: Social Networks Analysis: Methods and Applications. Cambridge University Press, Cambridge (1994)
175. Watkins, S.C.: Social Networks. In: Demeny, P., McNicoll, G. (eds.) Encyclopedia of Population, rev. ed., pp. 909–910. Macmillan Reference, New York (2003)
176. Watts, D.: Small Worlds: The Dynamics of Networks between Order and Randomness. Princeton University Press, Princeton (2003)
177. Watts, D.: Six Degrees: The Science of a Connected Age. W. W. Norton & Company (2004)
178. Wellman, B.: Networks in the Global Village. Westview Press, Boulder (1999)
179. Wellman, B.: Physical Place and Cyber-Place: Changing Portals and the Rise of Networked Individualism. International Journal for Urban and Regional Research 25(2), 227–252 (2001)

180. Wellman, B., Berkowitz, S.D.: Social Structures: A Network Approach. Cambridge University Press, Cambridge (1988)
181. Weng, M.: A Multimedia Social-Networking Community for Mobile Devices Interactive Telecommunications Program. Tisch School of the Arts/ New York University (2007)
182. White, H., Boorman, S., Breiger, R.: Social Structure from Multiple Networks: I Blockmodels of Roles and Positions. American Journal of Sociology 81, 730–780 (1976)

# Review Study on Fuzzy Cognitive Maps
# and Their Applications during the Last Decade

Elpiniki I. Papageorgiou

Dept of Informatics and Computer Technology
Technological Educational Institute of Lamia
Lamia, Greece
epapageorgiou@teilam.gr

**Abstract.** This survey work tries to review the most recent applications and trends on fuzzy cognitive maps (FCMs) at the last ten years. FCMs are inference networks, using cyclic directed graphs, for knowledge representation and reasoning. In the past decade, FCMs have gained considerable research interest and are widely used to analyze causal systems such as system control, decision making, management, risk analysis, text categorization, prediction etc. Some example application domains, such as engineering, social and political sciences, business, information technology, medicine and environment, where the FCMs emerged a considerable degree of applicability were selected Their dynamic characteristics and learning methodologies make them essential for modeling, analysis, prediction and decision making tasks as they improve the performance of these systems. A survey on FCM studies concentrated on FCM applications on diverse scientific fields is elaborated during the last decade.

**Keywords:** fuzzy cognitive maps, review, applications, styling, insert (key words).

## 1 Introduction

This study tries to gather the recent advances and trends on FCMs, their dynamic capabilities and application characteristics in diverse scientific areas. FCM represents a system in a form that corresponds closely to the way humans perceive it. Experts of each scientific area are used to express their knowledge by drawing weighted causal digraphs. The developed model is easily understandable, even by a non-technical audience and each parameter has a perceivable meaning [1,2].

In a graphical form, the FCMs are typically signed fuzzy weighted graphs, usually involving feedbacks, consisting of nodes and directed links connecting them. The nodes represent descriptive behavioral concepts of the system and the links represent cause-effect relations between the concepts. In the context of FCM theory, the fuzzy value of a concept denotes the degree in which the specific

M. Glykas (Ed.): Business Process Management, SCI 444, pp. 281–298.
springerlink.com          © Springer-Verlag Berlin Heidelberg 2013

concept is active in the general system, usually bounded in a normalized range of [0, 1]. Furthermore, the weights of the system's interrelations reflect the degree of causal influence between two concepts and they are usually assigned linguistically by experts [3]. They work by capturing and representing cause and effect relationships.

According to Codara (1998) [4], FCMs can be used for various purposes, including:

- to reconstruct the premises behind the behavior of a given agent, to understand the reasons for their decisions and for the actions the take, highlighting any distortions and limits in their representation of the situation (explanatory function);
- to predict future decisions and actions, or the reasons that a given agent will use to justify any new occurrences (prediction function);
- to help decision-makers ponder over their representation of a given situation in order to ascertain its adequacy and possibly prompt the introduction of any necessary changes (reflective function);
- to generate a more accurate description of a difficult situation (strategic function).

Essentially, a Fuzzy Cognitive Map is developed by integrating the existing experience and knowledge regarding a system. This can be achieved by using a group of human experts to describe the system's structure and behavior in different conditions. With FCM it is usually easy to find which factor should be modified and how and by this sense, an FCM is a dynamic modeling tool in which the resolution of the system representation can be increased by applying further mapping [5,6]. As simple mathematical models, FCMs represent the structured causality knowledge for qualitative and quantitative inference. The resulting fuzzy model can be used to analyze, simulate, and test the influence of parameters and predict the behavior of the system.

The main reasons someone uses the FCM approach are [7]: easy of use, easy to construct and parameterize, flexibility in representation (as more concepts/phenomena can be added and interact), low time performing, easily understandable/transparent to non-experts and lay people [8], handle with complex issues related to knowledge elicitation and management, handle with dynamic effects due to the feedback structure of the modeled system.

Furthermore, individual FCMs pertaining to a particular domain can be combined mathematically [1,2]. This means that FCMs allow for different experts and/or stakeholder views to be incorporated [9], and can provide a useful mechanism for combining information drawn from many sources to create a rich body of knowledge [10-12]. Finally, vector-matrix operations allow an FCM to model dynamic systems [1,13], allowing for the dynamic aspect of system behaviour to be captured [14]. Thus, FCMs have gained considerable research interest and accepted as useful methodology in many diverse scientific areas from knowledge modeling and decision making.

This work presents a survey on FCMs applications and trends in diverse scientific areas during the last decade exploring some of the most representative for each application study. The main aim of this study is to give an outline on how the FCMs increase their applications and more methodological efforts made by other researchers to enhance their applicability in different domains. It is difficult to present all the representative applications in each domain, as the number of FCM papers was extremely increased the last three years. Thus, we attempt to figure out only some of the most representative works of each domain, during the last decade.

## 2  Fuzzy Cognitive Maps and Applications

FCM is an efficient inference engine for modeling complex causal relationships easily, both qualitatively and quantitatively. Dickerson and Kosko (1994) used the FCM in a virtual world to model how sharks and fish hunt [15]. Parenthoen and his colleagues [16] successfully used the FCM to model the intentions and movements of a sheep dog and sheep in a virtual world.

During the past decade, FCMs played a vital role in the applications of diverse scientific areas, such as social and political sciences, engineering, information technology, robotics, expert systems, medicine, education, prediction, environment etc. Aguilar (2005) was the first who tried to gather the FCM applications in different scientific domains till 2004 [17]. In his work, most of the FCM applications were referred: administrative sciences, information analysis, popular political developments, engineering and technology management, prediction, education, cooperative man–machines, decision making and support, environmental management etc. After 2004, a large number of research studies related to methodologies for constructing or enhancing FCMs as well as innovative applications of FCMs were emerged. It is pinpointed that the number of research papers at 2010 is almost the double of the research papers presented till 2004 (see Table 1). The recent applications are focused not only to the previous referred domains but to more others such as telecommunications, game theory, e-learning, virtual environments, ambient intelligence, collaborative systems. New methodologies engaged with dynamic construction of FCMs, learning procedures, fuzzy inference structures were explored to improve the performance of them.

In this work, we concentrate our effort on FCM research studies after 2000 and especially on the presentation of the main categories of them at the last ten years (2001 to 2010). In particular, we describe recent research studies after 2008 for most of the application domains.

From the number of FCM studies opposed in Table 1 (last access in scopus at 20 February 2011), it is observed that during the last decade (we consider the years 2000 till 2010), there is a large increase on the number of research papers related to FCMs.

**Table 1** Research Studies of FCMs found in Scopus

| Year | Number of FCM-related studies from scopus | | |
| --- | --- | --- | --- |
| | *Studies* | *Journals & Book chapters* | *Conferences* |
| 2000 | 18 | 8 | 10 |
| 2001 | 22 | 2 | 20 |
| 2002 | 7 | 4 | 3 |
| 2003 | 19 | 9 | 10 |
| 2004 | 53 | 15 | 38 |
| 2005 | 35 | 16 | 19 |
| 2006 | 50 | 11 | 39 |
| 2007 | 56 | 19 | 37 |
| 2008 | 81 | 21 | 60 |
| 2009 | 84 | 26 | 58 |
| 2010 | (91+11 journals published in 2011) 102 | 50+11 Journals at 2011 | 41 |

Also, it is clearly shown that the last two years (2009-2010) the number of published papers in FCMs has been quadruple from the number of papers at the first years of decade (2000, 2001), and doubled from the papers published at 2004. Most of the research papers are related to FCMs applications and methodologies. 517 studies were accomplished during the past decade. A collection of papers with applications in various disciplines is presented in the book "Fuzzy Cognitive Maps: Advances in Theory, Methodologies, Applications and Tools edited by M. Glykas and prefaced by B. Kosko [6]. This is a significant index on FCM acceptability and applicability in research studies. Bar-chart in Figure 1 illustrates the FCM research studies for the last twenty years, including conference and journal papers, book chapters and technical reports, respectively for each year. It is clear from the chart that the FCMs have gained a considerable interest in research, which extremely increased during the last years. Furthermore, some types of typical problems solved by FCMs are modeling, prediction, interpreting, monitoring, decision making, classification, management. Paradigms of typical problems solved by FCMs are depicted in Table 2.

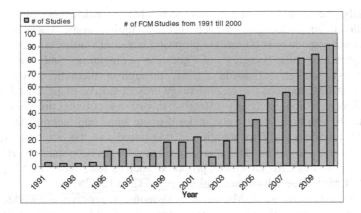

**Fig. 1** FCM papers during the last twenty years

**Table 2** Paradigms of typical problems solved by fcms

| Paradigm | Typical problems solved by FCMs |
|---|---|
| | *Description* |
| Control | Prediction interpreting, monitoring |
| Business | Planning, management, decision making, inference |
| medicine | Decision support, modeling, prediction, classification |
| robotics | Navigation, learning, prediction |
| environment | Knowledge representation, reasoning, stakeholders' analysis, policy making |
| Information technology | Modeling, analysis |

## 2.1  FCM Methodologies

In this section, a very short review is attempted to present the FCM methodologies and theories, depicting the estimation of their causal weights, the design and development process, the inference process etc.

As a generic model, the FCM relies on several assumptions. For example, the concepts' activation values are updated simultaneously at the same rate, and the causalities among the concepts are always in effect. However, these assumptions might not always hold, and the FCM isn't powerful or robust enough to model a dynamic, evolving virtual world. One other restriction is that use only simple monotonic and symmetric causal relations between concepts. But many real world causal relations are neither symmetric nor monotonic.

To solve these and other shortcomings and thus to improve the performance of FCMs, several methodologies were explored. Extensions to the FCMs theory are more than anything needed because of the feeble mathematical structure of FCMs and mostly the desire to assign advanced characteristics not met in other computational methodologies. Under this standpoint, some core issues are discussed and respective solutions are proposed in recent studies [18-25]; Pedryez (2010) presented the synergy of granular computing and evolutionary optimization to design efficiently FCMs, through a theoretic analysis [18]. Salmeron (2010) proposed the case of fuzzy grey cognitive maps, based on grey system theory, which is a very effective theory for solving problems within environments with high uncertainty, under discrete small and incomplete data sets [19]. Next, Iakovidis and Papageorgiou (2011) introduced the intuitionistic fuzzy sets and reasoning to handle the experts' hesitancy for decision making [20]. Also, more extensions of FCMs have been proposed, such as Rule-based FCMs [21,22], temporal FCMs [23], evolutionary multilayered FCMs [24,25] etc.

A dynamic version of FCMs, namely Dynamic Cognitive Networks, and a transformation process of cognitive maps, were proposed by Miao et al. [5,26] in order to explore the dynamic nature of concepts and to determine the strength of the cause and the cause-effect relationships as much as the degree of influence. Also, Fuzzy Cognitive Networks (FCNs) proposed as an operational extension of FCMs [27] that always reach equilibrium points in their operations especially to control unknown plants.

As another core task to design the FCMs, Song and his colleagues (2010) proposed a fuzzy neural network to enhance the learning ability of FCMs so that the automatic determination of membership functions and quantification of causalities can be incorporated with the inference mechanism of conventional FCMs. They employed mutual subsethood to define and describe the causalities in FCMs. It provides more explicit interpretation for causalities in FCMs and makes the inference process easier to understand. In this manner, FCM models of the investigated systems can be automatically constructed from data, and therefore are independent of the experts [28]. Next, Papageorgiou (2011) proposed a new methodology to design augmented FCMs combining knowledge from experts and knowledge from different data sources in the form of fuzzy rules [12].

It is worth noting that another important methodology for improving the performance of FCMs is learning algorithms. Learning methodologies for FCMs have been developed in order to update the initial knowledge of human experts and/or include any knowledge from historical data to produce learned weight matrices. The adaptive Hebbian-based learning algorithms, the evolutionary-based such as genetic algorithms and the hybrid approaches composed of Hebbian type and genetic algorithm were established to handle the task of FCM training [29,30-33]. These algorithms are the most efficient and widely used for training FCMs according to the existing literature [29,34].

## 2.2   Example Application Areas

Some example application areas were selected to present the way FCMs were applied and depicted in what follows. In Figure 2, the number of research studies accomplished in each one of the most common application domains during the last decade is depicted. At 2010, 13 application studies of FCMs were emerged in business, 12 in engineering and control, 8 in information technology, 5 in medical domain, and 6 in environment and agriculture.

## 1) Political and Social Sciences

FCMs emerged as a technique for modeling political and strategic issues situations and supporting the decision-making process in view of an imminent crisis. The research group of Andreou et al. proposed the use of the genetically evolved certainty neuron fuzzy cognitive map (GECNFCM) as an extension of CNFCMs aiming at overcoming the main weaknesses of the latter, namely the recalculation of the weights corresponding to each concept every time a new strategy is adopted [35]. That novel technique combined CNFCMs with genetic algorithms (GAs), the advantage of which lies with their ability to offer the optimal solution without a problem-solving strategy, once the requirements are defined. Using a multiple scenario analysis the value of the hybrid technique was demonstrated in the context of a model that reflects the political and strategic complexity of the Cyprus issue, as well as the uncertainties involved in it. Later, the same research group [24,36], presented the evolutionary FCMs for crisis management of the political problem of Cyprus.

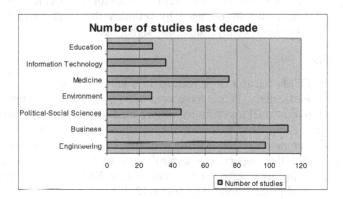

**Fig. 2** Number of FCM studies during the last decade (approximately +/- 5 studies per category)

An Ambient Intelligence (AmI) system, integrating aspects of psychology and social sciences, can be considered as a distributed cognitive framework composed by a collection of intelligent entities capable of modifying their behaviors by taking into account the user's cognitive status in a given time. Acampora et al. (2007), introduced a novel methodology of AmI systems' design that exploits multi-agent paradigm and a novel extension of FCMs theory benefiting on the theory of Timed Automata in order to create a collection of dynamical intelligent agents that use cognitive computing to define actions' patterns able to maximize environmental parameters as, for instance, user's comfort or energy saving [37].

Carvalho (2010) in his recent study, discussed the structure, the semantics and the possible use of FCM as tools to model and simulate complex social, economic and political systems, while clarifying some issues that have been recurrent in published FCM papers [38].

## 2) Medicine

At the last decade, FCMs have found important applicability in medical diagnosis and decision support [12, 39-45. In medical domain and in particular for medical decision support tasks, FCM based decision methodologies include an integrated structure for treatment planning management in radiotherapy [39], a model for specific language impairment [40], models for bladder and brain tumor characterization [41], an approach for the pneumonia severity assessment [42], and a model for the management of urinary tract infections [43]. Stylios et al. proposed FCM architectures for decision support in medicine [44].

Papakostas et al. (2007) implemented FCMs for pattern recognition tasks [45]. Froelich and Deja, 2009, proposed an FCM approach for mining temporal medical data [46]. Rodin et al. (2009) developed a fuzzy influence graph to model cell behavior in systems biology through the intracellular biochemical pathway [47]. Next, this model can be integrated in agents representing cells. Results indicate that despite individual variations, the average behavior of MAPK pathway in a cells group is close to results obtained by ordinary differential equations. The model was applied in multiple myeloma cells signaling.

## 3) Engineering

In this domain, FCMs found a large number of applications, especially in control and prediction. Particularly, FCMs have been used to model and support a plant control system, to construct a system for failure modes and effect analysis, to fine tune fuzzy logic controllers, to model the supervisor of a control system etc. Stylios and Groumpos, [2004] investigated the FCM for modeling complex systems and controlling supervisory control systems [48]. Papageorgiou et al. implemented learning approaches based on non-linear Hebbian rule to train FCMs that model industrial process control problems [29].

Recently, an integration of a cognitive map and a fuzzy inference engine was presented, as a cognitive–fuzzy model, targeting online fuzzy logic controller (FLC) design and self fine-tuning [49]. The proposed model was different than previous proposed FCMs in that it presents a hierarchical architecture in which the FCM process, available plant, and control objective data on represented knowledge to generate a complete FLC architecture and parameters description. Simulation results demonstrate model interpretability, which suggests that the model is scalable and offers robust capability to generate near optimal controller.

At 2010, Kottas et al. [27] presented basic theoretical results related to the existence and uniqueness of equilibrium points in FCN, the adaptive weight estimation based on system operation data, the fuzzy rule storage mechanism and the use of the entire framework to control unknown plants. The results are validated using well known control benchmarks. The same research team, at the same year, used FCN to construct a maximum power point tracker (MPPT), which may operate in cooperation with a fuzzy MPPT controller [50]. The proposed scheme outperforms other existing MPPT schemes of the literature giving very good maximum power operation of any PV array under different conditions such as changing insolation and temperature.

Beeson et al. (2010) proposed a factored approach to mobile robot map-building that handles qualitatively different types of uncertainty by combining the strengths of topological and metrical approaches [51]. This framework is based on a computational model of the human cognitive map and allows robust navigation and communication within several different spatial ontologies.

*4) Business*

In business, FCMs have found a great applicability. They can be used for product planning, for analysis and decision support. Some interested applications and worthy to be referred are illustrated. Jetter et al. (2006) used the concept of fuzzy front end for ideation, concept development and concept evaluation of new product development. This concept helped various problems managers who faced difficulty in early product development, as well as to systematic approaches to deal with them [52]. This approach helped in identification of market needs and technology potentials, detection and exploitation of idea sources, early stage assessment of ideas and product concepts, and successful management styles.

Yaman & Polat (2009) proposed the use of fuzzy cognitive maps (FCMs) as a technique for supporting the decision-making process in effect-based planning. With adequate consideration of the problem features and the constraints governing the method used, an FCM is developed to model effect-based operations (EBOs) [53]. In the mentioned study, the model was applied to an illustrative scenario involving military planning. Wei et al (2008) investigated the use of fuzzy cognitive time maps for modeling and evaluating trust dynamics in the virtual enterprises [54]. Bueno & Salmeron (2008) proposed a tool for fuzzy modeling Enterprise Resource Planning selection [55]. Salmeron (2009) proposed the augmented fuzzy cognitive maps for modeling LMS critical success factors [56].

Kim and his coworkers (2008) developed a hybrid qualitative and quantitative approach, using FCMs and GAs, to evaluate forward-backward analysis of RFID supply chain [57]. The research group of Trappey et al (2010) used FCMs to model and evaluates the performance of RFID-enabled reverse logistic operations [58]. Radio frequency identification (RFID) complying with the EPCglobal (2004) Network architecture, i.e., a hardware- and software-integrated cross-platform IT framework, is adopted to better enable data collection and transmission in reverse logistic management. Inference analysis using genetic algorithms contributes to the performance forecasting and decision support for improving reverse logistic efficiency [58]. The study provided a method to predict future logistic operation states and to construct a decision support model to manage system performance based on the forecast.

Büyüközkan et al. (2009) proposed a systematic way of analyzing collaborative planning, forecasting and replenishment (CPFR) supporting factors using FCM approach [59]. Through their study, it was verified the application of FCM where interrelated variables like decision variables and uncontrollable variables were used. Xirogiannis et al. (2010) addressed the problem of designing an "intelligent" decision support methodology tool to act as a back end to financial planning [60].

One of the challenges in Risk Analysis and Management (RAM) is identifying the relationships between risk factors and risks. Lazzerini et al. (2010), proposed Extended Fuzzy Cognitive Maps (E-FCMs) to analyze the relationships between risk factors and risks [61]. The main differences between E-FCMs and conventional FCMs are the following: E-FCMs have non-linear membership functions, conditional weights, and time delay weights. Therefore E-FCMs are suitable for risk analysis as all features of E-FCMs are more informative and can fit the needs of Risk Analysis. Particularly, the work explores the Software Project Management (SPM) and discusses risk analysis of SPM applying E-FCMs.

*5) Production Systems*

FCM can provide an interesting solution to the issue of assessing the factors which are considered to affect the operator's reliability and can be investigated for human reliability in production systems [13]. Bertolini & Bevilacqua (2010) investigated the human reliability in production systems which act as an excellent means to study a production process and obtain useful indications on the consequences which can be determined by the variation of one or more variables in the system examined.

Lo Storto et al., (2010) presented a methodological framework to explore the cognitive processes implemented by members of a software development team to manage ambiguous situations at the stage of product requirements definition [62]. FCMs were used in the framework to elicit cognitive schemes and developed a measure of individual ambiguity tolerance. Moreover, FCMs were used to design game-based learning systems because it has the excellent ability of concept representation and reasoning [63]. A novel game-based learning model which includes a teacher submodel, a learner submodel and a set of learning mechanisms was established.

In computer vision, which is a new emerging area, there are demanding solutions for solving different problems. The data to be processed are bi-dimensional (2D) images captured from the tri-dimensional (3D) scene. The objects in 3D are generally composed of related parts that joined form the whole object. Fortunately, the relations in 3D are preserved in 2D. Hence, there are necessary ingredients to build a structure under the FCMs paradigm. FCMs have been satisfactorily used in several areas of computer vision including: pattern recognition, image change detection or stereo vision matching [64]. Pajares (2010) established a general framework of FCMs in the context of 2D images and described the performance of three applications in the three mentioned areas of computer vision.

*6) Environment and Agriculture*

FCMs were applied in ecology and environmental management for: modeling a generic shallow lake ecosystem by augmenting the individual cognitive maps [65], assessing local knowledge use in agroforestry management [66], modeling of interactions among sustainability components of an agro-ecosystem using local knowledge [67], predicting modeling a New Zealand dryland ecosystem to anticipate pest management outcomes [68], semi-quantitative scenario, with an example from Brazil [69]. Recently, van Vliet et al., (2010) used FCMs as a communication and learning tool for linking stakeholders and modelers in scenario studies [7]. In their recent work demonstrated the potential use of a highly participatory scenario development framework that involves a mix of qualitative, semi-quantitative and quantitative methods. Giordano, et al. (2010) proposed a methodology based on a FCM to support the elicitation and the analysis of stakeholders' perceptions of drought, and the analysis of potential conflicts [70]. The same year, Kafetzis et al. (2010) investigated two separate case studies concerned with water use and water use policy [71]. The documentation and analysis of such stakeholders' models will presumably offer insights into the use and limitations of local knowledge and management, while concurrently providing a current approach to developing appropriate strategies for process-oriented problem solving and decision making in an environmental pollution context.

In agricultrure, FCMs used to represent knowledge and assess cotton yield prediction in precision farming by connecting yield defining parameters with yield in Cotton Crop Production in Central Greece as a basis for a decision support system [72].

## 7) Information Technology (IT)

In information technology (IT) project management, a FCM-based methodology helps to success modeling. Current methodologies and tools used for identifying, classifying and evaluating the indicators of success in IT projects have several limitations. These could be overwhelmed by employing the FCMs for mapping success, modeling Critical Success Factors (CSFs) perceptions and the relations between them [8]. Rodriguez-Repiso et al (2007) demonstrated the applicability of the FCM methodology through a case study based on a new project idea, the Mobile Payment System (MPS) project, related to the fast evolving world of mobile telecommunications.

Xiangwei et al. (2009) analyzed and summarized common software's usability quality character system in order to find a software usability malfunction discovers and improve problems [73]. They used FCM to describe the software quality character relationship and give an integrated training arithmetic, syntax pruning arithmetic, semantic pruning arithmetic and quality relationship analysis arithmetic to the method.

An interesting tool for FCM development was presented in [74] where the FCM is defined by concepts and relationships that can change during the execution time. Using the tool, someone can design a FCM, follow the evolution of a given FCM, change FCM defined previously, etc. Furfaro et al. (2010) presented a novel method for the identification and interpretation of sites that yield the highest potential of cryovolcanic activity in Titan and introduced the theory of FCMs for the analysis of remotely collected data in planetary exploration [75].

In telecommunications, FCMs applied for distributed wireless P2P networks [76]. Peer-to-peer (P2P) technologies have raised great research interest due to a number of successful applications in wired networks. Popular commercial applications such as Skype and Napster have attracted millions of users worldwide. A novel team-centric peer selection scheme based on FCMs, which simultaneously considers multiple selection criteria in wireless P2P networks, was proposed. The main influential factors and their complex relationships for peer selection in wireless P2P networks were investigated.

Table 3 FCM Applications at the first two months of 2011

| Research works published in 2011 | FCM applications at the first two months of 2011 | |
| --- | --- | --- |
| | Application area | Problem solving |
| Kannappan et al. [80] | medicine | Classification, prediction |
| Papageorgiou [12] | medicine | Knowledge representation, decision making |
| Beena, & Ganguli [79] | Structural damage detection | learning |
| Song et al. [78] | Business | Classification and prediction |
| Hanafizadeh, P., Aliehyaei, R. [81] | Soft system | Modeling, analysis |
| Iakovidis & Papageorgiou [20] | medicine | Decision making, reasoning |
| Lee, N., Bae, J.K., Koo, C. [82] | Sales assessment | reasoning |
| Chytas, P., Glykas, M., Valiris, G. [834] | Business | Planning, analysis |
| Jetter, A., Schweinfort, W. [84] | Solar energy | Modeling, policy scenarios |
| A. Baykasoglu, Z.D.U. Durmusoglu, V. Kaplanoglu, [71] | Industial process control | Learning, control |

Moreover, the participation of agents in FCM process [77] intends to a non-centralized detection of FCM stable state and to a modification process using asynchronous calculations. Stula et al. (2010) developed an agent-based fuzzy cognitive map (ABFCM) for injecting the concept of multi-agent system (MAS) into the FCM and the different inference algorithms in each node enabled the simulation of systems with diverse behavior concepts.

Of course, it was difficult in this study to present all the innovative and useful applications performed by FCMs and their extensions. We attempted to figure out some of the most mentioned applications emerged in the literature and to give the recent research directions of FCMs.

# 3   FCM Applications at 2011

During the first two months of year 2011, ten journal papers on FCM applications and modeling in different scientific fields were emerged. Table 3 includes these studies with the related application areas. So, the FCM has gradually emerged as a powerful paradigm for knowledge representation and a simulation mechanism that is applicable to numerous research and application fields. For example, Song et al. (2011) proposed a fuzzy neural network to enhance the learning ability of FCMs and incorporated the inference mechanism of conventional FCMs with the determination of membership functions. The effectiveness of this approach lies in handling the prediction of time series [78]. Beena et al. (2011) developed a new algorithmic approach for structural damage detection based on the use of FCM and Hebbian-based learning [79] and Baykasoglu et al. (2011) proposed a new training algorithm for FCMs, the Extended Great Deluge Algorithm (EGDA) [71]. Hanafizadeh and Aliehyaei applied FCMs in soft system methodology [81], Lee et al applied FCMs to sales opportunity assessment [82], Chytas et al applied FCMs to address the problems of proactive balanced scorecards [83], and Jetter et al. applied FCMs for scenario planning of solar energy [84]. Kanappan et al. [80] and Papageorgiou [12,20] applied FCMs in medical domain for classification and decision making.

# 4 Conclusion

The Fuzzy Cognitive Map methodology has been proven through the literature a very useful approach and cognition tool to model and analyze complex dynamical systems. There is a considerable number of application studies of FCMs in different domains and the application of FCMs emerges a rapid continue. They are helpful to the decision-makers ponder over their representation of a given situation in order to ascertain its adequacy and possibly prompt the introduction of any necessary changes. It is a research challenge for applying FCMs in diverse scientific areas especially when efficient methods to quantify causalities, to adapt and learn FCMs, are proposed which might be handle with the complex tasks of each domain, thus to improve the FCM performance.

**Acknowledgment.** The DebugIT project (http://www.debugit.eu/) is receiving funding from the European Community's Seventh Framework Programme under grant agreement n° FP7–217139, which is gratefully acknowledged. The information in this document reflects solely the views of the authors and no guarantee or warranty is given that it is fit for any particular purpose.

# References

[1] Kosko, B.: Fuzzy cognitive maps. International Journal of Man-Machine Studies 24(1), 65–75 (1986)
[2] Kosko, B.: Adaptive inference in fuzzy knowledge networks. In: Dubois, D., Prade, H., Yager, R.R. (eds.) Readings in Fuzzy Sets for Intelligent Systems. Morgan Kaufman, San Mateo (1993)
[3] Kosko, B.: Fuzzy Thinking (1993/1995) ISBN 0-7868-8021-X, (Chapter 12: Adaptive Fuzzy Systems)
[4] Codara, L.: Le mappe cognitive. Carrocci Editore, Roma (1998)

[5] Miao, Y., Liu, Z.Q., Siew, C.K., Miao, C.Y.: Dynamical cognitive network - an extension of fuzzy cognitive map. IEEE Transactions on Fuzzy Systems 9, 760–770 (2001)

[6] Glykas, G.: Fuzzy Cognitive Maps: Theory, Methodologies, Tools and Applications. Springer (July 2010)

[7] van Vliet, M., Kok, K., Veldkamp, T.: Linking stakeholders and modellers in scenario studies: The use of Fuzzy Cognitive Maps as a communication and learning tool. Futures 42(1), 1–14 (2010)

[8] Rodriguez-Repiso, L., Setchi, R., Salmeron, J.L.: Modelling IT Projects success with Fuzzy Cognitive Maps. Expert Systems with Applications 32(2), 543–559 (2007)

[9] Stach, W., Kurgan, L.A.: Expert-based and Computational Methods for Developing Fuzzy Cognitive Maps. In: Glykas, M. (ed.) Fuzzy Cognitive Maps: Advances in Theory, Methodologies, Tools and Applications. Springer (2010) ISBN-10: 36-42032-19-2

[10] Papageorgiou, E.I., Papandrianos, N.I., Apostolopoulos, D., Vassilakos, P.J.: Complementary use of Fuzzy Decision Trees and Augmented FCMs for Decision Making in Medical Informatics. In: Proc. of the 1st BMEI 2008, art. no. 4548799, Sanya, China, May 28-30, pp. 888–892 (2008)

[11] Papageorgiou, E.I., Stylios, C.D., Groumpos, P.P.: Novel architecture for supporting medical decision making of different data types based on Fuzzy Cognitive Map Framework. In: Proc. 28th IEEE EMBS, Conference 2007, Lyon, France, August 21-23, pp. 1192–1195 (2007)

[12] Papageorgiou, E.I.: A new methodology for Decisions in Medical Informatics using Fuzzy Cognitive Maps based on Fuzzy Rule-Extraction techniques. Applied Soft Computing 11, 500–513 (2011)

[13] Bertolini, M., Bevilacqua, M.: Fuzzy Cognitive Maps for Human Reliability Analysis in Production Systems. In: Kahraman, C., Yavuz, M. (eds.) Production Engineering and Management under Fuzziness. STUDFUZZ, vol. 252, pp. 381–415. Springer, Heidelberg (2010)

[14] Miao, Y., Miao, C., Tao, X., Shen, Z., Liu, Z.: Transformation of cognitive maps. IEEE Transactions on Fuzzy Systems 18(1), art. no. 5340662, 114–124 (2010)

[15] Dickerson, A., Kosko, B.: Virtual Worlds as Fuzzy Cognitive Maps. Presence 3(2), 173–189 (1994)

[16] Parenthoen, M., Reignier, P., Tisseau, J.: Put Fuzzy Cognitive Maps to Work in Virtual Worlds. In: Proc. 10th IEEE Int'l Conf. Fuzzy Systems, vol. 1, p. 38. IEEE CS Press (2001)

[17] Aguilar, J.: A survey about fuzzy cognitive maps papers. International Journal of Computational Cognition 3, 27–33 (2005)

[18] Pedrycz, W.: The design of cognitive maps: A study in synergy of granular computing and evolutionary optimization. Expert Systems with Applications (2010) (in press)

[19] Salmeron, J.: Modeling grey uncertainty with Fuzzy Grey Cognitive Maps. Expert Systems with Applications 37, 7581–7588 (2010)

[20] Iakovidis, D.K., Papageorgiou, E.: Intuitionistic Fuzzy Cognitive Maps for Medical Decision Making. IEEE Transactions on Information Technology in Biomedicine 15(1) (2011)

[21] Carvalho, J.P., Tome, J.A.B.: Rule Based Fuzzy Cognitive Maps in Socio-Economic Systems. In: Proc. of IFSA-Eusflat (2009)

[22] Carvalho, J.P., Tomé, J.A.: Rule Based Fuzzy Cognitive Maps - Expressing Time in Qualitative System Dynamics. In: Proceedings of the 2001 FUZZ-IEEE, Melbourne, Australia (2001)

[23] Zhong, H., Miao, C., Feng, Z.S.Y.: Temporal Fuzzy Cognitive Maps. In: 2008 IEEE World Congress on Computational Intelligence, Hong Kong, June 1-6, pp. 1831–1840 (2008)

[24] Andreou, A.S., Mateou, N.H., Zombanakis, G.A.: Evolutionary Fuzzy Cognitive Maps: A Hybrid System for Crisis Management and Political Decision-Making. In: Proc. Computational Intelligent for Modeling, Control & Automation CIMCA, Vienna, pp. 732–743 (2003)

[25] Andreou, A.S., Mateou, N.H., Zombanakis, G.A.: Soft computing for crisis management and political decision making: the use of genetically evolved fuzzy cognitive maps. Soft Computing Journal 9(3), 194–210 (2005), doi:10.1007/s00500-004-0344-0

[26] Miao, Y., Miao, C., Tao, X., Shen, Z., Liu, Z.: Transformation of cognitive maps. IEEE Transactions on Fuzzy Systems 18(1), art. no. 5340662, 114–124 (2010)

[27] Kottas, T.L., Boutalis, Y.S., Christodoulou, M.A.: Fuzzy Cognitive Networks: Adaptive Network Estimation and Control Paradigms. In: Glykas, M. (ed.) Fuzzy Cognitive Maps. STUDFUZZ, vol. 247, pp. 89–134. Springer, Heidelberg (2010)

[28] Song, H., Miao, C., Roel, W., Shen, Z., Catthoor, F.: Implementation of fuzzy cognitive maps based on fuzzy neural network and application in prediction of time series. IEEE Transactions on Fuzzy Systems 18(2), art. no. 5352265, 233–250 (2010)

[29] Papageorgiou, E.I., Stylios, C.D., Groumpos, P.P.: Unsupervised learning techniques for fine-tuning FCM causal links. Intern. Journal of Human-Computer Studies 64, 727–743 (2006)

[30] Papageorgiou, E.I., Groumpos, P.P.: A new hybrid learning algorithm for Fuzzy Cognitive Maps learning. Applied Soft Computing 5, 409–431 (2005b)

[31] Froelich, W., Juszczuk, P.: Predictive Capabilities of Adaptive and Evolutionary Fuzzy Cognitive Maps - A Comparative Study. In: Nguyen, N.T., Szczerbicki, E. (eds.) Intelligent Systems for Knowledge Management. SCI, vol. 252, pp. 153–174. Springer, Heidelberg (2009b)

[32] Stach, W., Kurgan, L.A., Pedrycz, W., Reformat, M.: Genetic learning of fuzzy cognitive maps. Fuzzy Sets and Systems 153(3), 371–401 (2005)

[33] Koulouriotis, D.E., Diakoulakis, I.E., Emiris, D.M., Zopounidis, C.D.: Development of dynamic cognitive networks as complex systems approximators: validation in financial time series. Applied Soft Computing 5, 157–179 (2005)

[34] Stach, W., Kurgan, L.A., Pedrycz, W.: A survey of fuzzy cognitive map learning methods. In: Grzegorzewski, P., Krawczak, M., Zadrozny, S. (eds.) Issues in Soft Computing: Theory and Applications (2005)

[35] Andreou, A., Mateou, N.H., Zombanakis, G.: The Cyprus Puzzle and the Greek-Turkish Arms Race: Forecasting Developments Using Genetically Evolved Fuzzy Cognitive Maps. Journal of Defence and Peace Making 14, 293–310 (2003)

[36] Andreou, A.S., Mateou, N.H., Zombanakis, G.A.: Soft computing for crisis management and political decision making: the use of genetically evolved fuzzy cognitive maps. Soft Computing Journal 9(3), 194–210 (2006)

[37] Acampora, G., Loia, V.: A Dynamical Cognitive Multi-Agent System for Enhancing Ambient Intelligence Scenarios. In: IEEE International Conference on Fuzzy Systems, art. no. 5277303, pp. 770–777

[38] Carvalho, J.P.: On the semantics and the use of Fuzzy Cognitive Maps in social sciences. In: Proc. 2010 IEEE World Congress on Computational Intelligence, WCCI 2010, art. no. 5584033 (2010)

[39] Papageorgiou, E.I., Stylios, C.D., Groumpos, P.P.: The Soft Computing Technique of Fuzzy Cognitive Maps for Decision Making in Radiotherapy. In: Haas, O., Burnham, K. (eds.) Intelligent and Adaptive Systems in Medicine, ch. 5, Taylor & Francis, LLC (2008)

[40] Georgopoulos, V.C., Malandraki, G.A., Stylios, C.D.: A fuzzy cognitive map approach to deferential diagnosis of specific language impairment. Artificial Intelligence in Medicine 29(3), 261–278 (2003)

[41] Papageorgiou, E.I., Spyridonos, P., Ravazoula, P., Stylios, C.D., Groumpos, P.P., Nikiforidis, G.: Advanced Soft Computing Diagnosis Method for Tumor Grading. Artificial Intelligence in Medicine 36(1), 59–70 (2006)

[42] Papageorgiou, E.I., Papandrianos, N.I., Karagianni, G., Kyriazopoulos, G., Sfyras, D.: A fuzzy cognitive map based tool for prediction of infectious diseases. In: Proceeding of FUZZ-IEEE 2009, World Congress, Korea, August 24-27, pp. 2094–2099 (2009b)

[43] Papageorgiou, E.I., Papadimitriou, C., Karkanis, S.: Management uncomplicated urinary tract infections using fuzzy cognitive maps. In: Proc. of the 9th ITAB 2009, Larnaca, Cyprus, November 5-7 (2009a) ISBN: 978-1-4244-5379-5

[44] Stylios, C.D., Georgopoulos, V.C.: Fuzzy Cognitive Maps Structure for Medical Decision Support Systems. In: Nikravesh, M., et al. (eds.) Forging the New Frontiers: Fuzzy Pioneers II. STUDFUZZ, vol. 218, pp. 151–174. Springer, Heidelberg (2008)

[45] Papakostas, G.A., Boutalis, Y.S., Koulouriotis, D.E., Mertzios, B.G.: Fuzzy cognitive maps for pattern recognition applications. International Journal of Pattern Recognition and Artificial Intelligence 22(8), 1461–1486 (2008)

[46] Froelich, W., Wakulicz-Deja, A.: Mining temporal medical data using adaptive fuzzy cognitive maps. In: 2009 Proceedings - 2009 2nd Conference on Human System Interactions, HSI 2009, art. no. 5090946, pp. 16–23 (2009)

[47] Rodin, V., Querrec, G., Ballet, P., Bataille, F., Desmeulles, G., Abgrall, J.–F.: Multi-Agents System to model cell signaling by using Fuzzy Cognitive Maps. Application to computer simulation of Multiple Myeloma. In: Proc. 2009 Ninth IEEE International Conference on Bioinformatics and Bioengineering, pp. 236–241 (2009)

[48] Stylios, C.D., Groumpos, P.P.: Modeling Complex Systems using Fuzzy Cognitive Maps. IEEE Transactions on Systems, Man and Cybernetics, Part A: Systems and Humans 34, 155–162 (2004)

[49] Gonzalez, J.L., Aguilar, L.T., Castillo, O.: A cognitive map and fuzzy inference engine model for online design and self fine-tuning of fuzzy logic controllers. International Journal of Intelligent Systems 24(11), 1134–1173 (2009)

[50] Kottas, T.L., Karlis, A.D., Boutalis, Y.S.: Fuzzy Cognitive Networks for Maximum Power Point Tracking in Photovoltaic Arrays. In: Glykas, M. (ed.) Fuzzy Cognitive Maps. STUDFUZZ, vol. 247, pp. 231–257. Springer, Heidelberg (2010b)

[51] Beeson, P., Modayil, J., Kuipers, B.: Factoring the mapping problem: Mobile robot map-building in the hybrid spatial semantic hierarchy. International Journal of Robotics Research 29(4), 428–459

[52] Jetter, A.J.M.: Fuzzy Cognitive Maps in engineering and technology management – what works in practice? In: Anderson, T., Daim, T., Kocaoglu, D. (eds.) Technology Management for the Global Future: Proceedings of PICMET 2006, Istanbul, Turkey, Portland, July 8-13 (2006)

[53] Yaman, D., Polat, S.: A fuzzy cognitive map approach for effect based operations: an illustrative case. Information Sciences 179(4), 382–403 (2009)

[54] Wei, Z., Lu, L., Yanchun, Z.: Using fuzzy cognitive time maps for modeling and evaluating trust dynamics in the virtual enterprises. Expert Systems with Applications 35(4), 1583–1592 (2008)

[55] Bueno, S., Salmeron, J.L.: Fuzzy modeling Enterprise Resource Planning tool selection. Computer Standards & Interfaces 30, 137–147 (2008)

[56] Salmeron, J.L.: Supporting decision makers with fuzzy cognitive maps: These extensions of cognitive maps can process uncertainty and hence improve decision making in R&D applications. Research Technology Management 52(3), 53–59 (2009)

[57] Kim, M.-C., Kim, C.O., Hong, S.R., Kwon, I.-H.: Forward-backward analysis of RFID-enabled supply chain using fuzzy cognitive map and genetic algorithm. Expert Systems with Applications 35(3), 1166–1176 (2008)

[58] Trappey, A.J.C., Trappey, C.V., Wub, C.-R.: Genetic algorithm dynamic performance evaluation for RFID reverse logistic management. Expert Systems with Applications: An International Journal 37(11), 7329–7335 (2010)

[59] Baykasoglu, A., Durmusoglu, Z.D.U., Kaplanoglu, V.: Training Fuzzy Cognitive Maps via Extended Great Deluge Algorithm with applications. Computers in Industry 62(2), 187–195 (2011)

[60] Xirogiannis, G., Glykas, M., Staikouras, C.: Fuzzy Cognitive Maps in Banking Business Process Performance Measurement. In: Glykas, M. (ed.) Fuzzy Cognitive Maps. STUDFUZZ, vol. 247, pp. 161–200. Springer, Heidelberg (2010)

[61] Lazzerini, B., Lusine, M.: Risk Analysis Using Extended Fuzzy Cognitive Maps. In: International Proc., ICICCI 2010, art. no. 5566004, pp. 179–182 (2010)

[62] Lo Storto, C.: Assessing ambiguity tolerance in staffing software development teams by analyzing cognitive maps of engineers and technical managers. In: 2nd Int. Conf. on Engineering System Management and Applications, ICESMA 2010, Sharjah (April 2010)

[63] Luo, X., Wei, X., Zhang, J.: Game-based Learning Model Using Fuzzy Cognitive Map Proceedings of the first ACM International Workshop on Multimedia Technologies for Distance Learning, Proceeding MTDL 2009 (2009) ISBN: 978-1-60558-757-8

[64] Pajares, G., Guijarro, M., Herrera, P.J., Ruz, J.J., de la Cruz, J.M.: Fuzzy Cognitive Maps Applied to Computer Vision Tasks. In: Glykas, M. (ed.) Fuzzy Cognitive Maps. STUDFUZZ, vol. 247, pp. 259–289. Springer, Heidelberg (2010)

[65] Tan, C.O., Ozesmi, U.: A generic shallow lake ecosystem model based on collective expert knowledge. Hydrobiologia 563, 125–142 (2006)

[66] Isaac, M.E., Dawoe, E., Sieciechowicz, K.: Assessing Local Knowledge Use in Agroforestry Management with Cognitive Maps. Environmental Management 43, 1321–1329 (2009)

[67] Ramsey, D., Norbury, G.L.: Predicting the unexpected: using a qualitative model of a New Zealand dryland ecosystem to anticipate pest management outcomes. Austral Ecology 34, 409–421 (2009)

[68] Rajaram, T., Das, A.: Modeling of interactions among sustainability components of an agro-ecosystem using local knowledge through cognitive mapping and fuzzy inference system. Expert Systems with Applications (2010) (in press)

[69] Kok, K.: The potential of Fuzzy Cognitive Maps for semi-quantitative scenario development, with an example from Brazil. Global Environmental Change 19, 122–133 (2009)

[70] Giordano, R., Vurro, M.: Fuzzy Cognitive Map to Support Conflict Analysis in Drought Management. In: Glykas, M. (ed.) Fuzzy Cognitive Maps. Studies in Fuzziness and Soft Computing, vol. 247, pp. 403–425. Springer, Heidelberg (2010)

[71] Kafetzis, A., McRoberts, N., Mouratiadou, I.: Using Fuzzy Cognitive Maps to Support the Analysis of Stakeholders' Views of Water Resource Use and Water Quality Policy. In: Glykas, M. (ed.) Fuzzy Cognitive Maps. STUDFUZZ, vol. 247, pp. 383–402. Springer, Heidelberg (2010)

[72] Papageorgiou, E.I., Markinos, A.T., Gemtos, T.A.: Soft Computing Technique of Fuzzy Cognitive Maps to Connect Yield Defining Parameters with Yield in Cotton Crop Production in Central Greece as a Basis for a Decision Support System for Precision Agriculture Application. In: Glykas, M. (ed.) Fuzzy Cognitive Maps. STUDFUZZ, vol. 247, pp. 325–362. Springer, Heidelberg (2010)

[73] Lai, X., Zhou, Y., Zhang, W.: Software Usability Improvement: Modeling, Training and Relativity Analysis. In: Proc. 2nd Int. Symp. on Information Science and Engineering, ISISE 2009, art. no. 5447282, pp. 472–475 (2009)

[74] Jose, A., Contreras, J.: The FCM Designer Tool. In: Glykas, M. (ed.) Fuzzy Cognitive Maps. SFSC, vol. 247, pp. 71–87. Springer, Heidelberg (2010)

[75] Furfaro, R., Kargel, J.S., Lunine, J.I., Fink, W., Bishop, M.P.: Identification of Cryovolcanism on Titan Using Fuzzy Cognitive Maps. Planetary and Space Science 5(5), 761–779 (2010)

[76] Li, X., Ji, H., Zheng, R., Li, Y., Yu, F.R.: A novel team-centric peer selection scheme for distributed wireless P2P networks. In: IEEE Wireless Communications and Networking Conference, WCNC, art. no. 4917532 (2009)

[77] Stula, M., Stipanicev, D., Bodrozic, L.: Intelligent Modeling with Agent-Based Fuzzy Cognitive Map. International Journal of Intelligent Systems 25(10), 981–1004 (2010)

[78] Song, H.J., Miao, C.Y., Wuyts, R., Shen, Z.Q., D'Hondt, M., Catthoor, F.: An extension to fuzzy cognitive maps for classification and prediction. IEEE Transactions on Fuzzy Systems 19(1), art. no. 5601761, 116–135 (2011)

[79] Beena, P., Ganguli, R.: Structural Damage Detection using Fuzzy Cognitive Maps and Hebbian Learning. Applied Soft Computing 11(1), 1014–1020 (2011)

[80] Arthi, K., Tamilarasi, A., Papageorgiou, E.I.: Analyzing the performance of fuzzy cognitive maps with non-linear hebbian learning algorithm in predicting autistic disorder. Expert Systems with Applications 38(3), 1282–1292 (2011)

[81] Hanafizadeh, P., Aliehyaei, R.: The Application of Fuzzy Cognitive Map in Soft System Methodology. Systemic Practice and Action Research, 1–30 (2011) (in press)

[82] Lee, N., Bae, J.K., Koo, C.: A case-based reasoning based multi-agent cognitive map inference mechanism: An application to sales opportunity assessment. Information Systems Frontiers, 1–16 (2011) (in press)

[83] Chytas, P., Glykas, M., Valiris, G.: A proactive balanced scorecard. International Journal of Information Management (2011) (article in Press)

[84] Jetter, A., Schweinfort, W.: Building scenarios with Fuzzy Cognitive Maps: An exploratory study of solar energy. Futures 43(1), 52–66 (2011)

# Mathematical Modelling of Decision Making Support Systems Using Fuzzy Cognitive Maps

Peter P. Groumpos and Ioannis E. Karagiannis

Laboratory for Automation and Robotics
Department of Electrical and Computer Engineering
University of Patras, 26500 Rio, Greece

**Abstract.** This chapter critically analyses the nature and state of Decision Support Systems (DSS) theories, research and applications. A thorough and extensive historical review of DSS is provided which focuses on the evolution of a number of sub-groupings of research and practice: personal decision support systems, group support systems, negotiation support systems, intelligent decision support systems, knowledge management- based DSS, executive information systems/ business intelligence, and data warehousing. The need for new DSS methodologies and tools is investigated. The DSS area has remained vital as technology has evolved and our understanding of Decision-Making process has deepened. DSS over the last twenty years has contributed both breadth and depth to DSS research. The challenge now is to make sense of it in "Decision Making" by planning it in understanding context and by searching new ways to utilize other advanced methodologies. The possibility of using Fuzzy Logic, Fuzzy Cognitive Maps and Intelligent Control in DSS is reviewed and analyzed. A new generic method for DSS is proposed, the Decision Making Support System (DMSS). Basic components of the new generic method are provided and fully analyzed. Case studies are given showing the usefulness of the proposed method.

**Keywords:** Decision Support Systems, Intelligent Control, Fuzzy Systems, Decision Making Support Systems, Fuzzy Cognitive Maps.

## 1 Introduction

One of the challenges of accepting the "operation" of any complex system is the ability to make Decisions so the system runs efficiently and cost effectively. However making Decisions concerning complex systems often strains our cognitive capabilities. Uncertainty and related concepts such as risk and ambiguity are prominent in the research and accompanied literature on Decision-Making. Uncertainty is a term used in subtly different ways in a number of scientific fields, including statistics, economics, finance, physics, psychology, engineering, medicine, energy, environment, biology, sociology, philosophy, insurance, geology, military

M. Glykas (Ed.): Business Process Management, SCI 444, pp. 299–337.
springerlink.com     © Springer-Verlag Berlin Heidelberg 2013

systems and Information and Communication Technologies (ICT). It applies to making decisions = predictions of future events, to physical measurements already made and/or computer generated data based on manmade "systems". This prominence is well deserved. Ubiquitous in realistic settings, uncertainty constitutes a major obstacle to effective Decision Making Process (DMP).

Currently, the prevalent view within many of the engineering, medical and human sciences, bestows the status of "scientific data" (physical and human produced data), mainly on those facts and propositions that stem directly from empirical and/or experimental work. In practice, experiment subjugates theory, leaving to theory the modest function of data interpretation. Nevertheless, "scientific data" are only isolated elements that must be interpreted and synthesized by holistic theory. In the absence of theory, the myriad arrays of "scientific data" turn into a heap of disparate material that is difficult to generalize, let alone correctly interpret.

Therefore in a complex system, there are a large number of "scientific data" been processed and a substantial amount of evidence that Decision Making (DM) and Human Intuitive Judgment (HIJ) can be far from "optimal" solutions. In almost all situations Decision Making (DM) has been a major focus of science throughout the human history. All these decisions prior to the development of the computer were made by people using various theories and methods and were recorded by hand.

These methods were originated from artificial intelligence, statistics, probability, cognitive psychology and information science. After the 1950, when the computer appeared for first time all these techniques were integrated in computing environments and thus enhancing the DM and the HIJ. Thus the concept of Decision Support systems (DSS) emerged after the 1950s as it will be seen in section 2. The concept of DSS since the 1980s is extremely broad and its definitions vary, depending on the scientist's or the researcher's point of view. Today the whole issue of defining DSS is still open. In this chapter a new systemic approach to the concept of "DSS" is undertaken.

Decision Support Systems (DSS) are defined as any interactive computer – based support system for making decisions in any complex system, when individuals and/or a team of people are trying to solve unstructured problems on an uncertain environment.

DSS have gained an increased popularity in various domains the last 10 years. There is no scientific field that, in one way or another, decisions are taken everyday using extensively advanced techniques of integrating digital computer systems. All scientific fields been mentioned earlier that encounter uncertainty are candidate for using one or another type of DSS.

DSS are especially valuable in situations in which, the amount of "scientific data" is prohibitive for the "human decision maker" without any aid to proceed in solving difficult problems faced by any complex system. Advanced DSS can aid human cognitive deficiencies by integrating various methodologies and tools utilizing a number of different information sources in order to reach "acceptable decisions". The benefits in using DSS are: increases efficiency, productivity, competiveness, cost effectiveness and high reliability. This gives business and other "systems" a comparative advantage over other competitors.

In this chapter the overall concept of Decision Support Systems is critically reviewed and analyzed. In section 2, a historical review starting from the ancient times till today is provided. In section 3 many but not all definitions and theories of DSS are analyzed. In section 4 the scientific areas of Fuzzy Logic and Intelligent Control are briefly reviewed as also their role in analyzing and modeling complex systems while basic theories of FCM are provided in section 5. The new generic DSS embedding "Decision Making" in the loop is presented and justified in section 6 while section 7 provides two illustrative case studies using the new proposed generic methodology. In section 8 a new Five Steps Approach to Success (5-SAS) which is further enabling to formulate more effective, flexible and cost effective DMSS. Future research topics are presented and certain specific directions are analyzed in section 9. Finally, section 10 provides a summary and closing remarks of the challenging issues been raised throughout the whole chapter.

## 2 A Historical Overview of Decision Support Systems (DSS)

Today it is still possible to reconstruct the history of computerized Decision Support Systems (DSS) from first-hand accounts and unpublished materials as well as published articles. History is both a guide to future activity in this field and a record of the ideas and actions of those who have helped advance our thinking and practice. In a technology field as diverse as DSS, history is not neat and linear. Different people and from different scientific fields have perceived the field of computerized DSS, from various points and so they report different accounts of what happened and what was important. Some of this can be sorted out, but more data gathering is necessary. For example in the field of Medicine, Clinical Decision Support Systems (CDSS) have been developed since the late 1960s and have played a very important role in providing health care for the patients.

Information Systems and Business researchers and technologists have built and investigated Decision Support Systems (DSS) for the last 60 years. Some researchers trace the origins of DSS to 1951 and the Lyons Tea Shops Business use of the LEO (Lyons Electronic Office I) _digital computer_. Now this is true but for computerized DSS ONLY. Decisions were made from the classical world and the time of Greek Civilization till today. From a strict historical point of view the Delphic Oracle could be considered as the first formal DSS. Ancient Delphi was a small City located in the Central Greece and was the focal point for intellectual enquiry, as well as an occasional meeting place for kings, leaders and intellectuals of the known ancient world. The Delphic Oracle extended considerable influence throughout the Ancient Greek world and it was consulted by everybody (from Kings to single citizens) before all major undertakings: wars, the founding colonies, legislating laws and so forth. It also was respected by the semi-Hellenic countries around the world such a Lydia, Caria, Egypt and even Persia. The

Delphic Oracle was using an extensive distributed system of "informal data points" throughout the cities of the Classical world. It had developed, on the Temple of Apollo, (ancient God) an extensive data base library for all events of that period. These information was used by the priests and Pythia before giving an answer-prophesy to the question been put in front of them. The first computerized DSS based on Distributed computer systems evolved in the early 1950s. However the term Decision Support Systems (DSS) was not used till the early 1970s.

In this chapter a starting point in collecting more firsthand accounts and in building a more complete mosaic of what was occurring in Universities, Research Institutes, software companies and in organizations to build and use computerized DSS in the last 60 years.

According to Keen [42]-[43], the concept of DSS has evolved from two main areas of research: the theoretical studies of organizational Decision Making (DM) done at the Carnegie Institute of Technology during the late 1950s and the technical work on interactive distributed systems mainly carried out at the Massachusetts Institute of Technology in the early 1960s. It is considered that the field of DSS became a scientific area of research and systemic studies in the early 1970s before gaining in intensity during the 1980s. in the 1980s , Executive Information Systems (EIS), Group Decision Support Systems (GDSS) and Organizational Decision Support Systems (ODSS) evolved from the single user and Model-Oriented DSS.

In the late 1960s, a new type of information system became practical – model-oriented DSS or Management Decision Systems (MDS). Two DSS pioneers, Peter Keen and Charles Stabell, claim the concept of decision support evolved from "the theoretical studies of organizational decision making done at the Carnegie Institute of Technology during the late 1950s and early '60s and the technical work on interactive computer systems, mainly carried out at the Massachusetts Institute of Technology in the 1960s [44]. Prior to 1965, it was very expensive to build large-scale information systems. At about this time, the development of the IBM System 360 and other more powerful mainframe systems made it more practical and cost-effective to develop Management Information Systems (MIS) in large companies. MIS focused on providing managers with structured, periodic reports. The goal of the first management information systems (MIS) was to make information in transaction processing systems available to management for decision-making purposes. Unfortunately, few MIS were successful [45]. Perhaps the major factor in their failure was that the IT professionals of the time misunderstood the nature of managerial work. The systems they developed tended to be large and inflexible and while the reports generated from managers' MIS were typically several dozen pages thick, unfortunately, they held little useful management information [45]-[46]. The title of Dearden's (1972) Harvard Business Review article, "MIS is a Mirage", summarized the feelings of the time.

The term "Decision Support Systems" first appeared in [47], although Andrew McCosh attributes the birth date of the field to 1965, when Michael Scott Morton's PhD topic, "Using a computer to support the decision-making of a manager" was accepted by the Harvard Business School (McCosh, 2004). Gorry and Scott Morton (1971) constructed a framework for improving management information systems using Anthony's categories of managerial activity [47] and Simon's taxonomy of decision types (Simon, 1960/1977). Gorry and Scott Morton conceived DSS as systems that support any managerial activity in decisions that are semi- structured or unstructured. Keen and Scott Morton [44] later narrowed the definition, or scope of practice, to semi-structured managerial decisions; a scope that survives to this day. The managerial nature of DSS was axiomatic in Gorry and Scott Morton [47], and this was reinforced in the field's four seminal books: Scott Morton [52], McCosh and Scott Morton [51], Keen and Scott Morton [44], and Sprague and Carlson [50].

Much of the early work on DSS was highly experimental. The aim of early DSS developers was to create an environment in which the human decision maker and the IT-based system worked together in an interactive fashion to solve problems; the human dealing with the complex unstructured parts of the problem, the information system providing assistance by automating the structured elements of the decision situation. The emphasis of this process was not to provide the user with a polished application program that efficiently solved the target problem. In fact, the problems addressed are by definition impossible, or inappropriate, for an IT-based system to solve completely. Rather, the purpose of the development of a DSS is an attempt to improve the effectiveness of the decision maker. In a real sense, DSS is a philosophy of information systems development and use and not a technology.

According to Sprague and Watson [48], around 1970 business journals started to publish articles on management decision systems, strategic planning systems and decision support systems. For example, Scott Morton and colleagues published a number of decision support articles in 1968. In 1969, Ferguson and Jones discussed a computer aided decision system in the journal Management Science. In 1971, Michael S. Scott Morton's ground breaking book **Management Decision Systems: Computer-Based Support for Decision Making** was published. In 1966-67 Scott Morton had studied how computers and analytical models could help managers make a key decision. He conducted an experiment in which managers actually used a Management Decision System (MDS). T.P. Gerrity, Jr. focused on DSS design issues in [49]. His system was designed to support investment managers in their daily administration of a clients' stock portfolio. DSS for portfolio management have become very sophisticated since Gerrity began his research. In 1974, Gordon Davis, a Professor at the University of Minnesota, published his influential text on Management Information Systems. He defined a Management Information System as "an integrated, man/machine system for providing information to support the operations, management, and decision-making functions in an organization." Davis's Chapter 12 titled "Information System Support for Decision Making" and Chapter 13 titled "Information System Support for Planning and Control" created the setting for the development of a broad foundation for DSS research and practice.

By 1975, J. D. C. Little was expanding the frontiers of computer-supported modeling. Little's DSS called BRANDAID and was designed to support product, promotion, pricing and advertising decisions. Also, Little (1970) in an earlier article identified criteria for designing models and systems to support management decision--making. His four criteria included: robustness, ease of control, simplicity, and completeness of relevant detail. All four criteria remain relevant in evaluating modern Decision Support Systems. Klein and Methlie (1995) note "A study of the origin of DSS has still to be written. It seems that the first DSS papers were published by PhD students or professors in business schools, who had access to the first time-sharing computer system: Project MAC at the Sloan School, the Dartmouth Time Sharing Systems at the Tuck School. In France, HEC was the first French business school to have a time-sharing system (installed in 1967), and the first DSS papers were published by professors of the School in 1970. The term SIAD ('Systèmes Interactif d'Aide à la Décision' the French term DSS) and the concept of DSS were developed independently in France, in several articles by professors of the HEC working on the SCARABEE project which started in 1969 and ended in 1974."

## 3 Decision Support Systems (DSS)

### 3.1 Definitions

The concept of a decision support system (DSS) is extremely broad and its definitions vary depending on the author's point of view. It can take many different forms and can be used in many different ways [53]. On the one hand, Finlay [54] and others define a DSS broadly as "a computer-based system that aids the process of decision making". In a more precise way, Turban [55] defines it as "an interactive, flexible, and adaptable computer-based information system, especially developed for supporting the solution of a non-structured management problem for improved decision making. It utilizes data, provides an easy-to-use interface, and allows for the decision maker's own insights." Other definitions fill the gap between these two extremes. For Keen and Scott Morton [44], DSS couple the intellectual resources of individuals with the capabilities of the computer to improve the quality of decisions. "DSS are computer-based support for management decision makers who are dealing with semi-structured problems." For Sprague and Carlson [50], DSS are "interactive computer based systems that help decision makers utilize data and models to solve unstructured problems "On the other hand, Schroff [56] quotes Keen [75] ("there can be no definition of DSS, only of Decision Support") to claim that it is impossible to give a precise definition including all the facets of the DSS. Nevertheless, according to Power [57], the term DSS remains a useful and inclusive term for many types of information systems that support decision making.

### 3.2 Basic Theories of Computerized Decision Support Systems

Typical application areas of DSSs are management and planning in business, health care, the military, and any area in which management will encounter complex decision situations. Decision support systems are typically used for strategic and tactical decisions faced by upper-level management—decisions with a reasonably low frequency and high potential consequences—in which the time taken for thinking through and modeling the problem pays off generously in the long run.

There are three fundamental components of DSSs.

- Database management system (DBMS). A DBMS serves as a data bank for the DSS. It stores large quantities of data that are relevant to the class of problems for which the DSS has been designed and provides logical data structures (as opposed to the physical data structures) with which the users interact. A DBMS separates the users from the physical aspects of the database structure and processing.
- Model-base management system (MBMS). The role of MBMS is analogous to that of a DBMS. Its primary function is providing independence between specific models that are used in a DSS from the applications that use them. The purpose of an MBMS is to transform data from the DBMS into information that is useful in decision making. Since many problems that the user of a DSS will cope with may be unstructured, the MBMS should also be capable of assisting the user in model building.
- Dialog generation and management system (DGMS). The main product of an interaction with a DSS is insight. As their users are often managers who are not computer-trained, DSSs need to be equipped with intuitive and easy-to-use interfaces. These interfaces aid in model building, but also in interaction with the model, such as gaining insight and recommendations from it.

While a variety of DSSs exists, the above three components can be found in many DSS architectures and play a prominent role in their structure.

Past practice and experience often guide computerized decision support development more than theory and general principles. Some developers say each situation is different so no fundamental theory is possible. Others argue that we have conducted insufficient research to develop theories. For these reasons, the theory of decision support and DSS has not been addressed extensively in the literature.

The following set of six propositions from the writings of the late Nobel Laureate Economist Herbert Simon form an initial theory of decision support. From [71] we draw three propositions.

Proposition 1: If information stored in computers is accessible when needed for making a decision, it can increase human rationality.

Proposition 2: Specialization of decision-making functions is largely dependent upon developing adequate channels of communication to and from decision centers.

Proposition 3: When a particular item of knowledge is needed repeatedly in decision making, an organization can anticipate this need and, by providing the individual with this knowledge prior to decision, can extend his or her area of rationality. Providing this knowledge is particularly important when there are time limits on decisions.

Now three additional propositions are identified:

Proposition 4: In the post-industrial society, the central problem is not how to organize to produce efficiently but how to organize to make decisions – that is, to process information. Improving efficiency will always remain an important consideration.

Proposition 5: From the information processing point of view, division of labor means factoring the total system of decisions that need to be made into relatively independent subsystems, each one of which can be designed with only minimal concern for its interactions with the others.

Proposition 6: The key to the successful design of information systems lies in matching the technology to the limits of the attention of users. In general, an additional component, person, or machine for an information-processing system will improve the system's performance when the following three conditions are true:

1. The component's output is small in comparison with its input so that it conserves attention instead of making additional demands on attention.
2. The component incorporates effective indexes of both passive and active kinds. Active indexes automatically select and filter information.
3. The component incorporates analytic and synthetic models that are capable of solving problems, evaluating solutions, and making decisions.

In summary, computerized decision support is potentially desirable and useful when there is a high likelihood of providing relevant, high quality information to decision makers when they need it and want it.

## 3.3  Theories for Modern Decision Support Systems

The modern era in DSS started in the late 1990s with the specification of HTML 2.0, the expansion of the World Wide Web in companies, and the introduction of handheld computing. Today, the Web 2.0 technologies, mobile-integrated communication and computing devices, and improved software development tools have revolutionized DSS user interfaces. Additionally, the decision support data store back-end is now capable of rapidly processing very large data sets. Modern DSS are more complex and more diverse in functionality than earlier DSS built prior to the widespread use of the World Wide Web. Today, we are seeing more decision automation with business rules and more Knowledge-Driven DSS. Current DSS are changing the mix of decision-making skills needed in organizations. Building better DSS may provide one of the "keys" to competing in a global business environment.

The following attributes are increasingly common in new and updated Decision Support Systems. Some attributes are more closely associated with one category of DSS, but sophisticated DSS often have multiple subsystems. Contemporary DSS include five attributes:

1. Multiple, remote users can collaborate in real-time using rich media.
2. Users can access DSS applications anywhere and anytime.
3. Users have fast access to historical data stored in very large data sets.
4. Users can view data and results visually with excellent graphs and charts.
5. Users can receive real-time data when needed.

## 3.4 Different Approaches to Decision Support Systems

DSS is not a homogenous field. There are a number of fundamentally different approaches to DSS and each has had a period of popularity in both research and practice. Each of these "DSS types" represents a different philosophy of support, system scale, level of investment, and potential organizational impact. They can use quite different technologies and may support different managerial constituencies. Another dimension to the evolution of DSS is improvement in technology, as the emergence of each of the DSS types has usually been associated with the deployment of new information technologies. The nature and development of four selected DSS types is discussed next.

### 3.4.1 Personal Decision Support Systems

Personal DSS (PDSS) are small-scale systems that are normally developed for one manager, or a small number of independent managers, for one decision task. PDSS are the oldest form of decision support system and for around a decade they were the only form of DSS in practice. They effectively replaced Management Information Systems (MIS) as the management support approach of choice. The world of MIS was that of the Cold War and the rise of the Multi-National Corporation. The focus of management in this environment was total integration, efficiency, and central control, and the large, inflexible MIS mirrored this organizational environment. The emergence of PDSS also mirrored its social and organizational environment. The 1960s and 1970s saw a radicalization of Western society, especially in response to the Vietnam War. The emphasis was on empowering individuals and a democratization of decision-making. PDSS followed this philosophy by supporting individual managers rather than attempting to support the more nebulous concept of "the organization".

The major contribution of PDSS to Information sciences (IS) theory is evolutionary systems development [58]. The notion that a DSS evolves through an iterative process of systems design and use has been central to the theory of decision support systems since the inception of the field. Evolutionary development in decision support was first hinted at [72] and [73] as part of their description of middle-out design. This was a response to the top-down versus bottom-up methodology debate of the time concerning the development of transaction processing systems. Courbon [74] provided the first general statement of DSS evolutionary development. In what they termed an "evolutive approach", development processes are not implemented in a linear or even in a parallel fashion, but in continuous action cycles that involve significant user participation. As each evolutive cycle is completed the system gets closer to its final or stabilised state. Keen [75], building on Courbon's work, developed a framework or model for understanding the dynamics of DSS evolution. The importance of this work was to give the concept a larger audience. Amongst other contributors to PDSS development theory, Sprague and Carlson [50] defined an evolutionary DSS development methodology, and Silver [76] extended Keen's approach by considering how PDSS restrict or limit decision-making processes.

### 3.4.2 Intelligent Decision Support Systems

Artificial Intelligence (AI) techniques have been applied to decision support and these systems are normally called intelligent DSS or IDSS [59] although the term knowledge-based DSS has also been used [60].Intelligent DSS can be classed into two generations: the first involves the use of rule-based expert systems and the second generation uses neural networks, genetic algorithms and fuzzy logic [61]. A fundamental tension exists between the aims of AI and DSS. AI has long had the objective of replacing human decision makers in important decisions, whereas DSS has the aim of supporting rather than replacing humans in the decision task. As a result the greatest impact of AI techniques in DSS has been embedded in the PDSS, GSS or EIS, and largely unknown to managerial users. This is particularly the case in data mining and customer relationship management.

### 3.4.3 Executive Information Systems and Business Intelligence

Executive Information Systems (EIS) are data-oriented DSS that provide reporting about the nature of an organization to management [62]. Despite the 'executive' title, they are used by all levels of management. EIS were enabled by technology improvements in the mid to late 1980s, especially client server architectures, stable and affordable networks, graphic user interfaces, and multidimensional data modeling. This coincided with economic downturn in many OECD countries that resulted in the downsizing phenomenon that decimated middle management. EIS were deployed to help try to manage the leaner reporting structures. The seminal EIS book [69] was titled ''Executive Support Systems'', reflecting the decision support heritage. Rockart had earlier contributed what became EIS's major theoretical contribution to general information systems theory, the notion of Critical Success Factors or CSF [63]. CSF are the small number of factors that must go right for an organization, business unit, or individual executive to prosper. If a manager notices from an EIS report that the business is not performing in any critical area, the EIS enables the manager to drill-down through a report hierarchy to discover the possible sources of the variance. The multidimensional view of data, institutionalized as the 'data cube'', was the foundation of early EIS vendor offerings like HOLOS and Cognos. This multidimensionality was later codified and described as OnLine Analytical Processing (OLAP) [70].

By the mid 1990s EIS had become main stream and was an integral component of the IT portfolio of any reasonably sized organization. The Business Intelligence (BI) movement of the late 1990s changed the direction or emphasis of EIS by focusing on enterprise-wide reporting systems although this organizational focus has yet to be widely realized in successful systems. Dashboard-style interfaces and web delivery changed the look and feel of EIS, and the broader measures of balanced score cards [64] displaced some, but not all, of the CSF framework of EIS reporting. Business Intelligence (BI) is a poorly defined term and its industry origin means that different software vendors and consulting organizations have defined it to suit their products; some even use 'BI' for the entire range of decision support approaches. Business Intelligence (BI) as the contemporary term for both model-oriented and data-oriented DSS that focus on management reporting, that is, BI is a contemporary term for EIS.

### 3.4.4 Data Warehouses

The development of large-scale EIS created the need for continuous high quality data about the operations of an organization. The bull market of the 1990s led to a plethora of mergers and acquisitions and an increasing globalization of the world economy. Large organizations were faced with significant challenges in maintaining an integrated view of their business. This was the environment of the birth of data warehousing. A data warehouse is simply a set of databases created to provide information to decision makers; they provide raw data for user-focused decision support through PDSS and EIS.

There are two fundamental approaches to data warehouses: enterprise level data warehouses [65] and division or department level data marts [66]. This architectural debate has raged since the mid 1990s and shows no signs of abating in practice. The major contribution of data warehousing to IS theory is dimensional modeling [67]. Using dimensional models very large data sets can be organized in ways that are meaningful to managers. They are also relatively easy to query and analyze. In this sense, data warehousing provides the large scale IT infrastructure for contemporary decision support. As a result data warehouse development is dominated by central IT departments that have little experience with decision support. A common theme in industry conferences and professional books is the rediscovery of fundamental DSS principles like evolutionary development [68]. An issue that needs to be addressed very serious and the appropriate attention.

## 3.5 Different Driven Types of Decision Support Systems

Another classification of DSS is from a structural driven point of view. Here we provide the most common types, without a lot of details. The interested reader can easily find the appropriate material on basic textbooks.

### Data Driven

These DSS has file drawer systems, data analysis systems, analysis information systems, data warehousing and emphasizes access to and manipulation of large databases of structured data.

**Model Driven**

The underlying model that drives the DSS can come from various disciplines or areas of specialty and might include accounting models, financial models, representation models, optimization models, etc. With model drive DSS the emphasize is on access to and manipulation of a model, rather than data, i.e. it uses data and parameters to aid decision makers in analyzing a situation. These systems usually are not data intensive and consequently are not linked to very large databases.

**Knowledge Driven**

These systems provide recommendation and/or suggestion schemes which aid the user in selecting an appropriate alternative to a problem at hand. Knowledge driven DSS are often referred to as management expert systems or intelligent decision support systems. They focus on knowledge and recommends actions to managers based on an analysis of a certain knowledge base. Moreover, it has special problem solving expertise and are closely related to data mining i.e. sifting through large amounts of data to produce contend relationships.

**Communication Driven**

This breed of DSS is often called group decision support systems (GDSS). They are a special type of hybrid DSS that emphasizes the use of communications and decision models intended to facilitate the solution of problems by decision makers working together as a group. GDSS supports electronic communication, scheduling, document sharing and other group productivity and decision enhancing activities and involves technologies such as two-way interactive video, bulletin boards, e-mail and others.

**Inter- and Intra-organization Driven**

These systems are driven by the rapid growth of Internet and other networking technologies such as broadband WAN's, LAN's, WIP and others. Inter-organization DSS are used to serve companies stakeholders (customers, suppliers, etc.), whereas intra-organization DSS are more directed towards individuals inside the company and specific user groups. The latter, because of their stricter control, are often stand-alone units inside the firm.

## 4   Fuzzy Logic and Intelligent Control

### 4.1   Fuzzy Logic Basic Theory

Fuzzy logic starts with and builds on a set of user-supplied human language rules. The fuzzy systems convert these rules to their mathematical equivalents. This simplifies the job of the system designer and the computer, and results in much more accurate representations of the way systems behave in the real world [28]-[30].

Additional benefits of fuzzy logic include its simplicity and its flexibility. Fuzzy logic can handle problems with imprecise and incomplete data, and it can model nonlinear functions of arbitrary complexity. "If you don't have a good plant model, or if the system is changing, then fuzzy will produce a better solution than conventional control techniques," says Bob Varley, a Senior Systems Engineer at Harris Corp., an aerospace company in Palm Bay, Florida.

In fuzzy logic, unlike standard conditional logic, the truth of any statement is a matter of degree. (How cold is it? How high should we set the heat?) We are familiar with inference rules of the form p -> q (p implies q). With fuzzy logic, it's possible to say (.5* p ) -> (.5 * q). For example, for the rule if (weather is cold) then (heat is on), both variables, cold and on, map to ranges of values. Fuzzy inference systems rely on membership functions to explain to the computer how to calculate the correct value between 0 and 1. The degree to which any fuzzy statement is true is denoted by a value between 0 and 1.

Not only do the rule-based approach and flexible membership function scheme make fuzzy systems straightforward to create, but they also simplify the design of systems and ensure that you can easily update and maintain the system over time.

The Mamdani Fuzzy Inference Systems (FIS) [25]-[26], was the first system proven in a practical way as universal approximator of functions. Later Kosko and Wang formally settled that any relationship among input and output variables can be approximated by means of FIS, built in linguistic terms with a high grade of accuracy [27] (universal approximator).

In Fuzzy Logic theory there are three outstanding definitions:

a)  Fuzzy Set is called any set that allows to its members having different degree of membership (Membership function) in universe [0,1].
b)  Universe of Discourse is the range of all possible values for an input to a fuzzy system
c)  Membership Function (MF) shows the degree that set x belongs to set A according to the following equation (Fig. 1):

$$\mu_A(x): X \to [0,1] \tag{4.1}$$

Fuzzy sets are usually represented from ordered pairs sets according to the following equation:

$$A = \int \{\mu_A(x)/x\} \text{ or } A = \sum \{\mu_A(x)/x\} \text{ for } x \in X \tag{4.2}$$

Fig. 2 shows the basic architecture of a fuzzy logic controller containing the rule-base, the inference mechanism and the defuzzification method.

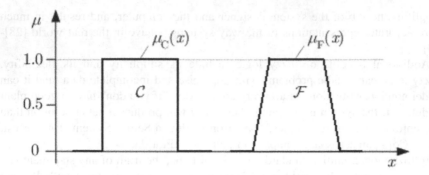

**Fig. 1** Membership function characteristics of a conventional/crisp set on the left and of a fuzzy one on the right

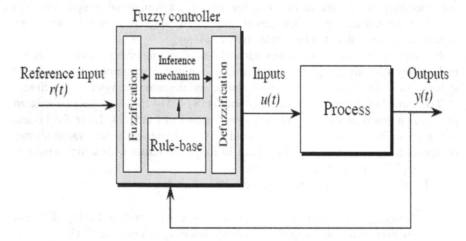

**Fig. 2** Fuzzy Controller Architecture

The fuzzy controller has four main components: (1) The rule-base" holds the knowledge, in the form of a set of rules, of how best to control the system. (2) The inference mechanism evaluates which control rules are relevant at the current time and then decides what the input to the plant should be. (3) The fuzzification interface simply modifies the inputs so that they can be interpreted and compared to the rules in the rule-base. And (4) the defuzzification interface converts the conclusions reached by the inference mechanism into the inputs to the plant.

Defuzzification refers to the way a crisp value is extracted from a fuzzy set as a representative value. In general, there are five methods [31]-[36] for defuzzifying a fuzzy set A of a universe of discourse Y. The adopted defuzzyfication strategy for this chapter is the Center of Area (COA):

$$\widehat{y_{COA}} = \frac{\int y_i \mu_A(y_i) y \, dy}{\int \mu_A(y_i) \, dy} \qquad (4.3)$$

where $\mu_A(y)$ is the aggregated output MF (i.e. Fig. 3) [35]. This is the most widely adopted defuzzyfication strategy, which is reminiscent of the calculation of expected values of probability distributions.

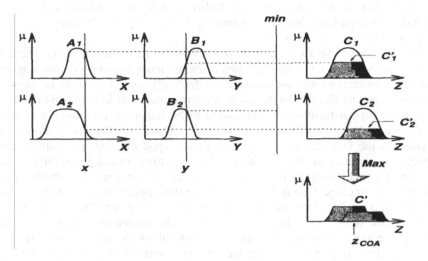

**Fig. 3** The Mamdani fuzzy inference system using min and max for T-norm and T-conorm operators, respectively

## 4.2 Intelligent Control and Intelligence

Intelligent control describes the discipline where control methods are developed that attempt to emulate important characteristics of human intelligence. These characteristics include adaptation and learning, planning under large uncertainty and coping with large amounts of data. Others describe as Intelligent Control, the discipline where control algorithms are developed by emulating certain characteristics of intelligent biological systems, is being fueled by recent advancements in computing technology and is emerging as a technology that may open avenues for significant technological advances. Today, the area of Intelligent Control tends to encompass everything that is not characterized as conventional control. The main difficulty in specifying exactly what is meant by the term Intelligent Control stems from the fact that there is no agreed upon definition of human intelligence and intelligent behavior and the centuries old debate of what constitutes intelligence is still continuing, nowadays among educators, psychologists, computer scientists and engineers.

It is appropriate at this point to briefly comment on the meaning of the word intelligent in "intelligent control". Note that the precise definition of "intelligence" has been eluding mankind for thousands of years. More recently, this issue has been addressed by disciplines such as psychology, philosophy, biology and of course by artificial intelligence (AI); note that AI is defined to be the study of mental faculties through the use of computational models. No consensus has emerged as yet of what constitutes intelligence. Intelligence is also considered as a very general mental capability that, among other things, involves the ability to reason, plan, solve problems, think abstractly, comprehend complex ideas, learn quickly and learn from experience. It is not merely book learning, a narrow academic skill, or test-taking smarts. Rather, it reflects a broader and deeper capability for comprehending our surroundings—"catching on," "making sense" of things, or "figuring out" what to do.

Intelligent controllers can be seen as machines which emulate human mental faculties such as adaptation and learning, planning under large uncertainty, coping with large amounts of data etc. in order to effectively control complex processes; and this is the justification for the use of the term intelligent in intelligent control, since these mental faculties are considered to be important attributes of human intelligence. An alternative term is "autonomous (intelligent) control"; it emphasizes the fact that an intelligent controller typically aims to attain higher degrees of autonomy in accomplishing and even setting control goals, rather than stressing the (intelligent) methodology that achieves those goals. We should keep in mind that "intelligent control" is only a name that appears to be useful today. In the same way the "modern control" of the 60's has now become "conventional (or traditional) control", as it has become part of the mainstream, what is called intelligent control today may be called just "control" in the not so distant future. What are more important than the terminology used are the concepts and the methodology, and whether or not the control area and intelligent control will be able to meet the ever increasing control needs of our technological society [37]-[40].

The term "intelligent control" has come to mean, particularly to those outside the control area, some form of control using fuzzy and/or neural network methodologies. Intelligent Control, however does not restrict itself only to those methodologies. In fact, according to some definitions of intelligent control not all neural/fuzzy controllers would be considered intelligent. The fact is that there are problems of control today, that cannot be formulated and studied in the conventional differential/difference equation mathematical framework using "conventional (or traditional) control" methodologies; these methodologies were developed in the past decades to control dynamical systems. To address these problems in a systematic way, a number of methods have been developed in recent years that are collectively known as "intelligent control" methodologies.

There are a number of areas related to the area of Intelligent Control. Intelligent Control is interdisciplinary as it combines and extends theories and methods from areas such as control, computer science and operations research. It uses theories from mathematics and seeks inspiration and ideas from biological systems. Intelligent control methodologies are being applied to robotics and automation, communications, manufacturing, traffic control, to mention but a few application areas. Neural networks, fuzzy control, genetic algorithms, planning systems, expert systems, and hybrid systems are all areas where related work is taking place. The areas of computer science and in particular artificial intelligence provide knowledge representation ideas, methodologies and tools such as semantic networks, frames, reasoning techniques and computer languages such as prolog. Concepts and algorithms developed in the areas of adaptive control and machine learning help intelligent controllers to adapt and learn. Advances in sensors, actuators, computation technology and communication networks help provide the necessary for implementation Intelligent Control hardware.

# 5 Basic Theories of FCM

## 5.1 Basic Theories

Fuzzy cognitive map is a soft computing technique that follows an approach similar to human reasoning and the human decision-making process. An FCM looks like a cognitive map, it consists of nodes (concepts) that illustrate the different aspects of the system's behavior. These nodes (concepts) interact with each other showing the dynamics of the model. Concepts may represent variables, states, events, trends, inputs and outputs, which are essential to model a system. The connection edges between concepts are directed and they indicate the direction of causal relationships while each weighted edge includes information on the type and the degree of the relationship between the interconnected concepts. Each connection is represented by a weight which has been inferred through a method based on fuzzy rules that describes the influence of one concept to another. This influence can be positive (a promoting effect) or negative (an inhibitory effect). The FCM development method is based on Fuzzy rules that can be either proposed by human experts and/or derived by knowledge extraction methods [1], in such a way that the accumulated experience and knowledge are integrated in the causal relationships between factors, characteristics and components of the process or system modeled [2].

## 5.2 Mathematical Representation of Fuzzy Cognitive Maps

The graphical illustration of an FCM is a signed directed graph with feedback, consisting of nodes and weighted arcs [15]. Nodes of the graph stand for the concepts that are used to describe the behavior of the system and they are connected by signed and weighted arcs representing the causal relationships that exist between the concepts (Fig. 4).

Each concept is characterized by a number $A_i$ that represents its value and it results from the transformation of the fuzzy real value of the system's variable, for which this concept stands, in the interval [0, 1]. Between concepts, there are three possible types of causal relationships that express the type of influence from a concept to the others. The weights of the arcs between concept $C_i$ and concept $C_j$ could be positive ($W_{ij} > 0$) which means that an increase in the value of concept $C_i$ leads to the increase of the value of concept $C_j$, and a decrease in the value of concept $C_i$ leads to the decrease of the value of concept $C_j$. Or there is negative causality ($W_{ij} < 0$) which means that an increase in the value of concept $C_i$ leads to the decrease of the value of concept $C_j$ and vice versa. The sign of $W_{ij}$ indicates whether the relationship between concepts $C_i$ and $C_j$ is direct or inverse.

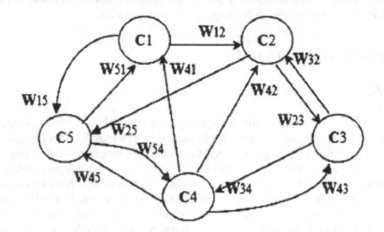

**Fig. 4** The fuzzy cognitive map model

The value $A_i$ of concept $C_i$ expresses the degree which corresponds to its physical value. At each simulation step, the value $A_i$ of a concept $C_i$ is calculated by computing the influence of the interconnected concepts $C_j$'s on the specific concept $C_i$ following the calculation rule:

$$A_i^{(k+1)} = f\left(A_i^{(k)} + \sum_{\substack{j \neq i \\ j=1}}^{N} A_j^{(k)} w_{ji}\right) \qquad (5.1)$$

where $A_i^{(k+1)}$ is the value of concept $C_i$ at simulation step $k + 1$, $A_i^{(k)}$ is the value of concept $C_j$ at simulation step $k$, $w_{ji}$ is the weight of the interconnection from concept $C_j$ to concept $C_i$ and $f$ is the sigmoid threshold function:

$$f = \frac{1}{1 + e^{-\lambda x}} \qquad (5.2)$$

where $\lambda > 0$ is a parameter determining its steepness. In this approach, the value $\lambda = 1$ has been used. This function is selected since the values $A_i$ of the concepts lie within [0, 1].

We briefly referred to FCMs as further information and details about Fuzzy Cognitive Maps and their theories are outlined analytically in [41].

# 6 A New DSS Methodology Using Decision Making Analysis and FCMs

## 6.1 A New Decision Making Support System (DMSS)

So far DSSs research has widened its focus to serve more and more different scientific disciplines. This can be seen from the many papers been published the last 10-15 years. Since the 2000s the DSS application in all scientific fields has exploded on a geometric way. The articles analyzed is DSS research published are numerous between 2000 and 2012 in 14 journals: Decision Sciences (DS); Decision Support Systems (DSS); European Journal of Information Systems (EJIS); Information and Management (I&M); Information and Organization (I&O), formerly Accounting, Management and Information Technologies; Information Systems Journal (ISJ); Information Systems Research (ISR); Journal of Information Technology (JIT); Journal of Management Information Systems (JMIS); Journal of Organizational Computing and Electronic Commerce (JOC&EC); Journal of Strategic Information Systems (JSIS); Group Decision and Negotiation (GD&N); Management Science (MS); and MIS Quarterly (MISQ).

It would take many pages to study, investigate and analyze all these papers in a systematic way. We should also mention that there are many other papers that are published in scientific journals of other fields, i.e. in Journals of the Medical field or the Manufacturing area or Fuzzy Systems. All these require a systematic research to analyze and formulate new generic methodologies for the challenging area of DSS.

Till now all different types or formulation of the "DSS" has a common denominator; **that decisions are made**. This generates the need to embed the "Decision Making" part as a necessary step in the overall integrated effort on taking decisions by humans been contronted by problems in studying and analyzing complex systems. In the present section we justify the need for the new terms of Decision Making Support Systems (DMSS) in which the experts play a major role. In addition on our DMSS approach the Fuzzy Cognitive Maps' theories will be utilized appropriately.

Today more than ever, modeling is rarely a one-shot process and good models, are usually refined and enhanced as their users gather practical experiences with the system recommendations. The generic approach of Decision Making Support System (DMSS) is shown in Fig. 5. Basic prerequisite for the smooth function of the proposed model is that we have a minimum number of experts ($N \geq 2$).

The new DMSS idea is an innovative approach because we define as a DMSS the relative box shown in Fig. 5 combining for the first time the following (in contrast with the conventional DSS):

a)  FCM model
b)  Experts ($N \geq 2$)
c)  Learning algorithms
d)  Decision Making Trees

The significant difference between the proposed generic DMSS and all other DSSs methodologies is that the four components a) to d) are present in any decision process. Please note that existing DSSs methodologies and tools are utilized. It is also very important to note that each one of the four components a) to d) has and plays a different role in the new generic DMSS. It is necessary to fully understand the potential of the FCMs models been combined with appropriate experts and utilizing learning techniques. In order to better understand the Decision Making process some theoretical remarks are provided next.

## 6.2  Decision Analysis and Fuzzy Cognitive Maps

Decision analysis is based on a number of quantitative methods that aid in choosing amongst alternatives. Traditional decision analysis is used to indicate decisions favoring good outcomes even though there is an uncertainty surrounding the decision itself. Furthermore, the value of each possible outcome of a decision, whether measured in costs and benefits or utility, usually varies [41].

Over the last years, several approaches have been investigated in the field of Decision Analysis, with the most popular one to be used that of Decision Trees (DT). Some methods combine DT with other machine learning techniques, such as Neural Networks [3] or Bayesian Networks [4]. However, very little work has been reported in combining a DT with FCMs. Some research work of this combination has been the literature the last ten years [5]-[8]. In this chapter the technique of combining a DT with an FCM model in Decision Analysis is presented.

The derived FCM model is subsequently trained using an unsupervised learning algorithm to achieve improved decision accuracy. In this chapter, the C4.5 has been chosen as a typical representative of the decision tree approach. Similarly, the Nonlinear Hebbian Learning (NHL) algorithm is chosen as a representative o unsupervised FCM training.

The generic approach of the DT-FCM's function is briefly outlined in Fig. 5. If there is a large number of input data, then the quantitative data are used to induce a Decision Tree and qualitative data (through experts' knowledge) are used to construct the FCM model. The FCM's flexibility is enriched by the fuzzification of the strict decision tests (derived fuzzy IF-THEN rules to assign weights direction and values). Finally, the derived FCM model (new weight setting and structure) is trained by the unsupervised NHL algorithm to achieve a decision [41].

This methodology can be used for three different circumstances, depending on the type of the initial input data: (1) when the initial data are quantitative, the DT generators are used and an inductive learning algorithm produce the fuzzy rules which then are used to update the FCM model construction; (2) when experts' knowledge is available, the FCM model is constructed and through unsupervised NHL algorithm is trained to calculate the target output concept responsible for the decision line; and (3) when both quantitative and qualitative data are available, the initial data are divided and each data type is used to construct the DTs and the FCMs separately. Then the fuzzy rules induced from the inductive learning restructure the FCM model enhancing it. At the enhanced FCM model the training algorithm is applied to help FCM model to reach a proper decision.

The new technique has three major advantages. First, the association rules derived from the decision trees have a simple and direct interpretation and introduced in the initial FCM model to update its operation and structure. For example, a produced rule can be: If the *variable 1* (input variable) has *feature A* Then the *variable 2* (output variable) has *feature B*.

Second, the procedure that introduces the Decision Tree rules into an FCM also specifies the weight assignment through the new cause-effect relationships among the FCM concepts. Third, as will be demonstrated through the experiments, this technique fares better than the best Decision Tree inductive learning technique and the FCM decision tool.

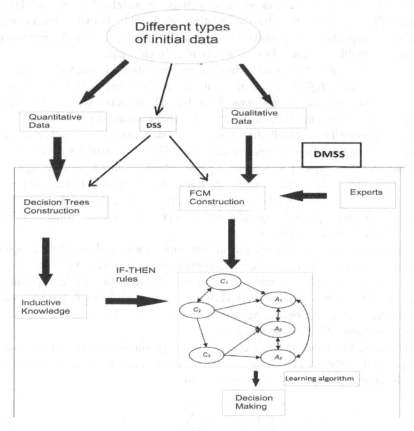

**Fig. 5** The proposed generic approach of Decision Making Support System

# 7 Case Studies

The method used to develop and construct a Decision Making Support System (DMSS) using FCMs has considerable importance in order to represent the policy decision procedure as accurately as possible. The methodology described here extracts the knowledge from the experts and exploits their experience of the process [9].

The appropriate experts, consisting in most cases of interdisciplinary teams, determine the number and kind of concepts that comprise the FCM models of the DMSS. Each expert from his/ her experience knows the main factors that contribute to the decision; each of these factors is represented by one concept of the FCM. The expert also understands potential influences and interactions between factors themselves or between factors and decisions, thus establishing the corresponding fuzzy degrees of causation between concepts. In this way, an expert's knowledge is transformed into a dynamic weighted graph, the DMSS using FCMs. Experts describe the existing relationship between the concepts firstly, as "negative" or "positive" and secondly, as a degree of influence using a linguistic variable, such as "low", "medium", "high", etc.

More specifically, the causal interrelationships among concepts are declared using the variable *Influence* which is interpreted as a linguistic variable taking values in the universe of discourse U = [-1, 1]. Its term set T(influence) is suggested to be comprised of eight variables. Using eight linguistic variables, an expert can describe in detail the influence of one concept on another and can discern between different degrees of influence. The nine variables used here are: T(influence) = {zero, very very low, very low, low, medium, high, very high, very very high, one}. The corresponding membership functions for these terms are shown in Fig. 6 and they are $\mu_z$, $\mu_{vvl}$, $\mu_{vl}$, $\mu_l$, $\mu_m$, $\mu_h$, $\mu_{vh}$, $\mu_{vvh}$ and $\mu_0$. A positive sign in front of the appropriate fuzzy value indicates positive causality while a negative sign indicates negative causality.

Once one expert describes each interconnection as above, then, all the proposed linguistic values for the same interconnection, suggested by experts, are aggregated using the SUM method and an overall linguistic weight is produced, which with the defuzzification method of center of area (COA), is transformed to a numerical weight $w_{ji}$, belonging to the interval [-1, 1]. A detailed description of the development of FCM model is given in [2].

Generally, the value of each concept at every simulation step is calculated, computing the influence of the interconnected concepts to the specific concept [9]-[10], by applying the following calculation rule:

$$A_i^{(k+1)} = f\left(k_2 A_i^{(k)} + k_1 \sum_{\substack{j \neq i \\ j=1}}^{N} A_j^{(k)} w_{ji}\right) \qquad (7.1)$$

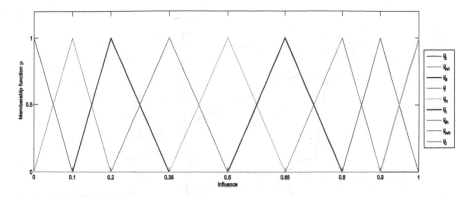

**Fig. 6** Membership functions of the linguistic variable *Influence*

where $A_i^{(k+1)}$ is the value of the concept $C_i$ at reputation step k+1, $A_i^{(k)}$ is the value of the concept $C_j$ at iteration step k, $w_{ij}$ is the weight of interconnection from concept $C_j$ to concept $C_i$ and $f$ is the sigmoid function. The $k_1$ expresses the influence of the interconnected concepts in the configuration of the new value of the concept $A_i$ and $k_2$ represents the proportion of the contribution of the previous value of the concept in the computation of the new value.

The sigmoid function f belongs to the family of squeezing functions, and usually the following function is used:

$$f = \frac{1}{1+e^{-\lambda x}} \qquad (7.2)$$

This is the unipolar sigmoid function, where $\lambda > 0$ determines the steepness of the continuous function $f(x)$. The following examples show how the FCMs lead to the proposed decision making approach strictly following the experts' knowledge.

## Example 7.1: Decision Making in Hybrid Renewable Energy System using FCMs

In this example it is considered that $k_1 = k_2 = 1$ , $\lambda = 1$ and an initial matrix $w^{initial} = [w^{ij}]$, i,j=1,...,N, with $w_{ii}=0$, i=1,...,N, is obtained.

In the current Decision Making Analysis model there are two decision concepts (outputs), i.e. the two renewable energy sources are studied: concept 4 PV-System and concept 5 Wind-Turbine-System. The factor concepts are considered as measurements (via special sensors) that determine how each RES will function in this model and they are:

- $C_1$ : insolation (kW$_p$/m$^2$)
- $C_2$ : temperature
- $C_3$ : wind
- $C_4$ : PV-System
- $C_5$ : Wind-Turbine-System

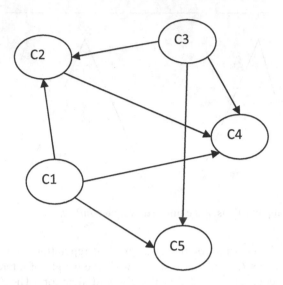

**Fig. 7** A conceptual model for Hybrid RES System

**Table 1** Weights between concepts for FCM for Hybrid RES System

|    | C1 | C2 | C3 | C4 | C5 |
|----|----|----|----|----|----|
| C1 | 0 | 0.15 | 0 | 0.82 | 0.1 |
| C2 | 0 | 0 | 0 | -0.24 | 0 |
| C3 | 0 | -0.15 | 0 | 0.26 | 0.76 |
| C4 | 0 | 0 | 0 | 0 | 0 |
| C5 | 0 | 0 | 0 | 0 | 0 |

In order to show how the crisp values of the weights created in Table_1, we are going to give a specific example for the calculation of the crisp value of a single weight describing the correlation between node C2 (temperature) and node C4 (PV-Systems' performance). Preferences of three experts on how they define this correlation follow:

**1st expert:**
If a small change happens in node $C_2$ then a very very low change is caused in node $C_4$
    Infer: Influence from concept $C_2$ to $C_4$ is negatively very very low

**2nd expert:**
If a small change happens in node $C_2$ then a very low change is caused in node $C_4$
    Infer: Influence from concept $C_2$ to $C_4$ is negatively very low

## 3$^{rd}$ expert:

If a small change happens in node $C_2$ then a low change is caused in node $C_4$

 Infer: Influence from concept $C_2$ to $C_4$ is negatively low

Fig 8 shows the three linguistic variables which are being proposed:

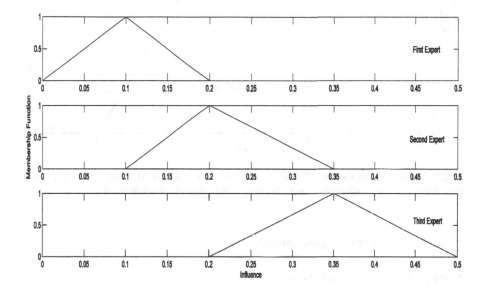

**Fig. 8** Example of the three linguistic variables proposed by three experts to describe the correlation among two concepts

These linguistic variables (very very low, very low, low) are aggregated and a total linguistic weight is produced, which is transformed into a crisp value w24=-0.2389 after the CoA defuzzyfication method (Fig. 9).

The same procedure was followed for the determination of the rest weights of the FCM model. A weight matrix $w^{initial}$ which contains the initial proposed weights of all interconnections among the concepts of the FCM model is shown in Table_1.

Detailed information for hybrid renewable energy systems are given in [11]-[14]. One case study from the literature is examined here concerning the decision making approach of hybrid renewable energy source system. In Table 2 the initial factors used by the model are presented. In addition, the degree of occurrence of each factor is denoted with qualitative degrees of very very high, very high, high, medium, low, very low, and 0 for insolation (C1), low, medium, high and very high for temperature (C2) and for wind (C3). Respectively for the output concepts C4, C5 the qualitative degrees are low, medium and high. C4 and C5 are considered a % percentage of the maximum performance at STC conditions.

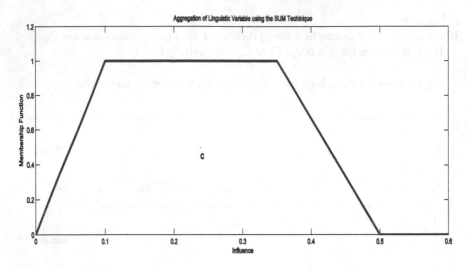

**Fig. 9** Aggregation of the three linguistic variables using the SUM method. The C point is the crisp value of the relative weight after the *CoA* defuzzyfication method

## Case_1 (without training algorithm)

**Table 2** Initial factor-concepts fuzzy value

| Factor-concepts | Case 1 |
|---|---|
| C1 | VVH |
| C2 | M |
| C3 | M |

The initial values of the outputs were set equal to zero.

**Table 3** Final decision-concepts

| Decision-concepts | Case 1 |
|---|---|
| C4 (PV-System) | **0.7937** |
| C5              (Wind-Turbine-System) | **0.7963** |

The iterative procedure is being terminated when the values of $C_i$ concepts has no difference between the latest two iterations. Considering $\lambda=1$ for the unipolar sigmoid function and after $N=10$ iteration steps the system reaches an equilibrium point.

We considered initial values for the concepts after COA defuzzyfication method [31]-[34]:

$$A^{(0)} = [0.90\ 0.3334\ 0.3334\ 0.839\ 0.459]$$

The fuzzy rule considered for the calculation of the initial conditions of the output concepts C4, C5 follows:

- *If* C1 is VVH *and* C2 is M *and* C3 is M   *Then* C4 is VVH *and* C5 is M;

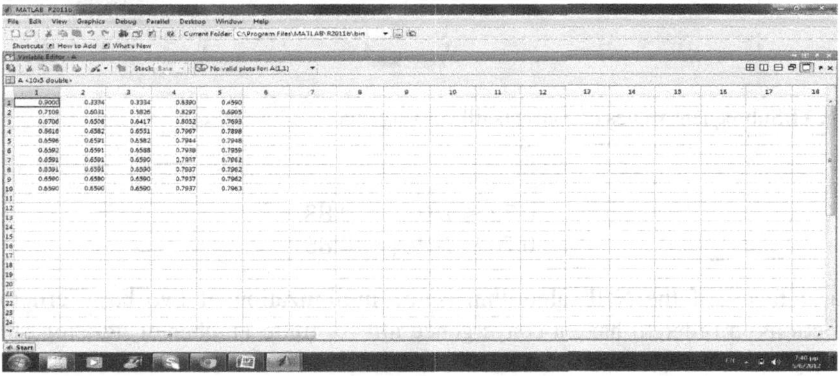

**Fig. 10** Value of concepts for each iteration step

It is observed that in the latest three iterations there is no difference between the values of concepts $C_i$. So after 10 iteration steps, the FCM reaches an equilibrium point where the values do not change any more from their previous ones, that is:

$$A^{(10)} = [0.6590\ 0.6590\ 0.6590\ \mathbf{0.7937}\ \mathbf{0.7963}]$$

Finally it is observed that the PV-System ($C_4$) and the Wind-Turbine-System ($C_5$) function under the 79.37% and 79.63% of their optimum performance in STC conditions respectively.

**Case_2 (with Nonlinear Hebbian Training algorithm)**

Firstly the experts suggested us a desired region where the decision output concepts (DOCs) should move. The desired regions for the output nodes reflect the prospered operation of the modeled system.

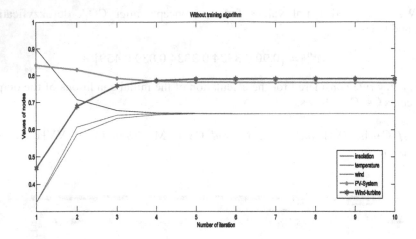

**Fig. 11** Subsequent values of concepts till convergence

$$0.72 \le DOC_4 \le 0.83$$
$$0.70 \le DOC_5 \le 0.85$$

Basic factor of the NHL algorithm is the minimization of two basic criterion functions in order to have a convergence after a finite number of iteration steps [16]-[17]:

$$F1 = J1 = \|DOC_i - T_i\|_2^2 < 0.005$$

and

$$F2 = J2 = \left|DOC_i^{(k)} - DOC_i^{(k-1)}\right| < 0.005$$

where $T_i$ is the hypothetic desired value of the output and it is usually the average of the defined range by the experts. Basic idea of the NHL algorithm is the adaption of the initial ($w_{ij}^{initial}$) matrix weights defined by the experts in a way that the Decision Output Concepts converge inside the desired region:

$$w_{ij}^{(k)} = \gamma * w_{ij}^{(k-1)} + \eta * A_j^{(k-1)}\left(A_i^{(k-1)} - sgn\left(w_{ij}\right) * w_{ij}^{(k-1)} * A_j^{(k-1)}\right) \quad (7.3)$$

Each non-zero element of the final weight matrix $w_{ij}^{final}$ has been improved and has been converged into an optimal value according to the specific criteria functions of the problem. We could define an acceptable range of change for these weights. If one weight takes a value out of the desired regions then we should consider it better whether experts' suggestion of the initial value of the weight is correct or not, i.e.:

if $\left| w_{ij}^{final} - w_{ij}^{initial} \right| > l$ then that means that concept $C_i$ has a different relationship (than the initial defined from the experts one) with concept $C_j$ and their correlation should be reevaluated.

The learning parameters $\gamma$, $\eta$ of the above equation are very important and they usually take values between $\gamma \ni [0.9, 1]$ and $\ni [0, 0.1]$.

Defining the initial values of the concepts:

$$A^{initial} = [0.90\ 0.3334\ 0.3334\ 0.839\ 0.459]$$

we take the following after 7 iterations:

$$A^{final} = [0.6592\ 0.6761\ 0.6588\ \mathbf{0.8003\ 0.7991}]$$

It is observed that the values of the concepts $C_4$, $C_5$ in the final state are inside the suggested desired regions.

## Example 7.2: Decision Making for the Stability of an Enterprise in a Crisis Period using FCMs

In this example it is considered that k1=k2=1, $\lambda$=1 and an initial matrix winitial=[wij], i,j=1,...,N, with wii=0, i=1,...,N, is obtained

In the current DMA model there is one decision concept (output), i.e. the stability of an enterprise in a crisis period is studied: concept_8. The factor concepts are considered as measurements (via special statistic rescarch) that determine how each measurement-concept will function in this model and they are:

- $C_1$ : sales
- $C_2$ : turnover
- $C_3$ : expenditures
- $C_4$ : debts & loans
- $C_5$ : research & innovation
- $C_6$ : investments
- $C_7$ : market share
- $C_9$ : present capital

- **$C_8$ : stability of enterprise (output of the system)**

At this point it should be noted that in economic systems we can't talk about causality but only for correlation between the defined factor-concepts of this problem. Experts noted that the acceptable-desired region for the final value of concept C8 is:

$$0.70 \leq C_8^{(final)} \leq 0.95$$

If $C_8^{(final)}$ is inside this region then we can say with great certainty that the enterprise is out of danger and the economic crisis period does not put at risk the stability and the smooth function of the enterprise.

Weights in table_4 are determined after defuzzifying (with COA method) the fuzzy values that were given from the experts (mostly economists) [18]-[24].

**Table 4** Weights between concepts for CFCM for Hybrid RES System

|     | C1  | C2  | C3  | C4   | C5   | C6   | C7  | C8   | C9   |
|-----|-----|-----|-----|------|------|------|-----|------|------|
| C1  | 0   | 0.6 | 0   | -0.4 | 0.2  | 0.3  | 0.6 | 0.8  | 0    |
| C2  | 0   | 0   | 0   | -0.2 | 0.2  | 0.5  | 0.1 | 0.3  | 0    |
| C3  | 0   | 0   | 0   | 0.4  | -0.5 | -0.4 | 0   | -0.6 | -0.5 |
| C4  | 0   | 0   | -0.4 | 0   | -0.7 | -0.8 | 0   | -0.7 | -0.4 |
| C5  | 0.2 | 0.3 | 0   | 0    | 0    | 0.5  | 0.3 | 0.2  | -0.2 |
| C6  | 0.3 | 0.2 | 0.6 | 0.5  | -0.3 | 0    | 0.3 | 0.3  | -0.4 |
| C7  | 0.4 | 0.3 | 0   | -0.2 | 0    | 0    | 0   | 0.4  | 0.5  |
| C8  | 0   | 0   | 0   | 0    | 0    | 0    | 0   | 0    | 0    |
| C9  | 0   | 0   | 0   | -0.3 | 0.2  | 0.4  | 0   | 0.2  | 0    |

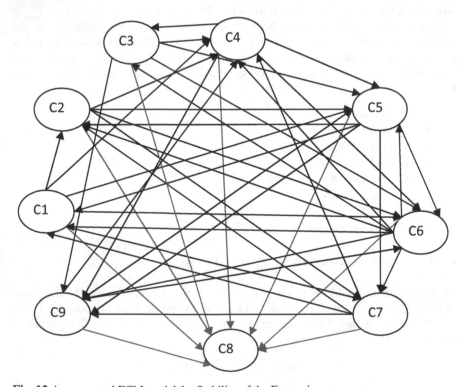

**Fig. 12** A conceptual FCM model for Stability of the Enterprise

In addition, the degree of occurrence of each input-concept factor is denoted with qualitative degrees of *high, medium,* and *low*. Respectively for the output concept C8 the qualitative degrees are *very low, low, medium, high* and *very high*.

**Table 5** Initial factor-concepts fuzzy value

| Factor-concepts | Case 1 |
|---|---|
| C1 | H |
| C2 | M |
| C3 | L |
| C4 | L |
| C5 | M |
| C6 | L |
| C7 | L |
| C9 | M |

The initial values of the outputs were set equal to zero.

**Table 6** Final decision-concepts

| Decision-concepts | Case 1 |
|---|---|
| C8 (Stability of the Enterprise) | **0.8391** |

The iterative procedure is being terminated when the values of $C_i$ concepts has no difference between the latest three iterations. Considering $\lambda=1$ for the unipolar sigmoid function and after 11 iteration steps the FCM reaches an equilibrium point.

We considered initial values for the concepts:

$$A^{(0)} = [0.8867\ 0.4667\ 0.0967\ 0.0967\ 0.4667\ 0.0967\ 0.0967\ 0.65\ 0.4667]$$

The fuzzy rule considered for the calculation of the initial condition of the output concept C8 follows:

- *If* C1 is H *and* C2 is M *and* C3 is L *and* C4 is L *and* C5 is M *and* C6 is and C7 is L *and* C9 is M *Then* C8 is VVH;

It is observed that in the latest three iterations there is no difference between the values of concepts $C_i$ . So after 11 iteration steps, the FCM reaches an equilibrium point where the values do not change any more from their previous ones, that is:

$$A^{(11)} = [0.8140\ 0.8708\ 0.7145\ 0.6121\ 0.4743\ 0.7462\ 0.8581\ \mathbf{0.8391}\ 0.4779]$$

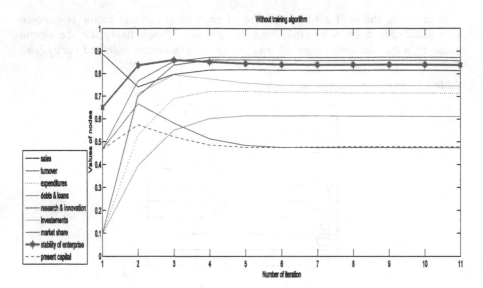

**Fig. 13** Subsequent values of concepts till convergence

Since the final value of $C_8^{(final)}$ is inside the acceptable region, defined by the experts, then we could assume with great certainty that the enterprise can survive the crisis period.

## 8 The Five Steps Approach to Success (5-SAS)

In order to achieve an optimal Decision Making Support System (DMSS) a new five steps approach is proposed:

1) Determine the final objective. The final objective should be realistic. Be optimistic and self-confident. You can reach no goal if you don't believe it. "**Think Different**" was an advertising slogan for Apple Computer in 1997. We borrowed the last words of this slogan and advise everyone with the following: " ...*while some see them as the crazy ones, we see genius. Because the people who are crazy enough to think they can change the world, are the ones who do.*"

2) Justify the reasons you want to reach your goal. If you want your reasons to meet the objective you have to choose realistic and reasonable ones. So *if* your reasons are reasonable *then* continue the procedure *else* redefine the final objective.

3) Define the initial conditions of the system. Specify analytically the present state of the system and possible reasons for justifying it. *If* the process is stochastic *then* continue the procedure *else* follow conventional methods.

4) Perform a systematic mathematical approach to solve the well defined problem. Then identify all possible/available methods and solutions. *If* there is at least one available solution *then* continue the procedure *else* try to investigate other solutions. *If* there is no method-solution *then* stop for the moment and start a research effort to generate a new method-solution. This might need alliance and/or close collaboration with other scientists.

5) Decide the optimal solution regarding specific criteria (cost-functions) according to the problem, i.e.: a) Realistic , b) Cost-effective, c) Executable, d) Reasonable and e)Time-effective.

Now applying the Five-Steps Approach to Success (5-SAS) into the DMSS, exploiting experts' knowledge and FCMs, will help to decide an acceptable solution.

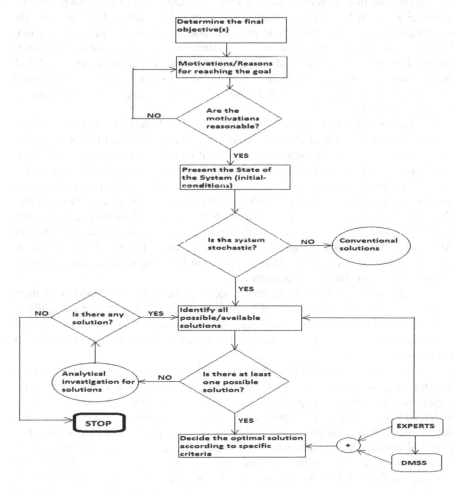

**Fig. 14** Flowchart of the Five Step Approach

This Five Step Approach (5-SAS) can and should be used when implementing DMSS methodologies. Especially this 5-SAS will be very useful when specific problems of complex systems (health, energy, agriculture, finance, environment, etc.) are considered, analyzed and investigated.

# 9 Future Research

By studying and analyzing most of DSS's research papers over the last 10 years as well as after studying the previous sections of this chapter a number of very interesting and challenging topics can be identified for further research.

There are a number of excellent theoretical and applied scholarly "results" on the broad concept of DSS. However there is no integral and unified theory behind all these techniques. Moreover, there is a sincere skepticism of even the possibility of developing such a theory. Many practitioners of human DMSSs note the significant gap between theoretical research and application. The proposed new DMSS in this chapter could be used as a starting point searching for such a unified theory. It is very interesting and promising the fact that in the proposed DMSS using experts, learning techniques and FCM's theory a new mathematical formulation is obtained and further research could enhance the capabilities of human decision makers.

It is known that, using the AHL algorithm, the FCM model is improved and the weights W are determined using a number of experts, so that the a new weight matrix can be used for the same initial values of concepts, but arriving on different final values. This gives new perspectives in searching for a more efficient and active DMSSs. The link of the AHL algorithms and their positive contributions to further, explore the potential of DMSS in solving problems of complex systems. This direction can be pursued in searching for new Evolutionary Computation Learning Algorithms for specific applications. It should be emphasized that the AHL is problem- dependent and although it starts using the initial weight matrix, but throughout the process of DM is independent from the initial conditions of the system. This needs further investigation

The new proposed DMSS algorithm needs to be fully developed on a more generic software platform. Then with appropriate changes it should be available for daily use when real-time data from a given application are provided. Special attention must be paid as to how the experts are selected and used on a real DMSS application.

The Five Steps Approach to Success (S-SAS) needs further mathematical development so it can be used on many real-life applications. Again the importance of using N-experts in the proposed methodology should not be underestimated. The most significant weakness of the new DMSS is their dependence on the experts beliefs been used to construct the FCMs and their potential convergence to undesired or unrealistic states of the complex system. This however can be further investigated by introduction new unsupervised learning methods for FCM training which then in return improves further the credibility and reliability of the new proposed DMSS methodology including the 5-SAS.

Another area for future research is to explore the current trends in mobile computing. Mobile computing is a facilitator that provides the means for the user to interact with existing systems, regardless of the location of either the user or the system. Mobile devices provide a new platform for DMSSs that challenges traditional approaches to DSSs. The size, speed, and reach of "scientific data" combined with continuously available support introduce a substantial technological advantage to Decision Making (DM). Mobile devices capture "scientific data" and allow for real-time monitoring or updating of data from the field, which, in turn, can be fed back into the decision loop. Mobile computing has complexities that will require some theoretical foundational work, however although the technology exists, connect users to resources can be difficult. Thus the need to explore both the technology and the use of the technology in real DMSSs. A new system is needed to coordinate collaborative intelligent systems using FCMs designed specifically for mobile applications.

## 10 Summary and Closing Remarks

Decision Support Systems (DSS) are powerful tools integrating methods from different scientific fields for supporting difficult decisions been made when problems for complex systems are investigated. DSS are gaining an increased popularity in many domains. More and more people have the need to use DSS software tools in their everyday life. They are especially valuable in situations in which the amount of available "information" or "scientific data" is prohibitive for the human mind to reach an "optimal" and/or "acceptable" decision. However in this process "precision" and "optimality" are of great importance to the decision maker.

In this chapter the historical commentary, starting from the Delphi Oracle to present DSS is less prescriptive than other works on other scientific fields. However it highlights the plethora of DSS theoretical research and their application to various scientific fields for the last 50 years or so. Nevertheless in this chapter a critical overview of the theoretical, research and application results been published in the broad field of DSS has been provided. It is very interesting to follow the development of the field of DSS, which in less than 20 (last) years has shown a great research interest, in order to embed to the whole procedure the philosophy of the "Decision Making" and not just the "Decision Support". Expansion of Internet and Communications is a promising environment for developing a Decision Making Support System (DMSS) tool/software for professionals and researchers in many scientific fields.

This chapter revealed the need of Decision Making (DM) and not only Decision Support Systems (DSS). The conventional DSS rarely contained the "Decision Making" part as an integrated step in the overall decision making loop. One of the major obstacles of the effectiveness of this process is uncertainty. Combination and collaboration of Fuzzy Cognitive Maps (FCMs), experts, data base, learning algorithms and the decision making process leaded us to the proposed Decision Making Support System (DMSS). This could provide a new step towards the minimization of the uncertainty.

Rapid growth of Internet and other networking technologies such as broadband WAN's, LAN's and WIP provide full and easy accessibility in data-base of many scientific fields. This way we don't just trust experts' knowledge but we can also compare it with past data-base and historical facts in order to make the experts selection procedure more reliable and interactive.

Given problems of complex systems in the presence of nonlinearities, uncertainties, impression or complexity can now be investigated in a new promising way through the proposed DMSS. Future direction of this research could be the development of a DMSS-FCM tool for daily use under real-time data. In this way the new FCM model will become more dynamic and flexible. The included nodes/concepts will tend to behave more and more like neurons of a real nervous system. This will further improve Decision Making of humans taking into consideration the various human factors of different experts but with one main objective: to improve the human performance.

However, there is an important difference: single man's nervous system provides only decisions in limited scientific fields and not always an acceptable and implementable one. The proposed DMSS-FCM model will provide decisions in any scientific field with certainties of acceptability and feasibility.

This chapter, therefore, can be useful to a broad spectrum of professionals, researchers and students in many scientific fields. Moreover it can be a starting point for a more interdisciplinary research work on all these different fields.

# References

[1] Stylios, C.D., Groumpos, P.P., Georgopoulos, V.C.: An fuzzy cognitive maps approach to process control systems. Journal of Advanced Computational Intelligence 3, 409–417 (1999)

[2] Stylios, C.D., Groumpos, P.P.: Modeling complex systems using fuzzy cognitive maps. IEEE Transactions on Systems, Man and Cybernetics, Part A: Systems and Humans 34, 155–162 (2004)

[3] D'alche-Buc, F., Zwierski, D., Nadal, J.: Trio learning: a new strategy for building hybrid neural trees. Neural Syst. 5(4), 255–274 (1994)

[4] Janssens, D., Wets, G., Brijs, T., Vanhoof, K., Arentze, T., Timmersmans, H.: Intergrating Bayesian networks and decision trees in a sequential rule-based transportation model. Europ. J. Operat. Research (2005)

[5] Podgorelec, V., Kokol, P., Tiglic, S.B., Rozman, I.: Decision Trees: An overview and their Use in Medicine. Journal of Medical Systems 5 (October 2002)

[6] Papageorgiou, E., Stylios, C., Groumpos, P.P.: An Intergrated Two-Level Hierarchical Decision Making System based on Fuzzy Cognitive Maps (FCMs). IEEE Trans. Biomed. Engin. 50(12), 1326–1339 (2003)

[7] Papageorgiou, E.I., Groumpos, P.P.: A weight adaption method for fine-tuning Fuzzy Cognitive Map casual links. Soft Computing Journal 9, 846–857 (2005)

[8] Papageorgiou, E., Stylios, C., Groumpos, P.P.: Unsupervised learning techniques for fine-tuning Fuzzy Cognitive Map casual links. Intern. Journal of Human-Computer Studies 64, 727–743 (2006)

[9] Groumpos, P.P., Stylios, C.D.: Modeling supervisory control systems using fuzzy cognitive maps. Chaos Solit Fract. 11(3), 329–336 (2000)

[10] Papageorgiou, E.I., Stylios, C.D., Groumpos, P.P.: Active Hebbian learning algorithm to train fuzzy cognitive maps. International Journal of Approximate Reasoning 37, 219–249 (2004)

[11] Hamrouni, N., Jraidi, M., Cherif, A.: Solar radiation and ambient temperature effects on the performances of a PV pumping system. Revue des Energies Renouvelables 11(1), 95–106 (2008)

[12] Anderson, P.M., Bose, A.: Stability Simulation of Wind Turbine Systems. IEEE Trans. on Power Apparatus and Systems PAS-102(12), 3791–3795 (1983)

[13] Nehrir, M.H., Lameres, B.J., Venkataramanan, G., Gerez, V., Alvarado, L.A.: An approach to evaluate the general performance of stand-alone wind/photovoltaic generating systems 15(4), 433–439 (2000)

[14] Yang, H., Zhoo, W., Lu, L., Fang, Z.: Optimal sizing method for stand-alone hybrid solar-wind system with LPSP technology by using genetic algorithm 82(4), 354–367 (2007)

[15] Kosko, B.: Fuzzy cognitive maps. International Journal of Man–Machine Studies 24, 65–75 (1986)

[16] Hyvarinen, A., Oja, E.: Independent component analysis by general nonlinear Hebbian-like learning rules. Signal Processing 64(3), 301–313 (1998)

[17] Karhunen, J., Joutsensalo, J.: Nonlinear Hebbian algorithm for sinusoidal frequency estimation. In: Aleksander, J., Taylor, J.G. (eds.) Artificial Neural Networks, vol. 2, pp. 1099–1102. North-Holland, Amsterdam (1992)

[18] Harris, A.J.: Lifeline: Call Centers and Crisis Management. Risk Management 55(5), 42–55 (2008)

[19] Moynihan, D.: Learning under uncertainty: Networks in Crisis Management. Public Administration Review 68(2), 350–365 (2008)

[20] Bolloju, N., Khalifa, M., Turban, E.: Intergrating knowledge management into enterprise environments for the next generation decision support. Decision Support Systems: Directions for the Next Decade 33(2), 163–176 (2002)

[21] Lam, W.: Investigating success factors in enterprise application: A case driven analysis. European Journal of Information Systems 14, 175–187 (2005)

[22] Al-Mashari, M., Al-Mudimigh, A., Zairi, M.: Enterprise resourse planning: A taxonomy of critical factors. European Journal of Operational Research 146(2), 352–364 (2003)

[23] Caves, R.: Multinational Enterprise and Economic Analysis, 3rd edn. (2007)

[24] Zhiyun, L.: Banking Structure and the Small-Medium-Sized Enterprise Financing. Economic Research Journal (2002)

[25] Mamdani, E., Gaines, B.: Fuzzy Reasoning and its applications. Publ. Academic Press, London (1981)

[26] Mamdani, E.H.: Application on Fuzzy Logic to approximate reasoning using linguistic synthesis. IEEE Trans. on computers C26. Dic., 1182–1191 (1977)

[27] Wang, L.: Fuzzy systems are universal approximators. In: Proc. of Int. Conf. on Fuzzy Engineering, pp. 471–496 (1992)

[28] McNeil, D., Freiberger, P.: Fuzzy Logic: The Revolutionary Computer Technology That is Changing Our World (April 1994)

[29] Zadeh, L.A.: Making computer think like people. IEEE Spectrum 21, 26–32 (1984)

[30] Haack, S.: Do we need fuzzy logic? Int. Jrnl. of Man-Mach. Stud. 11, 437–445 (1979)

[31] Runkler, T.A., Glesner, M.: Defuzzification and ranking in the context of membership value semantics, rule modality, and measurements theory. In: European Congress on Fuzzy and Intelligent Technologies, Aachen (September 1994)

[32] Pfluger, N., Yen, J., Langari, R.: A defuzzyfication strategy for a fuzzy logic controller employing prohibitive information in command formulation. In: Proceedings of IEEE International Conference on Fuzzy Systems, San Diego, pp. 717–723 (1992)

[33] Runkler, T.A.: Selection of appropriate defuzzyfication methods using application specific properties. IEEE Transaction on Fuzzy Systems 5(1), 72–79 (1997)

[34] Saade, J.J.: A unifying approach to defuzzification and comparison of the outputs of fuzzy controller. IEEE Trans. Fuzzy Syst. 4, 227–237 (1996)

[35] Jang, J.-S.R.: Fuzzy Inference Systems. ch. 4, 73–91 (1997)

[36] Mamdani, E.H., Assilian, S.: An experiment in linguistic synthesis with a fuzzy logic controller. Intenational Journal of Man-Machine Studies 7(1), 1–13 (1975)

[37] Albus, J.S.: Outline for a Theory of Intelligence. IEEE Transactions on Systems, Man and Cybernetics 21(3), 432–509 (1991)

[38] Saridis, G.N., Valavanis, K.P.: Analytical Design of Intelligent Machines. Automatica 24(2), 123–133 (1988)

[39] White, D.A., Sofge, D.A. (eds.): Handbook of Intelligent Control Neural, Fuzzy, and Adaptive Approaches. Van Nostrand Reinhold, New York (1992)

[40] Gupta, M.M., Sinha, N.K. (eds.): Intelligent Control: Theory and Practice. IEEE Press, Piscataway (1994)

[41] Glykas, M.: Fuzzy Cognitive Maps: Advances in Theory, Methodologies, Tools and Applications. Springer, Heidelberg (2010)

[42] Keen, P.G., Morton, S., Michael, S.: Decision Support Systems: An organizational perspective. Addison-Wesley Pub. Co., Reading (1978)

[43] Keen, P.G.: Decision Support Systems: the next decade. Decision Support Systems 3(3), 253–265 (1987)

[44] Keen, P.G.W., Scott Morton, M.S.: Decision support systems: an organizational perspective. Addison-Wesley (1978)

[45] Ackoff, R.L.: Management Misinformation Systems. Management Science 14(4), 147–157 (1967)

[46] Mintzberg, H.: Making Management Information Useful. Management Review 64(5), 34–38 (1975)

[47] Gorry, G.A., Scott-Morton, M.S.: A framework for management information systems. Sloan Management Review 13(1), 50–70 (1971)

[48] Sprague, R.H., Watson, H.J.: Bit by Bit: Toward Decision Support Systems. California Management Review 22(1), 60–68 (1979)

[49] Gerrity Jr, T.P.: Design of Man-Machine Decision Systems: An Application to Portfolio Management. Sloan Management Review 12(2), 59–75 (1971)

[50] Sprague, J. R.H., Carlson, E.D.: Building effective decision support systems. Prentice-Hall, Inc., Englewood Cliffs (1982)

[51] McCosh, A.M., Scott Morton, M.S.: Management decision support systems. Wiley, New York (1978)

[52] Scott Morton, M.S.: Management Decision Systems. Harvard Business School Press, Boston (1971)

[53] Alter, S.L.: Decision support systems: Current practice and continuing challenges. Addison-Wesley, Phillipines (1980)

[54] Finlay, P.: Introducing Decision Support Systems, 2nd edn. NCC/Blackwell, Malden (1994)

[55] Turban, E.: Decision Support Systems and Expert Systems. Prentice Hall (1995)

[56] Schroff, A.: An Approach to User Oriented Decision Support Systems, Inaugural-Dissertation Nr. 1208, Druckerei Horn, Bruchsal (1998)

[57] Power, D.J.: "What is a DSS?" originally published in DSStar. The On-Line Executive Journal for Data Intensive Decision Support 1(3) (1997)

[58] Arnott, D.: Decision support systems evolution: Framework, case study and research agenda. European Journal of Information Systems 13(4), 247–259 (2004)

[59] Bidgoli, H.: Intelligent Management Support Systems, Greenwood, Westport CT (1998)

[60] Doukidis, G.I., Land, F., Miller, G.: Knowledge Based Management Support Systems. Ellis Horwood, Chichester (1989)

[61] Turban, E., Aronson, J.E., Liang, T.-P.: Decision Support Systems and Intelligent Systems, 7th edn. Pearson Education, Upper Saddle River (2005)

[62] Fitzgerald, G.: Executive information systems and their development in the U.K.: A research study. International Information Systems 1(2), 1–35 (1992)

[63] Rockart, J.F.: Chief executives define their own data needs. Harvard Business Review 57, 81–93 (1979)

[64] Kaplan, R.S., Norton, D.P.: The Balanced Scorecard: Translating Strategy into Action. Harvard Business School Press, Cambridge (1996)

[65] Inmon, W., Hackathorn, R.: Using the Data Warehouse. John Wiley and Sons, New York (1994)

[66] Kimball, R., Reeves, L., Ross, M., Thornwaite, W.: The Data Warehousing Lifecycle Toolkit. John Wiley and Sons, Chichester (1998)

[67] Kimball, R.: The Data Warehousing Toolkit. John Wiley and Sons, Chichester (1996)

[68] Keen, P.G.W.: Let's focus on action not information: Information is a misleading and damaging IS term. Computerworld 31(46) (1997)

[69] Rockart, J.F., DeLong, D.W.: Executive Support Systems: The Emergence of Top Management Computer Use. Dow Jones-Irwin, Illinois (1988)

[70] Codd, E.F., Codd, S.B., Salley, C.T.: Providing on-line analytical processing (OLAP) touser-analysts: An IT mandate. E.F. Codd and Associates (1993) (unpublished manuscript)

[71] Simon, H.: Administrative Behavior. Free Press, Glencoe

[72] Meador, C.L., Ness, D.N.: Decision support systems: An approach to corporate planning. Sloan Management Review 15(2), 51–68 (1974)

[73] Ness, D.N.: Decision Support Systems: Theories of Design. Presented at the Wharton Office of Naval Research Conference on Decision Support Systems, Philadelphia, Pennsylvania, November 4-7 (1975)

[74] Courbon, J.C., Grajew, J., Tolovi, J.: Design and Implementation of Interactive Decision Support Systems: An Evolutive Approach, Technical Report, Institute d'Administration des Enterprises, Grenoble, France (1978)

[75] Keen, P.G.W.: Adaptive Design for DSS. Database 12(1-2), 15–25 (1980)

[76] Silver, M.S.: Systems that Support Decision Makers: Description and Analysis. John Wiley and Sons, New York (1991)

# Fuzzy Cognitive Strategic Maps in Business Process Performance Measurement

Michael Glykas[1,2,3]

[1] Faculty of Business and Management, University of Wallangong, Dubai
[2] University of Aegean, Department of Financial and Management Engineering,
   31, Fostini Street, Chios, 82 100, Greece
   mglikas@aegean.gr
[3] Aegean Technopolis, The Technology Park of the Aegean Region, Chios, Greece

**Abstract.** This paper describes a methodology for the development of a Proactive Balanced Scorecard (PBSCM). The Balanced Scorecard is one of the most popular approaches developed in the field of performance measurement. However, in spite of its reputation, there are issues that require further research. The present research addresses the problems of the Balanced Scorecard by utilizing the soft computing characteristics of Fuzzy Cognitive Maps (FCMs). By using FCMs, the proposed methodology generates a dynamic network of interconnected Key Performance Indicators (KPI), simulates each KPI with imprecise relationships and quantifies the impact of each KPI to other KPIs in order to adjust targets of performance.

**Keywords:** Performance Measurement, Balanced Scorecard, Fuzzy Cognitive Maps, Simulation.

## 1 Introduction

Today, companies are evolving in turbulent and equivocal environments (Drucker 1993; Kelly 1998; Grove 1999). This requires companies to be alert and watchful so as to detect weaknesses (Ansoff 1975) and discontinuities in regard to emerging threats and opportunities and to initiate further probing based on such detections (Walls, Widmeyer et al. 1992). The strategic role of performance measurement systems has been widely stressed in management literature. These systems provide managers with useful tools to understand how well their organisation is performing and to assist them in deciding what they should do next (Neely 1998; Waggoner, Neely et al. 1999).

Performance measurement systems have grown in use and popularity over the last twenty years. Organisations adopted performance measurement systems for a variety of reasons, but mainly to achieve control over the organisation in ways that traditional accounting systems do not permit (Kellen 2003). A review of the literature shows that traditional performance measurement systems (based on

M. Glykas (Ed.): Business Process Management, SCI 444, pp. 339–364.
springerlink.com

financial measures) have failed to identify and integrate all those factors that are critical in contributing to business excellence (Kaplan 1983; Kaplan 1984; Hayes 1988; Eccles 1991; Fisher 1992; Maskell 1992).

During the last decade, a number of frameworks, that help in designing and implementing performance measurement systems, has been identified in the literature, such as the Balanced Scorecard(Kaplan and Norton 1992), the Performance Prism (Kennerley and Neely 2000), the Performance Measurement Matrix (Keegan 1989), the Results and Determinants framework(Fitzgerald, Johnston et al. 1991), and the SMART pyramid (Lynch and Cross 1991). These frameworks aim to assist organisations in defining a set of measures that reflects their objectives and assesses their performance appropriately. The frameworks are multidimensional, explicitly balancing financial and non-financial measures (Kennerley and Neely 2002). Furthermore, a number of researchers have proposed a wide range of criteria for designing performance measurement systems (Globerson 1985; Maskell 1992; Morris 2002).

Despite, the existence of numerous approaches (frameworks, criteria, etc.) it is evident, from the literature, that the need for a broader research in the field of performance measurement is required. The criticism about the *static* nature of performance measurement systems as well as the *relationships* and *trade-offs* that exist among different measures is the catalyst for this research. Furthermore, the software applications that have been developed so far lack of an analytic capability and they can't do *predictive modelling* (Morris 2002). Despite the many attempts in this area (EIS, Decision Support tools), it is claiming that these tools do not necessarily advance the decision-making process.

The main objective of this research is to propose a methodology (not a new performance measurement framework) that will support existing measurement framework(s) during the process of performance measurement systems' design, implementation and use, and to advance the decision-making process. Conforming to the most favoured approach, we have adopted the Balanced Scorecard, to explore the existence of *trade-offs* among measures within the *dynamic* nature of performance measurement systems that provide *predictive modelling* capabilities. The use of FCMs in the development of a Balanced Scorecard, will allow prospective decision-makers to incorporate their insights into the model. They may select the most preferable measures, add new ones, test the relationships between them, and visualise holistic outcomes.

This paper consists of 5 sections. Section 2 provides a literature review and research background; section 3 presents the proposed methodology. Section 4 discusses the applicability of the proposed methodology. Finally, section 5 concludes this paper.

## 2 Literature Review

(Senge 1992) argues that, in today's complex business world, organisations must be able to learn how to cope with continuous change in order to be successful. In this changing environment, the need for adequate design, implementation and use of performance measurement systems, is greater than ever. (Eccles 1991) claims that it will become increasingly necessary for all major businesses to evaluate and modify their performance measures in order to adapt to the rapidly changing and highly competitive business environment.

The introduction of a performance measurement system is based on a three-stage process: design, implementation and use. Failing to implement any of these stages will result into a non-robust performance measurement system. When attempting to improve organisational performance by utilising performance measurement systems a critical point is the selection of appropriate measures. Anticipating this, several approaches have been introduced (frameworks, criteria, etc.). However, in spite the availability of such approaches, the need to further research the area of performance measurement is necessary.

Several authors have recognised that much more has to be done in order to identify the relationships among measures (Flapper, Fortuin et al. 1996; Neely 1999; Bititci, Turner et al. 2000). Kaplan, when interviewed by de Waal (de Waal 2003), argued that cause-and-effect relationships should be tested further. Nevertheless, in almost all cases, organisations ignore the dynamic interdependencies and trade-offs among measures. Furthermore, criticism exists regarding performance measurement systems and their static nature. According to (Kennerley and Neely 2002), consideration is being given to what should be measured today, but little attention is being paid to the question of what should be measured tomorrow. They suggest that measurement systems should be dynamic and must be modified as circumstances change. A radical rethink of performance measurement is now necessary more than ever (Corrigan 1998; Takikonda and Takikonda 1998). In an attempt to describe and test cause-and-effect relationships, (Kaplan and Norton 2001), proposed the use of strategy maps. However, the causal relationships that strategy maps claim to model are not always linear and one-way (Kaplan and Norton refer only to linear and one-way cause and effect chain), but mostly a fuzzy mess of interactions and interdependencies.

(Kellen 2003), argues that in the area of executive management only 6 in 10 executives place confidence in the data presented to them. He points out that one of the main factors that prevent measurement is the fuzzy objectives. By the same token, (Suwignjo, Bititci et al. 2000), explains that in a performance measurement system a large number of multidimensional factors can affect performance. Integrating those multidimensional effects into a single unit can only be done through subjective, individual or group judgement. It is impossible to have an objective measurement and scale system for each different dimension of measurement that can facilitate objective value trade-off between different measures. They argue that techniques, which are suited to fuzzy paradigms, should be considered.

Identifying the relationships and trade-offs that exist among measures will be a great step towards the design of a robust performance measurement system. However, the robustness of the performance measurement system is also based on its successful implementation and use. According to (Neely, Mills et al. 2000), implementation is not a straightforward task due to fear/resistance, politics and subversion. (Dumond 1994) claims that the main problems in the implementation of performance measurement systems are raised due to the lack of communication and dissemination of performance information. According to (De Geus 1994) even a simplified but credible (causal) model can be a powerful communication and learning tool. In the same token, (Morecroft 1994) argues that models are more effective when they become integral parts of management debate, communication, dialogue and experimentation. It is possible for managers to gain insights about how their actions might affect outcomes if they work with models. Furthermore, experimentation with models creates a cycle of increased learning and improved models.

Finally, further to all the aforementioned issues, (Morris 2002), argues that software applications that have been developed so far, lack of an analytic capability and they cannot carry out predictive modelling. Despite the many attempts in this area (EIS, Decision Support tools), it is claimed that these tools don't necessarily advance the decision-making process.

## 2.1 Balanced Scorecard

According to (Kaplan and Norton 1996), the balanced scorecard supplements traditional financial measures with criteria that measure performance from three additional perspectives – those of customers, internal business processes, and learning and growth (Fig. 1).

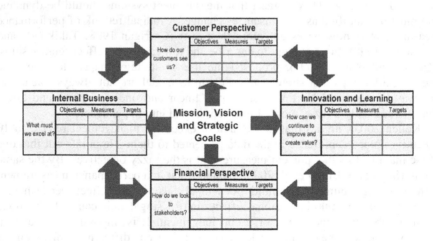

**Fig. 1** The Balanced Scorecard

♦ **Customer Perspective**

Since companies create value through customers, understanding how they view performance becomes a major aspect of performance measurement.

♦ **Internal Business Process Perspective**

According to (Kaplan and Norton 2000), in the internal-business-process perspective, executives identify the critical internal processes in which the organisation must excel.

♦ **Learning and Growth Perspective**
According to (Kaplan and Norton 2000), this perspective of the balanced scorecard identifies the infrastructure that the organisation must build to create long- term growth and improvement. Learning and growth come from three principal sources: 1. People; 2. Systems; and 3. Organisational procedures.

♦ **Financial Perspective**
Within the balanced scorecard, financial measures remain an important dimension. Financial performance measures indicate whether a company's strategy, implementation, and execution are contributing to bottom-line improvement.

### 2.1.1 Limitations of the Balanced Scorecard

Balanced Scorecard (Kaplan and Norton 1992), briefly described previously, is the most popular framework in the area of performance measurement. The introduction of the Balanced Scorecard was mainly based on a transition from the traditional financial performance measurement systems towards a more balanced approach (financial and non-financial measures) that includes several measures in a multi-dimensional structure. In spite of its "reputation", there are several issues related to the Balanced Scorecard, which need further research. More particularly:

❑ **Cause and effect consider to be one-way in nature**
The cause and effect concept is a very important element to consider in an attempt to construct a Balanced Scorecard. However, the way cause and effect is illustrated is rather problematic. Measures in the Balanced Scorecard are placed in a cause and effect chain rather a systemic approach. (Kaplan and Norton 1996a), argue that '*the financial objectives serve as the focus for the objectives and measures in all the other scorecard perspectives*'. This statement ignores any feedback loops that might exist.

❑ **Trade-offs among measures and among the four perspectives are ignored**
Ignoring the trade-offs among measures as well as among the four perspectives is rather not an efficient approach. By doing so, the *communication of strategy* and *dissemination of performance information* is restricted because users are not in position to identify and learn why and how certain things have occurred.

❑ **Measures are equally weighted**
All the measures in Balanced Scorecard are given the same weighting. This is not what happens in reality. Some measures may be more important and have greater impact compared to others. Weighting the measures among each other is critical on decision-making.

❑ **Design techniques used for the development of a Balanced Scorecard are rather poor in illustrating the dynamics of a system** (absence of feedback loops)
Two of the most usual design techniques used for the development of the Balanced Scorecard are the bubble diagram and the Generic Value Chain model (Fig. 2 (a & b)). Recently, (Kaplan and Norton 2001), introduced a new model; the strategy maps (Fig. 2 (c)). However, as it has been observed, these models

lack the ability of representing feedback loops. This is not very suitable for communicating strategy as well as exploring the interrelationships among measures and in turn objectives. Ignoring the feedback loops (two-way cause and effect) at the design stage of a performance measurement system will lead to a non-effective representation of the organisation and the dynamics that are involved. Introducing new measures in this way restricts the possibility to identify the consequences that might be raised in the whole system.

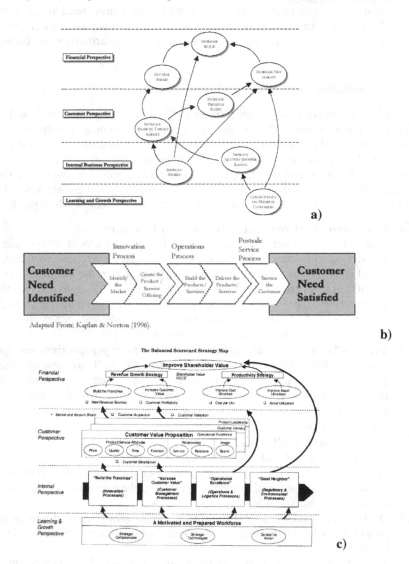

**Fig. 2** Design techniques for the development of a Balanced Scorecard

## 2.2  A Fuzzy Logic View - FCMs

Fuzzy logic was introduced in 1965 by Zadeh as a means of representing data and manipulating data that was not precise, but rather fuzzy. The theory of fuzzy logic provides a mathematical strength to capture the uncertainties associated with human cognitive processes, such as thinking and reasoning. Since its first appearance, fuzzy logic has been used in a variety of applications, such as image detection of edges, signal estimation, classification and clustering. A fuzzy logic technique represents an alternative solution to the design of intelligent engineering systems. Thus, fuzzy rule-based experts systems are widely applied nowadays, this being supported by the fact that fuzzy logic is linguistic rather than numerical, something which makes it similar to human thinking and hence simpler to understand and put into practice. It is not within the scope of this paper to present an overview of fuzzy logic and the reader is directed to the seminal work on the subject by (Zadeh, 1997) and in the more recent non-mathematical text by (Kosko, 1998). In this paper, the concept of an FCM is used to define the state of a sct of variables/objectives.

FCMs are soft computing tools which combine elements of fuzzy logic and neural networks. They are fuzzy signed directed graphs with feedback loops, in which the set of concepts (each concept represents a charactcristic, state or variable of the system/model; concepts stand for events, actions, goals, values and/or trends of the system being modelled as an FCM), and the set of causal relationships is modelled by directed arcs (Fig. 3). FCM theory developed recently (Kosko 1986) as an expansion of cognitive maps that had been employed to reprcscnt social scientific knowledge (Axelrod 1976), to make decision analysis (Zhang, Chen et al. 1989) and to analyse extend-graph thcoretic behaviour(Zhang and Chen 1988).

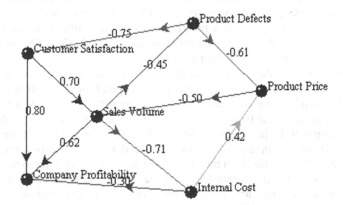

**Fig. 3** A Simple FCM

Figure 3, illustrates an FCM which is used to simulate the behaviour of *Company Profitability* in terms of other factors that positively or negatively affect its state (behaviour). In the figure above, *Company Profitability* is directly affected by the following factors: *Customer Satisfaction* (positive effect), *Sales Volume* (positive effect) and *Internal Cost* (negative effect). Directed, signed and weighted arcs, which represent the causal relationships that exist between the concepts, interconnect the FCM concepts. For example, in figure 3, there is a strong positive relationship from the *Customer Satisfaction* concept to the *Company Profitability* concept. Each concept is characterised by a numeric value that represents a quantitative measure of the concept's presence in the model. A high numeric value indicates the strong presence of a concept. The numeric value results from the transformation of the real value of the system's variable, for which this concept stands, to the interval [0,1]. All the values in the graph are fuzzy, so weights of the arcs are described with linguistic values (such as: "strong", "weak", etc) that can be "defuzzified" and transformed to the interval [-1,1].

Studying this graphical representation, one can conclude which concept influences other concepts and their interconnections. This representation makes updating the graph structure easy, as new information becomes available or as more experts are asked. This can be done, for example, by the addition or deletion of an interconnection or a concept.

Between concepts, there are three possible types of causal relationships expressing the type of influence of one concept on another. The weight of an interconnection, $W_{ij}$, for the arc from concept $C_i$ to concept $C_j$, can be positive ($W_{ij}>0$), which means that an increase in the value of concept $C_i$ leads to the increase of the value of concept $C_j$, and a decrease in the value of concept $C_i$ leads to the decrease of the value of concept $C_j$. Or there is a negative causality ($W_{ij}<0$), which means that an increase of the value of concept $C_i$ leads to the decrease of the value of concept $C_j$ and vice versa. When there is no relationship from concept $C_i$ to concept $C_j$, then ($W_{ij}=0$). In figure 3, the weight of the interconnection between the concepts, *Company Profitability* and *Sales Volume* is positive (represented by a blue arc and a positive value) and is illustrated as follows: if *Sales Volume* is high then *Company profitability* will be high – if *Sales Volume* is low then *Company Profitability* will be low.

An expert defines the main concepts that represent the model of the system, based on his knowledge and experience on the operation of the system. At first, the expert determines the concepts that best describe the system. He knows which factors are crucial for the modelling of the system and he represents each one by a concept. Moreover, he has observed which elements of the system influence other elements and for the corresponding concepts he determines the positive, negative or zero effect of one concept on the others. He describes each interconnection with a linguistic value that represents the fuzzy degree of causality between concepts. The linguistic weights are transformed into numerical weights using the methodology proposed by (Stylios, Groumpos et al. 1999).

When the FCM starts to model the system, concepts take their initial values and then the system is simulated. At each step, the value of each concept is determined by the influence of the interconnected concepts on the corresponding weights:

$$a_i^{t+1} = f(\sum_{j=1, j \neq i}^{n} w_{ji} a_j^t) \tag{1}$$

where $a_i^{t+1}$ is the value of concept Ci at step t+1, $a_j^t$ the value of the interconnected concept Cj at step t, Wji the weighted arc from concept Cj to Ci, and f a threshold function. Three threshold functions have been identified in the literature (Kosko 1998) and are described below:

$$S_i(x_i) = 0, x_i \leq 0$$
$$S_i(x_i) = 1, x_i > 0$$

**bivalent**

$$S_i(x_i) = -1, x_i \leq -0.5$$
$$S_i(x_i) = 0, -0.5 < x_i < 0.5$$
$$S_i(x_i) = 1, x_i \geq 0.5$$

**trivalent**

$$S_i(x_i) = \frac{1}{1 + e^{-cx_i}}$$

**logistic signal, c = 5**

# 3  A Proactive Balanced Scorecard Methodology (PBSCM)

## 3.1  *Successful Execution of Strategy: A New Component*

According to Kaplan and Norton (Kaplan and Norton 2004) successful execution of a strategy (Breakthrough Results) requires two components:

{**Breakthrough Results**}

=

{**Describe the Strategy**}

+

{**Manage the Strategy**}                    (2)

The philosophy of the two components is simple:

- You can't manage (second component) what you can't measure (first component)
- You can't measure what you can't describe (Breakthrough Results)

According to (Kaplan and Norton 2004), their first book, *The Balanced Scorecard*, has addressed the first component by showing how to measure strategic objectives in multiple perspectives. It also presented the early ideas regarding the second component, how to manage the strategy. Their second book, *The Strategy-Focused Organisation*, has provided a more comprehensive approach for how to manage the strategy. It has also introduced strategy maps for the first component, how to describe the strategy. Their third book, Strategy Maps, goes into much more detail on this aspect, using linked objectives in strategy maps to describe and visualize the strategy. They rewrite the above "equation" as follows:

$$\{\textbf{Breakthrough Results}\}$$
$$=$$
$$\{\textbf{Strategy Maps}\} \rightarrow [\textit{Describe}]$$
$$+$$
$$\{\textbf{Balanced Scorecard}\} \rightarrow [\textit{Measure}]$$
$$+$$
$$\{\textbf{Strategy-Focused Organisation}\} \rightarrow [\textit{Manage}] \quad (3)$$

However, it is our belief that both "equations" ((2), (3)) omit an important component: **Simulate the Strategy**. Hence, we rewrote the "equation" (2) as follows:

$$\{\textbf{Breakthrough Results}\}$$
$$=$$
$$\{\underline{\textbf{Simulate the Strategy}}\}$$
$$+$$
$$\{\textbf{Describe the Strategy}\}$$
$$+$$
$$\{\textbf{Manage the Strategy}\} \quad (4)$$

By incorporating this new component (Simulate the Strategy) in the above "equation" we aim to overcome all the limitations identified in the literature review (in particularly in section 2.2.1) and view performance measurement and in particularly the Balanced Scorecard within a systemic approach. In order to address this new component, we suggest the use of Fuzzy Cognitive Maps (FCMs). As it was described previously, FCMs are fuzzy signed directed graphs with feedback loops, in which the set of objects is modelled by the nodes, and the set of causal relationships is modelled by directed arcs. The FCM theory, was developed recently (Kosko 1986) as an expansion of the cognitive maps that had been employed to represent social scientific knowledge (Axelrod 1976), to make decision analysis (Zhang, Chen et al. 1989) and to analyse extend-graph theoretic behaviour (Zhang and Chen 1988). FCMs combine the strengths of cognitive maps with fuzzy logic. By representing human knowledge in a form more representative of natural human language than traditional concept mapping techniques, FCMs ease knowledge engineering and increase knowledge-source concurrence. The characteristics and the structure of FCMs allow us to re-write "equations" (3) and (4) as follows:

$$\{\textbf{Breakthrough Results}\}$$
$$=$$
$$\{\textbf{FCMs}\} \rightarrow [\textit{Simulate}]$$
$$+$$
$$\{\textbf{FCMs}\} \rightarrow [\textit{Describe}]$$
$$+$$
$$\{\textbf{Balanced Scorecard}\} \rightarrow [\textit{Measure}]$$
$$+$$
$$\{\textbf{Strategy-Focused Organisation}\} \rightarrow [\textit{Manage}] \quad (5)$$

In the above "equation", in the first instance (simulate), we use the simulation characteristics of the FCMs theory. The FCM approach involves forward-chaining (what-if analysis). The forward-chaining provides business domain experts with the capability to reason about the map they have constructed (nodes, relationships and weights) and examine different scenarios. In the second instance (describe), we utilise the representation capabilities of the FCMs theory. FCMs are illustrated as causal-loop diagrams. This is very suitable for communicating strategy as well as exploring the interrelationships among measures and in turn objectives.

## 3.2 Overview of the PBSCM

The methodology for the development of a Proactive Balanced Scorecard is depicted in the figure below (Fig. 4). PBSCM is capable of illustrating non-linear interactions and feedback loops through the use of FCMs as a causal-loop diagram and performing what-if scenarios through the use of FCMs simulation.

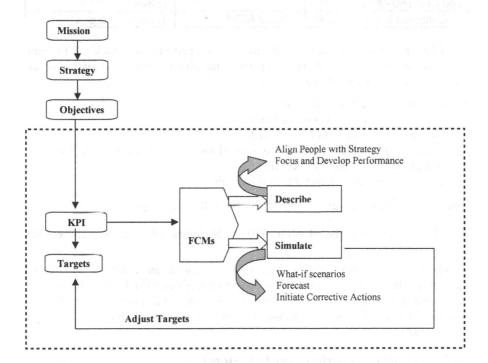

**Fig. 4** The PBSCM

The PBSCM goes through a series of stages that involve: (1) inputs to be provided, and (2) outcomes to be generated. Business domain experts and/or professionals of Performance Measurement/Balanced Scorecard are people with specific business expertise that contribute towards providing the business knowledge for the PBSCM. The following table (Table 1) indicates the stages of a PBSCM together with the inputs and outcomes of each stage:

**Table 1** Inputs and Outcomes of the PBSCM

| STAGE | INPUT | OUTCOME |
|---|---|---|
| 1. Establishing the Mission, Vision, Strategic Objectives, Perspectives and Critical Success Factors (CSF) | 1. Interviews with Middle and Top Management<br>2. Internal Company Data | 1. Mission<br>2. Vision<br>3. Strategic Objectives<br>4. Perspectives<br>5. CSF |
| 2. Identify Key performance Indicators (KPI) | 1. CSF | 1. KPI in each perspective |
| 3. Establish Targets | 1. KPI | 1. Target for each KPI |
| 4. Define relationships among the identified KPI | 1. KPI | 1. FCM with no weights |
| 5. Assign Linguistic Variables to Weights and Concepts-(KPI) | 1. FCM with no weights | 1. Final FCM with Weights and Concept values |
| 6. Continuous Improvement | 1. Final FCM | 1. Adjust Targets |

Before proceeding to each of the aforementioned stages a kick-off meeting takes place between the domain experts. The aim of this meeting is for all participants to contribute towards:

- Establishing the PBSCM team
- Clarifying the objectives of the team
- Identifying the context of the PBSCM
- Selecting the reference material to be used for the construction of the PBSCM
- Anticipating possible user benefits
- Preparing further actions for the participants

The PBSCM methodology is composed of the following stages:

### 🗁 Establishing the Mission, Vision, Strategic Objectives, Perspectives and CSF

In this stage the focus is on understanding the organisation's strategy, culture and capabilities in order to specify the Strategic Objectives (which state the specific goals/directions the organisation aims to achieve), Perspectives and Critical Success Factors (things the organisation must do well to achieve its strategic objectives).

### 🗎 Identify Key performance Indicators (KPI)

This stage aims to narrow down the list of all possible measures into a shortest one that provides the KPIs, which will be used in each perspective.

### 🗎 Establish Targets

Measurement alone is not good enough. We must drive behavioural changes within the organisation if we expect to execute strategy. This requires establishing a target for each KPI within the Balanced Scorecard. Targets are designed to drive and push the organisation as to meet its strategic objectives. Targets need to be realistic so that people feel comfortable about trying to execute on the target.

**📖✍ Define relationships among the identified KPI**
As the KPIs constituting the different perspectives have been derived, the relationships between these KPIs (KPIs are represented as concepts in the FCM) have to be defined. An edge connecting two KPIs represents a relationship. The direction of the relationship (i.e. which KPI affects the other) is denoted by the direction of the arrow on this edge. The FCM that has been constructed (Fig. 5) using the method mentioned above does not contain any information except that there are relationships between abstract concepts (KPI). The next step is to enrich the map with numerical values, which are assigned to the concepts and relationships.

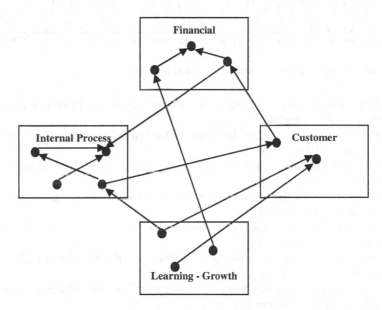

**Fig. 5** Define Relationships Among the Identified KPIs

**📖✍ Assign Linguistic Variables to Weights and Concepts (KPI)**

Knowledge on the behaviour of a system is rather subjective and in order to construct a model of the system it is proposed to utilise the experience of experts. Experts are asked to describe the causality among concepts using linguistic notions (A fuzzy logic perspective). They will determine the influence of one concept to the other as "negative" or "positive" and then they will describe the grade of the influence with a linguistic variable such as "strong", "weak", etc. *Influence* of one concept over another, is interpreted as a linguistic variable in the interval [-1,1]. Its term set **T(influence)** is: **T(influence)** = {negatively very-very high, negatively very high, negatively high, negatively medium, negatively low, negatively very low, negatively very-very low, zero, positively very-very low, positively very low, positively low, positively medium, positively high, positively very high, positively very-very high}.

We propose a semantic rule M to be defined at this point. The above-mentioned terms are characterized by the fuzzy sets whose membership functions µ are shown in Figure 6.

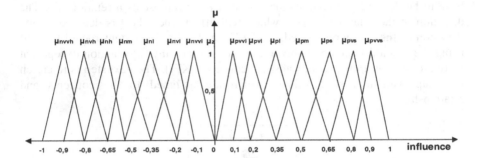

**Fig. 6** Membership functions of linguistic variable influence

- M(negatively very-very high)= the fuzzy set for "an influence close to -90%" with membership function $\mu_{nvvh}$
- M(negatively very high)= the fuzzy set for "an influence close to 80%" with membership function $\mu_{nvh}$
- M(negatively high)= the fuzzy set for "an influence close to 65%" with membership function $\mu_{nh}$
- M(negatively medium)= the fuzzy set for "an influence close to -50%" with membership function $\mu_{nm}$
- M(negatively low)= the fuzzy set for "an influence close to -35%" with membership function $\mu_{nl}$
- M(negatively very low)= the fuzzy set for "an influence close to -20%" with membership function $\mu_{nvl}$
- M(negatively very-very low)= the fuzzy set for "an influence close to -10%" with membership function $\mu_{nvvl}$
- M(zero)= the fuzzy set for "an influence close to 0" with membership function $\mu_z$
- M(positively very-very low)= the fuzzy set for "an influence close to 10%" with membership function $\mu_{pvvl}$
- M(positively very low)= the fuzzy set for "an influence close to 20%" with membership function $\mu_{pvl}$
- M(positively low)= the fuzzy set for "an influence close to 35%" with membership function $\mu_{pl}$
- M(positively medium)= the fuzzy set for "an influence close to 50%" with membership function $\mu_{pm}$

- M(positively high)= the fuzzy set for "an influence close to 65%" with membership function $\mu_{ph}$
- M(positively very high)= the fuzzy set for "an influence close to 80%" with membership function $\mu_{pvh}$
- M(positively very-very high)= the fuzzy set for "an influence close to 90%" with membership function $\mu_{pvvh}$

The membership functions are not of the same size since it is desirable to have finer distinction between grades in the lower and higher end of the influence scale. As an example, three experts propose different linguistic weights for the interconnection $W_{ij}$ from concept $C_i$ to concept $C_j$: (a) positively high (b) positively very high (c) positively very-very high. The three suggested linguistics are integrated using a sum combination method and then the "defuzzification" method of the centre of gravity (CoG) is used to produce a weight $W_{ij}=0,73$ in the interval [-1,1]. This approach has the advantage that experts do not have to assign numerical causality weights but to describe the degree of causality among concepts.

A similar methodology can be used to assign values to concepts. The experts are also asked to describe the measurement of each concept using linguistic notions once again. Measurement of a concept is also interpreted as a linguistic variable with values in the interval [-1,1]. Its term sct **T(Measurement) = T(Influence)**. A new semantic rule M2 (analogous to M) is also defined and these terms are characterized by the fuzzy sets whose membership functions $\mu_2$ are analogous to membership functions $\mu$.

### 3.4   Continuous Improvement

The purpose of this phase is to continuously update the usability of the FCMs in order to provide improved user support. The continuous improvement cycle requires the users to run a simulation exercise on the FCM (using weight and concept values defined in the previous stage) and test its effectiveness in response to the targets defined previously. The adjustment will be based on the behaviour of the FCM during simulation and on the results it delivers.

## 4   Using the FCM Modeler Tool

### 4.1   Designing Fuzzy Cognitive Maps

Concepts represent a business metric or a decision variable in a certain domain. As the concepts constituting the different model categories are derived, the relationships between these concepts have to be defined. A relationship is represented by a line connecting the concepts.

The direction of the relationship (i.e. which concept affects the other) is denoted by the direction of the arrow on this line. This is depicted in **Figure 7**.

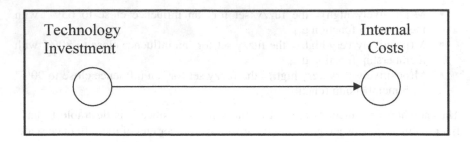

**Fig. 7** Example of FCM Concepts and Links

In **Figure 7**, therefore, the Technology Investment concept *somehow* affects the Internal Costs concept.

The FCMs that have been constructed using the method above do not contain any information except that there are relationships between abstract concepts.

The next step is to enrich the maps with numerical values, which are assigned to the concepts in order to denote their state at a certain time. The concept's state can change based on the FCM Modeler algorithm presented in appendix D. FCM Modeler concept states have three distinct values, namely: -1,0,1 in order to enable business domain experts to express their negative (-1), neutral or do not know (0), or positive (1) belief about the state of a concept. This is depicted in **Figure 8**.

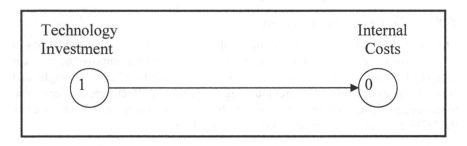

**Fig. 8** FCM node values example

**Figure 8** shows that, at this particular point in time, the user believes strongly that his or her company is investing in technology and that internal costs are neither high nor low. The question of what will be the next states of the two nodes will be answered once the relationship weights have been assigned.

Prior to the assignment of specific weights, the user must decide on the type of relationship that exists between concepts. In FCM Modeler there are two types of relationships:

- *Positive causal relationships* where the value of the effect concept increases when the state value of the cause concept increases and the value of the effect concept decreases when the state value of the cause value decreases.
- *Negative causal relationships* where the value of the effect concept increases when the state value of the cause concept decreases and the value of the effect concept decreases when the state value of the cause concept increases.

The graph in **Figure 9** shows schematically the assignment of relationship weights to our example.

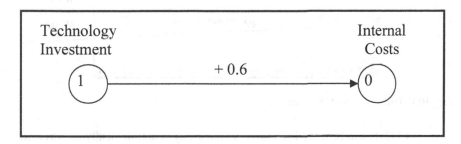

**Fig. 9** Positive Causal Relationship Example

**Figure 9** shows that technology investment and the positive causal relationship results in increased internal costs, i.e the next state value of internal costs will become one (1).

The degree of belief or certainty that a relationship exists is numerically denoted with a number in the bipolar interval of [-1,..,1]. The negative values in that range denote a negative causal relationship, while the positive values denote a positive causal relationship.

The value of zero (0) denotes that there is no relationship between the cause and effect concepts. In the case of a zero relationship, it is not necessary for the user to draw an arrowed line representing the relationship.

The calculation of the next state of an effect node is performed by adding the multiplications and of the nodes of the cause nodes by the weights of the causal relationships. For example, if the concepts "technology investment" and "training costs" influence the concept "internal costs" with positive causal relationships with weights of 0.4 and 0.3, respectively, the calculation formula of the next state of internal costs is:

value fo technology investment state * 0.4 + value of training costs state * 0.3 = value of internal costs state.

Users of the FCM Modeler tool can create an FCM by inserting concepts, relationships and weights.

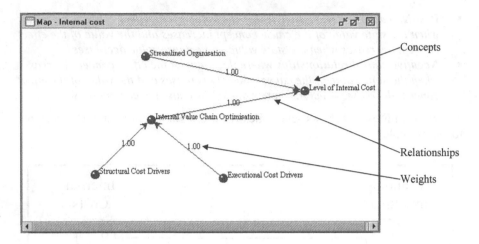

**Fig. 10** Relationships and Weights

Nodes of FCMs can be linked. The project hierarchy is automatically constructed.

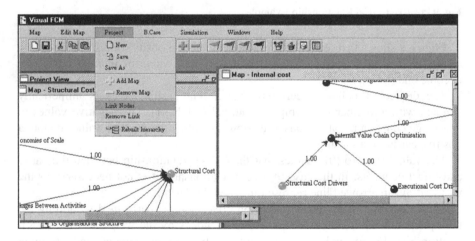

**Fig. 11** Project Hierarchy

Concepts and weights in the  FCMs that belong to a project have been assigned default values. In the business case module of the tool concept and weight  values can be assigned and stored. Many different business cases can be assigned to a project thus creating different business scenarios-cases. The changes in the FCM that belongs into a Business Case will not affect the same FCM in the Project View or in the Repository view. The FCMs in the two later  views will remain as they were (i.e. with their default values).

In the business case view the Structural Cost Driver FCM will have concept and weight values as shown in the following figure.

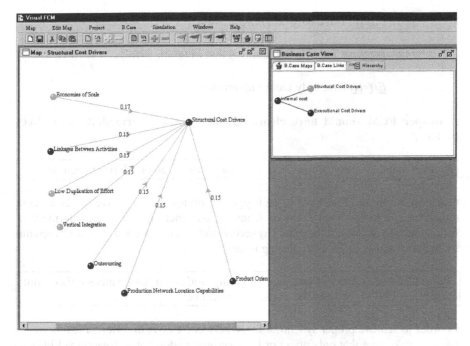

**Fig. 12** Business Case Example Map

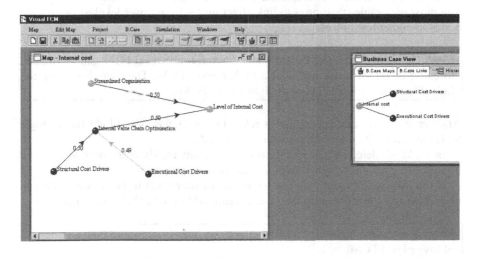

**Fig. 13** Business Case Example Map Continued

## 4.2  FCM Design Principles for Successful Simulation

For proper FCM design that will lead to successful simulation we classify FCM nodes into three categories:

☐  *__Cause__*: Have only outgoing arrows

☐  *__Intermediate__*: Have incoming and outgoing arrows

☐  *__Effect__*: have only incoming arrows

**A proper FCM should have clearly defined cause intermediate and effect nodes**.

| *__Rule__*: A map has to contain at Least Two nodes and a relationship between them |
| --- |

Cause nodes in an FCM have to be triggering nodes, i.e they have to get a value either from a lower level FCM or from the external environment. Intermediate nodes are also allowed to become triggering nodes but is a good standard to create FCMs with cause nodes as triggering nodes

| *__Rule__*: Cause and intermediate  are triggering nodes that take values either from lower level FCMs or from the external environment. |
| --- |

In order to achieve proper synchronisation between FCMs in different levels we have to make sure that only effect or intermediate nodes from a lower level FCM can be linked to cause or intermediate nodes of a higher level FCM.
The most preferable situation is to link effect nodes from lower level to cause nodes in higher level FCMs.

| *__Rule__*: Only effect or intermediate nodes from a lower level FCM can be linked to cause or intermediate nodes of a higher level FCM. The user should try to link effect from lower to cause in higher level FCMs. |
| --- |

Another synchronisation constraint is related with the number of links that a single node can have with nodes from lower level FCMs.

In FCM Modeler we allow only one node from FCMs at a lower level to be linked with a node at a higher level for consistency purposes.

If the user wants more than one nodes at a lower level to be  linked with one node at a higher level then the user must create additional FCMs to cater for this.

| *__Rule__*: A node in a higher level FCM cannot be linked with more than one nodes of lower level FCMs |
| --- |

### *Simulation Results*

Simulation results are presented in various diagrams and table results.  The figure below presents the simulation report of Internal Cost business case and contains a table presenting the linked nodes in higher and lower level FCMs as well as the Business Case Hierarchy.

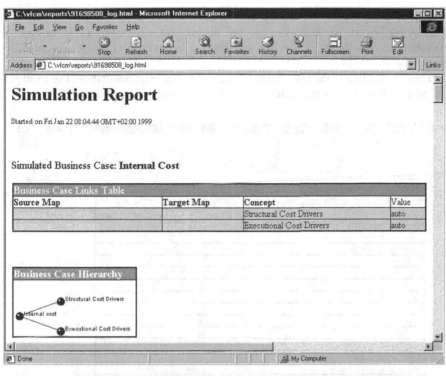

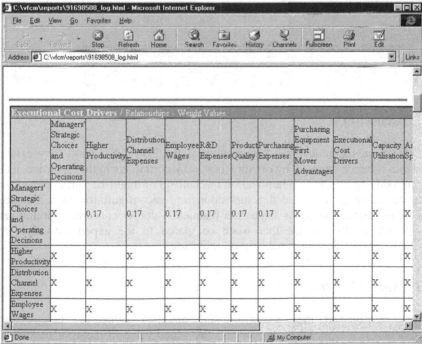

FCM weight values are presented in table format as shown in the following figure. An **X** symbol in a cell denotes the absence of a relationship. A **number** denotes the existence of a relationship between two nodes, weighted by this value.

Other tables depict the node values before and after implementation. For example the table bellow depicts the initial values of the executional cost drivers FCM

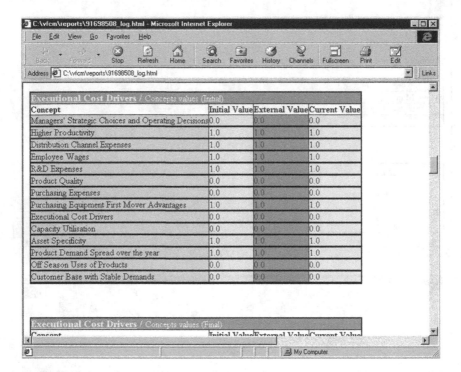

## 5 Discussion

As far as the theoretical value is concerned, the PBSCM methodology extends previous research attempts by (a) allowing fuzzy definitions in the cognitive maps, (b) introducing a specific interpretation mechanism of linguistic variables to fuzzy sets, (c) allowing dynamic decomposition and reconfiguration of a Balanced Scorecard strategy-map. As far as the practical value is concerned, preliminary evaluation results indicate that when compared to the expert estimates, the methodology provides reasonably good activities.

### 5.1  Added Value

Having established the theoretical and practical value of the proposed methodology, it is useful to discuss also the added value of incorporating such a methodology during the development of the Balanced Scorecard. It is presumed in this paper that the resulting methodology provides real value to the principle beneficiaries and stakeholders of Balanced Scorecard projects. For example:

- The methodology eases significantly the complexity of deriving expert decisions concerning the Balanced Scorecard development.
- The proposed methodology serves as a back-end to provide holistic strategic performance evaluation and management.

## 5.2 Preliminary Usability Evaluation

Senior managers of two major IT enterprises have evaluated the usability of the proposed methodology and have identified a number of benefits that can be achieved by the utilization of the proposed mechanism as a methodology for Balanced Scorecard development. A summary of major business benefits (as identified by senior managers) is provided to improve the autonomy of this paper:

- *Shared Goals*
  - ❖ Concept-driven simulation pulls individuals together by providing a shared direction and determination for strategic change.
  - ❖ Shared performance measurement enables business units to realize how they fit into the overall business model of the enterprise and what their actual contribution is.
  - ❖ Senior management receives valuable input from the business units (or the individual employees) who really comprehend the weaknesses of the current strategic model as well as the opportunities for performance change.
- *Shared Culture*
  - ❖ All business units at the enterprise feel that their individual contribution is taken under consideration and provide valuable input.
  - ❖ All business units and individuals at the enterprise feel confident and optimistic; they realize that they will be the ultimate beneficiaries of the Balanced Scorecard exercise.
  - ❖ The information sharing culture supports the enterprise's competitive strategy and provides the energy to sustain this by exploiting group and individual potential to its fullest.
- *Shared Learning*
  - ❖ The enterprise realizes a high return on its commitment to human resources.
  - ❖ There is a constant stream of improvement within the enterprise.
  - ❖ The entire enterprise becomes increasingly receptive to strategic changes, since the benefits can be easily demonstrated to individual business units.
- *Shared Information*
  - ❖ All business units and individuals have the necessary information needed to set clear objectives and priorities.
  - ❖ Senior management can effectively control all aspects of the strategic process
  - ❖ The enterprise reacts rapidly to threats and opportunities.

# 6 Conclusions

This paper proposes a **P**roactive **B**alanced **S**corecard **M**ethodology (**PBCSM**). The proposed decision aid may serve as a back end to Balanced Scorecard development and implementation. By using FCMs, the proposed methodology draws a causal representation of KPIs; it simulates the KPIs of each perspective with imprecise relationships and quantifies the impact of each KPI to other KPIs in order to adjust performance targets. The underlying research addressed the problems of the current Balanced Scorecard development process. The main objective of this research is to propose a methodology (not a new performance measurement framework) that will support existing measurement framework(s) during the process of performance measurement systems' design, implementation and use, and to advance the decision-making process. Future research will focus on conducting in depth studies to test and promote the usability of the methodology and to identify potential pitfalls.

# References

Ansoff, H.I.: Managing strategic surprise by response to weak signals. California Management Review XXVIII(2), 21–33 (1975)

Axelrod, R.: Structure of decision: The cognitive maps of political elites. Princeton University Press, Princeton (1976)

Bititci, U.S., Turner, T., et al.: Dynamics of performance measurement systems. International Journal of Operations & Production Management 20(6), 692–704 (2000)

Corrigan, J.: Performance Measurement: Knowing the Dynamics. Australian Accounting 68(9), 30–31 (1998)

De Geus, A.: Modeling to predict or to learn? Productivity Press, Portland (1994)

de Waal, A.A.: The future of the balanced scorecard: An interview with Professor Dr. Robert S. Kaplan. Measuring Business Excellence 7(1), 3 (2003)

Drucker, P.F.: Managing in Turbulent Times. Harper Collins Publishers (1993)

Dumond, E.J.: Making best use of performance measures and information. International Journal of Operations & Production Management 14(9), 16–31 (1994)

Eccles, R.G.: Performance measurement manifesto. Harvard Business Review 69, 131–137 (1991)

Fisher, J.: Use of non-financial performance measures. Journal of Cost Management 6, 31–38 (1992)

Fitzgerald, L., Johnston, R., et al.: Performance Measurement in Service Businesses. The Chartered Institute of Management Accountants, London (1991)

Flapper, S.D., Fortuin, L., et al.: Towards consistent performance management systems. International Journal of Operations & Production Management 16(7), 27–37 (1996)

Globerson, S.: Issues in developing a performance criteria system for an organisation. International Journal of Production Research 23(4), 639–646 (1985)

Grove, A.: Only the paranoid survive: How to Exploit the Crisis Points that Challenge Every Company. Bantam Books (1999)

Hayes, R.H., Wheelwright, S.C., Clark, K.B.: Dynamic Manufacturing: Creating the Learning Organization. Free Press, New York (1988)

Kaplan, R.S.: Measuring manufacturing performance: A new challenge for managerial accounting research. The Accounting Review 58(4), 686–705 (1983)

Kaplan, R.S.: Yesterday's accounting undermines production. Harvard Business Review 62, 95–101 (1984)

Kaplan, R.S., Norton, D.P.: The balanced scorecard - measures that drive performance. Harvard Business Review, 71–92 (January-February 1992)

Kaplan, R.S., Norton, D.P.: Using the Balanced Scorecard as a Strategic Management System. Harvard Business Review (January-February 1996)

Kaplan, R.S., Norton, D.P.: The Balanced Scorecard - Translating Strategy into Action. Harvard Business School Press, Boston (1996a)

Kaplan, R.S., Norton, D.P.: Why does business need a Balanced Scorecard? (2000), http://www.corpfinance.riag.com/

Kaplan, R.S., Norton, D.P.: Translating the Balanced Scorecard from Performance measurement to Strategic Management: Part 1. Accounting Horizons 15(1), 87–10 (2001)

Kaplan, R.S., Norton, D.P.: Strategy Maps: Converting Intangible Assets into 19Tangible Outcomes. Harvard Business School Press, Boston (2004)

Keegan, D.P., Eiler, R.G., Jones, C.R.: Are your performance measures obsolete? Management Accounting (US) 70(12), 45–50 (1989)

Kellen, V.: Business Performance Measurement, At the Crossroads of Strategy, Decision-Making, Learning and Information Visualization (2003), http://www.kellen.net/bpm.html

Kelly, K.: New rules for the new economy: ten ways the network economy is changing everything. Fourth Estate, London (1998)

Kennerley, M.P., Neely, A.D.: Performance measurement frameworks – a review. In: 2nd International Conference on Performance Measurement, Cambridge (2000)

Kennerley, M.P., Neely, A.D.: A framework of the factors affecting the evolution of performance measurement frameworks. International Journal of Operations and Production Management 22(11), 1222–1245 (2002)

Kosko, B.: Fuzzy Cognitive Maps. International Journal of Man-Machine Studies 24, 65–75 (1986)

Kosko, B.: Hidden patterns in combined and adaptive knowledge networks. International Journal of Approximate Reasoning 2, 377–393 (1998)

Lynch, R.L., Cross, K.F.: Measure Up - The Essential Guide to Measuring Business Performance, London, Mandarin (1991)

Maskell, B.H.: Performance Measurement for World Class Manufacturing: A Model for American Companies. Productivity Press, Cambridge (1992)

Morecroft, J.: Executive knowledge, models, and learning. Productivity Press, Portland (1994)

Morris, H.: Balanced Scorecard Report: Insight, experience and ideas for strategy focused organisations. Harvard Business School Publishing 4(1), 1–17 (2002)

Neely, A.: Measuring Business Performance - Why, What and How. Economist Books, London (1998)

Neely, A.: The performance measurement revolution: why now and what next? International Journal of Operations & Production Management 19(2), 205 (1999)

Neely, A., Mills, J., et al.: Performance measurement system design: developing and testing a process-based approach. International Journal of Operations & Production Management 20(10), 1119–1145 (2000)

Niven, P.R.: Balanced Scorecard Step-by-Step. John Wiley (2002)

Senge, P.N.: The Fifth Discipline: The Art and Practice of the Learning Organization. Century Business Press, London (1992)

Stylios, C.D., Groumpos, P.P., et al.: A fuzzy cognitive map approach to process control systems. J. Adv. Comput. Intell. 3, 1–9 (1999)

Suwignjo, P., Bititci, U.S., et al.: Quantitative models for performance measurement system. Int. J. Production Economics 64, 231–241 (2000)

Takikonda, L., Takikonda, R.: We Need Dynamic Performance Measures. Management Accounting 80(3), 49–51 (1998)

Waggoner, D.B., Neely, A.D., et al.: The forces that shape organisational performance measurement systems: an interdisciplinary review. International Journal of Production Economics 60-61, 53–60 (1999)

Walls, J.G., Widmeyer, G.R., et al.: Building an information system design theory for vigilant EIS. Information System Research 3(1), 36–59 (1992)

Zhang, W.R., Chen, S.S.: A logical architecture for cognitive maps. In: Second IEEE International Conference on Neural Networks, San Diego, CA (1988)

Zhang, W.R., Chen, S.S., et al.: Pool 2: a generic System for cognitive map development and decision analysis. IEEE Transactions on Systems, Man and Cybernetics 19, 31–39 (1989)

Zadeh, L.A.: The Roles of Fuzzy Logic and Soft Computing in the Conception, Design and Deployment of Intelligent Systems. In: Nwana, H.S., Azarmi, N. (eds.) Software Agents and Soft Computing: Towards Enhancing Machine Intelligence. LNCS, vol. 1198, pp. 83–190. Springer, Heidelberg (1997)

# Capturing Domain-Imposed Requirements Based on Basic Research Findings

Siaw Ming Wong[1], Jean-Yves Lafaye[2], and Patrice Boursier[3]

[1] Open University Malaysia / Université de La Rochelle
Jalan Tun Ismail, 50480 Kuala Lumpur, Malaysia
wsming@themis.com.my
[2] Laboratoire d'Informatique, Image et Interaction,
Université de La Rochelle, La Rochelle, France
jylafaye@univ-lr.fr
[3] College of Computer Studies
AMA International University, Bahrain
pboursier@amaiu.edu.bh

**Abstract.** The current means of obtaining domain-imposed requirements through users or domain experts are often suboptimal especially for the relatively new area of interest. This paper suggests the use of basic research findings as the more objective source and proposes an approach that translates research findings into a UML model based on which the domain-imposed requirements can be extracted. By using business project management as the domain of interest, it outlines the steps in the said approach and describes the use of the resulting domain model during the requirement specification for a Project Management Information System that caters specially to the needs of business projects. Theoretically, the same method can be applied to the other areas of management and an enterprise domain model could be developed in a similar way. Given equity access to all software developers, it is envisaged that meeting standard domain-imposed requirements would become a pre-requisite for competing enterprise systems in the future.

**Keywords:** Domain modeling, Knowledge specification, Unified modelling language, Requirement specifications.

## 1 Introduction

System requirements may be divided into two sub-types namely user-defined requirements which reflect the wishes of users on how the system should be used; and domain-imposed requirements which are facts and laws of the real-world domain [36]. Thus while user-defined requirements are most accurately gathered from the users, domain-imposed requirements could be obtained from various sources. One of the most common approaches is to draw them out of the domain

M. Glykas (Ed.): Business Process Management, SCI 444, pp. 365–388.
springerlink.com      © Springer-Verlag Berlin Heidelberg 2013

experts within the organization, the industry or the academicians specializing in the field. Alternatively, these requirements could be derived through domain engineering effort on existing application systems to work out the ontology of the domain [35].

For requirements gathering process that involves an extensive amount of human interactions, inaccuracy attributed to subjective and/or misinterpretations has been a known issue. The success of the domain engineering effort on the other hand, is highly dependent on the quality of the analyzed systems and consequently, the result is not necessarily a comprehensive representation of the problem domain. Furthermore, since a same subject matter may be approached from many different perspectives, a complete and honest description of the domain knowledge can only be obtained after compiling, digesting and consolidating input from various sources. In this regard, recognized standards developed for the domain help to reduce the effort required to perform this task and serves as a credible basis for domain-imposed requirements to be identified and captured. However, standards which are essentially formalized norms of the domain are likely to be established only for the more mature knowledge areas rather than the relatively green fields of interest. Relying solely on the conventional approach to gather system requirements for the latter in this case, is likely to obtain the usage requirements rather than a complete set of invariant functionalities which must be in place to support the fundamentals. Since the field is still being researched upon, the more objective source for acquiring the domain-imposed requirements would be none other than the basic research finding itself. The challenge is how to capture the descriptive form of basic research finding that addresses the problem space, in such a way that they could be leveraged for the requirement specifications of its supporting application system in the solution space.

This paper proposes a method that uses an UML domain model as the intermediary; by first translating the basic research findings into the domain model, and the domain model in turn becomes the basis where the domain-imposed requirements of the supporting application systems are extracted. The first half of the paper explains the rationale behind why UML model was adopted and how the method was developed. The second half of the paper put the method into action by illustrating how it has been successfully applied in the requirement study of a Project Managing Information System (P.M.I.S) based on the findings of a basic research on business project management. This is followed by the last section which concludes the paper, discusses the limitations and highlights the positive implications if the method is deployed on a larger scale in a coordinated fashion.

## 2    The Method and Its Underlying Thoughts

### 2.1    *UML Domain Model as the Desired Specification Format*

The first consideration of the method is to identify the most suitable format so that the domain knowledge acquired from the basic research can be captured and expressed in an unambiguous manner. Ontology was chosen as the starting point due to its origin as the centre piece in the Knowledge Representation (KR) paradigm ase "an explicit specification of conceptualization" [18]. A number of researches have already been conducted to elicit domain-imposed requirements based on domain ontologies [30]; but formal ontology encoded in ontology language is not the ideal format since basic researches are likely to involve non-technical researchers. In which case, the creation of a conceptual model at the end of the "Conceptualization" step in the ontology development process as depicted in Fig. 1 [16, 28] becomes a more appealing option.

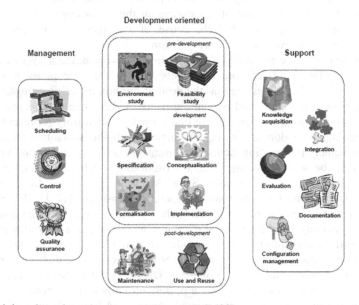

**Fig. 1** Activity of ontology development process [16],[28]

The first advantage conceptual model has over formal ontology is its "friendlier" format which could be better comprehended by both domain experts and the ontology engineers. But more importantly, the use of conceptual model allows the targeted domain to be represented without being bounded by the limitation of the ontology languages which tend to compromise expressiveness in favour of efficiency [28]. The preferred modelling language on the other hand, should be one that has the ability to capture the domain knowledge more explicitly than natural language descriptions; in addition, these expressions in semi-formal notations must be able to facilitate the specifications of systems requirements. Based on these considerations, Unified Modelling Language (UML) emerges as the key candidate as it was designed for the purpose of systems modelling, but has since been used as a general-purpose modelling language including modelling of enterprises [29],[17].

Managed and created by Object Management Group (OMG), the use of UML in ontology engineering was first suggested by Cranefield [11] given the connections between the standard UML and ontology concepts i.e. classes, relations, properties, inheritance. Although not all concepts are fully compatible with each other, various mapping techniques to bridge the gap have already been developed [32]. Moreover, UML is the key modelling language supporting the principle of Model Driven Architecture or MDA, an alternative software engineering approach where software systems can be developed through progressive transformation of models, by introducing more platform dependent specifications into the model in stages [5]. This means that the resulting UML model could be fed directly into an application system development cycle, satisfying not only the intent of leveraging on the knowledge specifications for requirements specifications, but the reuse of the these specifications in the development of the eventual software solution.

Let's now validate the ability of UML in specifying domain knowledge. It is said that are two kinds of domain models namely the concept models and the process models [2], suggesting that the desired domain model is complete only if it captures both the static and dynamic aspects of the domain. Praxeme Institute shares a similar point of view stating that a model which truly represents a basic understanding of a domain is one that captures all the semantics related to its concepts namely information, actions and transformations. In addition, it pinpoints that this can be achieved using UML's definitions for Attributes & Operations to capture information and action; State Machine Diagrams to capture the transformations; and Class to connect these 3 dimensions [41]. UML version 2.2 in which case, meets the requirements perfectly with 14 types of diagrams in two categories namely (1) the structural diagrams which emphasize on describing the things in the system/domain being modelled; and (2) the behavioural diagrams that capture transition of the internal states of objects and collaborations among objects. That being said, the use of Class Diagram and State Machine Diagram alone are insufficient as representation of domain knowledge in reality is far more complex and involves many inter-related concepts. Additional types of diagrams have to be introduced to support the specification as a result namely (1) Package Diagram to supplement the Class Diagram by organizing the relevant Classes into logical groups so that the resulting model is more structured and comprehensible; (2) Communication Diagrams to provide snapshots of the interactions among Objects for each of the states defined in the State Machine Diagram.

In the case of using UML in specifying system requirements, Use Case Diagram which is designed to show a purpose that an actor can use the system for, has long been used to support the practice of capturing requirements as use case descriptions [37]. Sequence Diagram which is designed to demonstrate communication among objects on the other hand, is ideal for defining the required exchange between the human and the system as well as the expected interactions among objects within the system in response to the human's request. The system requirements in which case, can be developed by using (1) Use Case Diagrams to capture the usage scenarios by illustrating the key users and the use cases; (2) Sequence Diagram to detail each of the uses cases.

In view of its adequacy in addressing the two vital aspects, a domain model specified using UML is endorsed as the intermediary.

## 2.2  Translating the Basic Research Findings into the UML Model

Given the similarity in purpose, the task of translating basic research findings into the UML model is expected to be similar to the "Conceptualization" step (in the ontology development process) of creating a model of the relevant domain knowledge at the knowledge level by expanding the key concepts following either the bottom up, top-down or middle-out strategy [16]. But a closer examination revealed that "Knowledge acquisition" and "Integration" related activities must be treated differently. With reference to Fig 1, these 2 sets of activities are positioned as "support" activities in the ontology development processes. This is understandable as conventional ontology engineering is motivated by the production of an ontology-based application system at the end of the process chain and thus, activities other than those directly related to the actual development of the ontology are classified as supporting or peripheral. The objective of our domain modelling on the other hand is to capture and specify knowledge as they are acquired. Thus, "Knowledge acquisition" is a key activity rather than a support activity; in fact, it is the starting point of this modelling exercise. Similarly, the junctures of "integrating" with existing work must also be explicitly stated and incorporated as part of the main stream so that the developed domain model could be automatically reused in the next cycle of knowledge acquisition and specification. Furthermore, in contrast to conventional ontology engineering process where the resulting model is fed into the next step of "formalization", the modelling exercise for our purpose is completed at this point. As such, testing and validation activities must also be incorporated to ensure that the resulting model is complete in representing the knowledge.

Based on these additional considerations, the domain modelling or knowledge specification process is devised as follows:

1) Define the modelling scope by expanding the key component in the domain knowledge with supporting concepts/terms using the prevailing standards of the domain;

    Specifically this refers to capturing the list of supporting concepts/terms into a "Specification Table" and the key relationships among them into the "Key-relationships Diagram" (a loose form of an Entity-Relationship diagram). The Key-relationships Diagram will not capture all the inter-relationships because its primary objective is only to provide an overview; otherwise the development efforts would be unjustly spent on building intermediaries. Furthermore by expressing the key inter-relationships in a graphical form, the format of the knowledge specification is a step closer to the UML's Class Diagram, making the transition to the next stage more intuitive.

2) Reuse similar existing work (which may be in the form of conceptual models or ontologies) by integrating them with the defined scope (i.e. output of step 1); and resolving the "conflicts" among the overlapping terms and definitions if any.

    In consideration that similar work may already exist and to ensure that the developed model will be leveraged to incorporate findings of future research, existing work that contains fundamental concepts of the targeted domain should be sourced to see if they could be reused and built upon. The basis of whether they will be adopted or rejected is based on 3 evaluation criteria namely "comprehensiveness", "suitability" and "expandability". "Comprehensiveness" refers to the completeness of the existing work in describing the domain; "Suitability" refers to whether its context (i.e. the perspective of which the domain knowledge is interpreted) is aligned with the current modelling exercise; and "Expandability" refers to the likelihood that the work will be expanded by its developer in the future, which is indicative by its popularity in the user community.

The ideal scenario would be where a similar UML model already existed, in which case, the task of the modelling exercise is just to identify the overlaps and extend the existing domain model with new terms and relationships. This may sound unlikely but it will be the exact scenario where the method is deployed to incorporate new research findings in the same domain. If the potentially reusable work is not in a similar format, the modelling exercise would have to extract the underlying list of terms and relationships accordingly. In any case, there are likely to be overlaps between the terms defined in the existing work and the content of the Specification Table. If these reusable existing work are selected based on the 3 predefined criteria as described earlier, the differences are likely a matter of choice of words rather than conceptualization. The recommended approach therefore, is to resolve the "conflict" by aligning the definitions of the affected terms in the Specification Table and the Key-Relationships Diagram to that of the adopted work. The rationale behind this "giving-in" attitude is simple. The aim of this domain modelling is to capture knowledge and thus, efforts should be spent in enriching the existing work rather than debating the superiority of vocabularies. In addition, changing definition of existing terms may impact the integrity of the adopted work thus potentially jeopardising the stability and validity of the final model.

The output of this step in summary are (1) the edited Specification Table; (2) the edited Key-relationships Diagram.

3)  Develop the UML specifications by translating the terms and relationships in the finalized scope into Class Diagrams, Package Diagrams, State Machine Diagrams and Communication Diagrams.

Creation of Package Diagram is the starting point so that a macro view containing all the logical groups is first obtained and can be used to guide subsequent work. Within each Package, sub-Packages are defined for the finer logical grouping of the terms. Creation of the Class Diagram is next and this can be achieved by populating the Package with Classes, where each Class is a direct translation of a term in the Specification Table. The next step is to define Associations among the Classes that justify their logical groupings within the Package, followed by definitions of Associations across Packages by working on a pair of Packages at a time. The final step of building the structural aspect of the model is to substantiate the Classes with more explicitness by defining Attributes, Constraints and Operations; based on the specifications in the adopted prevailing standards.

For the behavioural aspect of the model, creation of State Machine Diagram is worked on first and the starting point is to identify the "ruling" Class. A "ruling" Class is a Class whose transition of state controls the behaviour of the model or interactions among a subset of Classes at any one time. Given the State Machine Diagram of the ruling Class, Communication Diagram(s) are created to illustrate each state by defining the expected interactions among Classes, as well as the sequence of their occurrences through invoking the predefined Operations. In addition, dependencies among the Classes are indicated. Last but not least, the State Machine Diagrams should also be used to define the more sophisticated Classes which could impact the behaviour of the model in a similar was as the "ruling" Class but on a smaller scale.

4)  Validate the resulting model by checking it against the defined scope and testing it against real-life scenario

The first aspect of checking completeness against the defined scope can be done manually by checking the resulting model against the list of terms in the Specifications Table and the relationships defined in the Key-relationships Diagram. If a term or relationship is missing or misrepresented, changes are made in Class Diagram first, followed by the State Machine Diagram and the Communication Diagram. It should also be noted that since Operations must be predefined in the Class so that they are reused during the definitions of interactions in the Communication Diagram, the integrity between the structural and behavioural specifications is in effect maintained throughout the construction of the model.

The second aspect of testing the validity of the model can be achieved by creating Object Diagram using real-life data. With reference to the Class Diagram, the Object Diagram can be developed by instantiating one Class at a time starting with the ruling Class. As a miss-out in the model is uncovered, for example if a piece of real-life data could not be mapped to any Class, details are recorded but changes should not be applied to the model immediately. The affected UML diagrams are to be corrected only at the end of the validation exercise in one go, so that the versioning of the model can be properly managed. Moreover, as the miss-outs may be related to each other, further rationalization must be performed before the model is altered. This step of instantiations and corrections are carried out repeatedly until the model is capable of producing a corresponding Object Diagram that well represents the real-life scenario.

## 2.3  Capturing the Domain-Imposed Requirements

As discussed in section 2.1, this refers to the creation of Use Case Diagrams and Sequence Diagrams. The process of developing Use Case Diagrams to capture the system requirements in terms of usage scenarios is no different from the normal practice. The desired connection between problem world and the usage world is established during the creation of the Sequence Diagram to denote the series of actions that must be carried out in support of the Use Case according to the rules and norms of the domain. Since these domain-imposed requirements have already been captured by the domain model, the Sequence Diagrams could be built by reusing them especially the interactions among Objects specified in the Communication Diagrams. In fact, the domain-imposed requirements stated in the Sequence Diagrams should just be a transcription of relevant segments in Communication Diagrams. This is because any fundamental differences between the two would imply that the captured system requirements are deviating from the standard practice. This also means that the Sequence Diagram could easily separate the two types of system requirements i.e. (1) the domain-imposed requirements which are extracted directly from the domain model; (2) the user-imposed requirements gathered from the user community of the system.

The output at the end of the requirement study exercise in this case, is a UML model that contains both the domain knowledge specifications and the system requirements specifications.

## 3    An Example – The Case of P.M.I.S and Business Project Management

This section details how the devised method was applied to develop the system requirements of a Project Management Information System (P.M.I.S); based on the findings of an empirical research on business project management, i.e. the underlying problem domain of the system. This empirical research was a doctorate research in business administration and given the interest of this paper in requirement engineering, only the essences of this research and summary of its results are presented here.

### 3.1    The Basic Research on Business Project Management

Business projects in the context of the research, refers to non-standard, cross-functional in-house initiatives that aims at effecting internal changes within the business organization. Since it is used as a template for operational & strategic redesign [9], business project is playing an increasingly important role in the organization today. As reported by PriceWaterhouseCoopers (PWC)'s cross sectors survey, 200 companies from 30 countries are running a total of 10,640 projects a year worth in excess of US $4.5 billion [24].

Given the inception of project as a means of delivering unique product and/or services to external clients deployed primarily by the construction and engineering industries, project management professional bodies have been in existence since the 1960s. Most of these professional bodies have developed some form of project management standards which are used as a basis to certificate the qualification of their member practitioners [10, 14, 20-21, 34]. But despite the availability of the project management standards, business project failure rate remains unacceptably high. Based on the survey conducted by PIPC (a global project management specialist firm), one in three projects fails to deliver on time or within budget; and 60% fails to deliver the benefits as set out in business case [33]. Standish Group's

3rd Quarter 2004 Research on the IT industry indicated that 18% of all surveyed projects have failed i.e. cancelled prior to completion or delivered and never used; while 53% are challenged i.e. late, over budget and/or with less than the required features and functions [12].

That being said, it is also noted that while most of the standards acknowledged that project management effectiveness can be affected by elements and conditions in the operating environment, not much interest has been placed on incorporating context-specific requirements into the standard project management practices and processes. From this perspective, none of the general project management standards today are found to have adequately addressed the subject of "managing business projects" with due considerations to the influencing organizational factors and their collective effects on the attainment of business project success. Literature review further pointed out that unit of analysis of previous research has always been "project". In addition, the nature of these studies are mostly uni-variant descriptive analysis [7, 22, 25, 27, 39] or bi-variant causal analysis between an independent variable and project success [6, 8, 13, 15, 44]; and just about all the identified variables are found to pose some effect on project success depending on scenario. Furthermore, there is still very limited empirical work that examines specifically the subject of business project management despite the earlier calls to focus on business projects as an area for future research [26, 42]. In essence, the current understanding of business project management is incomplete.

Motivated by the gap in knowledge as described above, the empirical research was conducted with the objective to obtain a better understanding of business project management as a subject matter in its own right, by investigating why business project fails from the organization's perspective. In particular, the research aimed at identifying the key components of an effective business project management practice; and delivering a theory that explains the phenomena of continued business project failure.

## 3.2   The Basic Research Findings

Backed by the findings of its case studies, the research concluded that business project success should be measured in terms of meeting both the project's and the organization's objectives. The key components of an effective business project management practice were identified to be (1) "Core business project management competencies" i.e. business project managers must be proficient in the area of integration management, time management, communication management and scope management; (2) "Organization Support" i.e. an integrated programme management function must exist to support the project managers by integrating project, programme and the parent organizations; and (3) "Integrated P.M.I.S", i.e. an information system that enables seamless information exchange between the projects, programmes and the parent organization throughout the project life cycle to support the project managers in delivering their work. More specifically, the success at the project level is directly dependent on the effective application of the core project management competencies; whilst success at the organization level needs to be further enabled by the integrated programme management function.

The areas requiring support from the "integrated P.M.I.S" on the other hand are (a) formalization of the project/programme during the initiation and planning stage; (b) tracking of project/programme status and performance during the execution stage, (c) consolidation of lessons learned during the closing stage as well as enabling project knowledge retrieval and reference as and when required. Based on these findings, a business project management theory was developed stating that "business project is likely to fail if it is not managed as an integral parts of business enterprise with equal emphasis as its business-as-usual operations". Otherwise said, the research is pointing out that more adjustments in the organization settings are required so that business project is truly positioned "inside" the parent organization before they are likely to succeed.

The essence of these research findings, in particular, the key components of an effective business project management practice can be summarized into a theoretical framework as depicted in Fig. 2 [39]. Given the domain's key components and their causal relationships, the development of the business project management (hereafter refers to as BProjM) domain model can now be kicked off.

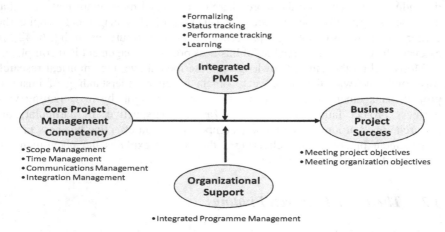

**Fig. 2** Business Project Management Theoretical Framework [39]

## 3.3   Creating the Business Project Management UML Domain Model

With reference to section 2.2, the first step was to define the modelling scope by expanding the key components of the domain with supporting terms based on prevailing standards if available. The Specification Table for BProjM domain model as depicted in Table 1, is a result of expanding the components in the theoretical framework by adopting two international project management standards developed by the Project Management Institute (PMI) namely (1) Project management Body of Knowledge or PMBOK [34]; and (2) The standards of program management [40].

**Table 1** Extract of the Specification Table

| Key terms (consolidated) | Definition |
|---|---|
| **Project Fundamentals** | |
| PROJECT | A group of people working together in order to achieve specific objective(s), given limited time and resources. |
| PROGRAMME | A group of projects organized to meet a set of organizational objectives. |
| **Core Business Project Management Functions** | |
| SCOPE MANAGEMENT | A subset of PM that includes processes to ensure that the project includes all the work required and only the work required, to complete the project successfully. |
| TIME MANAGEMENT | A subset of PM that includes processes to ensure timely completion of project. |
| COMMUNICATIONS MANAGEMENT | A subset of PM that includes processes to ensure timely and appropriate generation, collection, dissemination, storage and ultimate disposition of project information. |
| INTEGRATION MANAGEMENT | A subset of PM that includes processes to ensure that the various elements of the project are properly coordinated. |
| **Integrated Programme Management** | |
| INTEGRATED PROGRAMME MANAGEMENT | An organizational practice that includes processes to enable on-going initiation, execution and completion of top-down driven business change programme which may consist of many sub-programmes defined with specific organizational objectives. |
| **Project Results** | |
| PROJECT OBJECTIVE | Purposes or goals of the project, the aims to be achieved and the desired end results. |
| PROGRAMME OBJECTIVE | Purposes or goals of the programme, the aims to be achieved and the desired end results. |
| ORGANIZATIONAL OBJECTIVE | Purpose of goals of the organizations, the aims to be achieved and the desired end results. |
| PROGRAMME KPIS | Quantitative measures of the programme objectives. |
| ORGANIZATIONAL KPIS | Quantitative measures of the organizational objectives. |
| PROGRESS | The completion of planned project activities. |
| OUTPUT (WORK RESULTS) | The direct output of project plan execution. |
| PERFORMANCE | The contribution of the project at the organizational level. |
| Documentation | |

The Key-relationships Diagram on the other hand, is a result of explicitly specifying the relationships among the supporting terms which are excavated from the narratives of the two PMI standards. As depicted in Fig. 3, it highlights all the key relationships that must be incorporated by the BProjM domain model and indicates that the modelling must cover (1) terms and relationships directly related to business project management i.e. the main scope of the modelling work; (2) the interfaces between these terms and those residing in the operating environment i.e. the programme as well as the business enterprise at large.

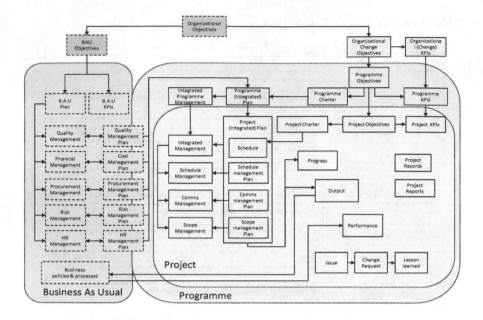

**Fig. 3** The Key-relationships Diagram for BProjM domain model

Step 2 of reusing existing work was carried out by injecting the resulting Specification Table and Key Relationship Diagram with supporting terms from (1) PROMONT (PROject Management ONTology) [1] which contains generic and fundamental terms of project; (2) OMG standards for defining Computational Independent Model (CIM) for enterprise namely (a) Business Motivation Model (BMM) [3] which contains a set of built-in concepts that define the elements of business plans; (b) Business Process Definition Meta-Model (BPDM) [4] which is a framework for understanding and specifying the processes of a community or organization; and (c) Organization Structure Meta-Model (OSM) [31] which provides the concepts behind the organization to serve as the basis for exchanging organizational models.

Given the finalized modelling scope, step 3 of developing the UML specifications translated the terms in the Specification Table into Classes, interpreted the relationships in the Key-relationship Diagram into appropriate types of Associations (either Standard, Inheritances, Composition or Aggregation); and constructed the series of Package Diagrams, Class Diagrams, State Machine Diagrams and the Communication Diagrams accordingly.

Executed in iterations with step 3 was step 4 of testing and validating the model for completeness at multiple levels. The definition of the model was manually checked against the defined scope, the technical integrity of the model was assured by the audit function of the UML tool Objecteering, while the validity of the model is tested against the real-life scenario by creating Object Diagram using the data collected during the case study of the empirical research. Below some examples of the discrepancies which were uncovered during the creation of the Object Diagram:

1) Before: PROGRAMME may be an Aggregation of PROJECTS

    **Findings:** Programme may be an Aggregation of projects and sub-programmes.

    **Amendments:** Add a new Association to PROGRAMME which points to itself.

2) Before: INITIATIVE produces OUTPUT, based on the PROMONT definition that OUTPUT is an outcome of an INITIATIVE, which can be a physical product, a service or a document.

    **Findings:** The outcome of an initiative may comprise multiple products, services or documents which are produced throughout the project rather than right at the end.

    **Amendments:** the Association "INITIATIVE produces OUTPUT" is replaced by "TASK produces OUTPUT".

3) Before: TASK is specified without ability to create Subtask.

    **Findings:** The 2nd findings triggered a check at the TASK's definition in the Specification Table where it is realized that Task should also be defined with Subtasks.

    **Amendments:** Add a new Association to TASK which points to itself.

The final result is a Business Project Management (BProjM) domain model comprising 10 Packages, 58 Classes, 10 State Machine Diagrams and 6 Communication Diagrams. An overview of the domain model is as depicted in the Package Diagram in Fig. 4.

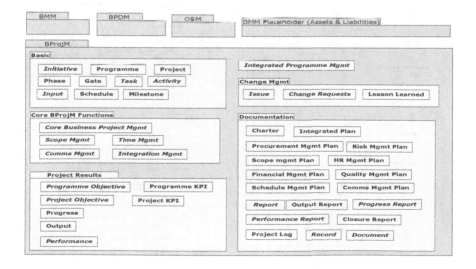

**Fig. 4** Business Project Management Domain Model - Package Diagram

With reference to Fig. 4, the core of the domain model is the "BProjM" Package which contains the essential concepts /components of an effective business project management practice. This Package contains 44 Classes organized into 5 Sub-Packages which correspond to the logical groupings in the Specification Table namely (1) "Basic" i.e. fundamental concepts of project; (2) "Core business project management" or i.e. the 4 business project management competencies which are essential for achieving delivery project success; (3) "Project Results" i.e. criteria that defines project success; (4) "Change Management" i.e. concepts related to change managements within projects; (5) "Documentation" i.e. various classes of documents produced by projects. The key research finding that the project must be positioned within the parent organization is supported by the definitions of project environment using OMG Packages as depicted at the top half of Fig. 3.

In terms of State Machine Diagram, the key ruling Class for the business project management domain is INITIATIVE; and thus the INITIATIVE's State Machine Diagram that paraphrases a standard programme/project life cycle was the first to be developed. As indicated at the top left hand corner of Fig. 5, business needs trigger the launch of a change initiative and assumes the first state of "Initiation". The initiative only transits into the next state of "Planning" if the initiative's Charter has been approved by the stakeholders. Similarly, it will only take on the "Execution" state if its Plan has been approved. Finally, only upon completion of all activities and acceptance of its output by the stakeholders will the initiative be closed.

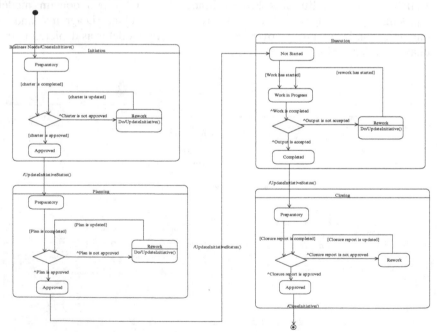

**Fig. 5** State Machine Diagram for INITIATIVE

The other 9 State Machine Diagrams defined for the more sophisticated Classes which captures the behaviour of a further subset of related Classes within the larger settings were:

(1)   PHASE -- which could take on the state of "Not started", "In Progress", "Cancelled", or "Completed"; depicting the phase of the project/programme is in;

(2)   TASK -- which could take on the state of "Initial", "Planned", "In Progress", "Cancelled" or "Completed"; depicting the status of a task;

(3)   ACTIVITY – which could take on the state of "Not started", "Cancelled", "In Progress" or "Completed"; depicting the status of an activity;

(4)   OUTPUT -- which could take on the state of "Planned", "In Progress", "Rework", "Cancelled", "Completed" or "Accepted"; depicting the stages of the project deliverable;

(5)   PROGRESS -- which could take on the state of "On time", "Behind schedule", "Ahead of schedule" or "No progress"; describing the overall status of the project/programme;

(6)   PERFORMANCE -- which could take on the state of "Not Met" or "Met"; describing whether the expected contributions of the project at the organizational level have been met;

(7)   DOCUMENT -- which could take on the state of "Draft", "Approved" or "Distributed"; describing the states which a document must go through before it is finally published and distributed;

(8)   ISSUE -- which could take on the state of "New", "Processing", or "Closed"; defining the state an issue faced by a programme/project is expected go through before it can be officially closed;

(9)   CHANGE REQUEST -- which could take on the state of "New", "Processing" or "Closed"; defining the states a Change request is expected go through before it can be officially closed.

Last but not least in support of the INITIATIVE's State Machine Diagram to define the behaviour of the BProjM domain model are 6 Communication Diagrams which correspond to the INITIATIVE's state and sub-state that the project could be in at any one time namely (1) "Initiation" state, (2) "Planning" state, (3) "Execution – status update and monitoring" state, (4) "Execution – change management" state; (5) "Execution – reporting" state and (6) "Closing" state.

Taking the state of "Execution – status update and monitoring" as an example, the Communication Diagram as depicted in Fig. 6 defines that project execution is monitored at two levels. At the project level in the lower half of the diagram, it is shown that update of activity status triggers the status update of other related Classes namely TASK, PHASE, INPUT and PROGRESS. Of particular importance is the update of progress status which will indicate if the project is on time, behind schedule or ahead of schedule. The update of task status on the other hand, triggers an update of the output status to reflect if the output of this task is in progress, completed, cancelled, reworked or accepted. Similarly, update of phase status triggers an update of milestone status. If this happens to be the juncture where a milestone is hit, the actual benefit realization date in PERFORMANCE will be updated accordingly.

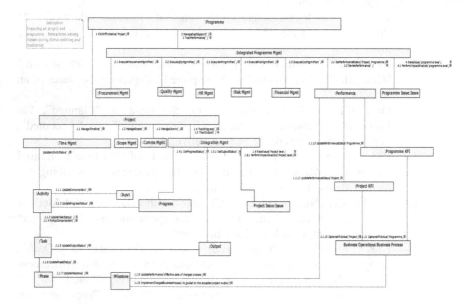

**Fig. 6** Communication Diagram for INITIATIVE's "Execution – status update and monitoring" state

In the centre of the diagram is "Integration Management", i.e. the last checkpoint at the project level where the overall progress of the project and the status of the output are reviewed. Issues if any will be raised and by doing so, the project transits into the "change management" mode in the "Executing" state. In the last segment on the right, integrated programme management reviews the performance of the projects as well as the programmes; and may also initiate the transition into the "change management" mode by raising cross-projects or programme level issues.

## 3.4    Capturing P.M.I.S Domain-Imposed Requirements from the UML Model

As highlighted by the empirical research, the "integrated P.M.I.S" required to support business project management is to enable effective exchange of information between projects, programmes and the parent organizations throughout the project life cycle. With this as the starting point, the first step was to develop the Use Case Diagrams to describe the usage scenarios i.e. how the system is expected to support each stage of the project life cycle. As presented in Table 2, the result is 4 Use Case Diagrams defined with (1) Use Cases i.e. the expected system functionalities of the P.M.I.S; and (2) the list of stakeholders who will be invoking them.

**Table 2** Overview of the P.M.I.S Requirements

| Use Case Diagram | Use Cases | Stakeholders |
|---|---|---|
| Initiation | • Initiate new programme<br>• Initiate new project<br>• Generate project charter<br>• Review and approve project charter<br>• Generate programme charter<br>• Review and approve programme charter | • Senior Management<br>• Programme Manager |
| Planning | • Initiate programme planning<br>• Develop project plan<br>• Review and approve project plan<br>• Complete programme plan<br>• Review and approve programme plan | • Senior Management<br>• Programme Manager<br>• Project Manager |
| Execution | • Retrieve work assignment<br>• Update work status<br>• Review project status<br>• Review project performance<br>• Review programme dashboard | • Senior Management<br>• Programme Manager<br>• Project Manager<br>• Project Team Members |
| Closing | • Prepare project closure report<br>• Review and close project<br>• Prepare programme closure report<br>• Review and close programme | • Senior Management<br>• Programme Manager<br>• Project Manager |

To elaborate further on this, let's take the Use Case Diagram created for the "Executing" state of the project as an example. As indicated in Fig. 7 [43], the integrated P.M.I.S is to allow the project team members to retrieve their work assignment and captures the work status upon completion of a day's work. In addition, the system is expected to retrieve the latest information about the project and/or the programme; presents and formats them according to the requirements of different stakeholders namely the Project manager (who reviews project status), Programme manager (who reviews project benefits) and senior management (who reviews programme dashboard).

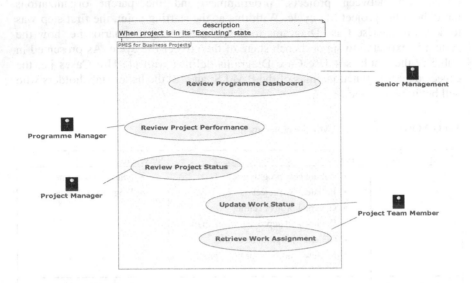

**Fig. 7.** Use Case Diagram for P.M.I.S – during project execution [43]

Each of the Use Case in this Use Case Diagram in turn, is accompanied by a Sequence Diagram that details the system functionality to be fulfilled by the integrated P.M.I.S. Using "Update Work Status" as an example, Fig. 8 is the Sequence Diagram which describes both user-defined requirements and the domain-imposed requirements.

Since the key objective here is to demonstrate how the domain model can be leveraged during the requirement study stage, only the domain-imposed requirements are specified in sequence of actions whilst the user-defined requirements are simply described using a note. The rationale behind this is so that the diagram remains compact and readable. Thus the ellipse on the left is the summary of the user-defined requirements which reflect users' goal, intention and wishes, while the ellipse on the right denotes the domain-imposed requirements which are essentially a transcript of the Communication Diagram for "Status Update and Monitoring" as described in section 3.3.

What appears to be straight forward here is not at all trivial in terms of impact. The system requirements for "Update work status" captured through the conventional means of interviewing project team members are likely to provide a false impression that that it is just a simple function that retrieves and updates only the activity's status. This "discrepancy" may not be uncovered until the requirements of management reporting are discussed with the project managers. But given the domain model in particular the Communication Diagram for "Status Update and Monitoring" in this case, the domain-imposed requirements for the function to also trigger the status update for the corresponding TASK, OUTPUT, PHASE and MILESTONE could be detected much earlier and captured into the

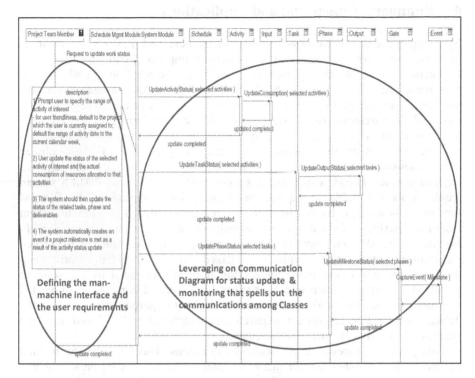

**Fig. 8** Sequence Diagram for P.M.I.S - during 'Update work status' [43]

system requirements accordingly. In addition, since business project management is a relatively new field of study, the domain model developed based on academic research findings can be used as a credible reference to converge the stakeholders' views. As uncovered during the empirical research, different project managers do have different views even on the fundamentals of how ACTIVITY, TASK, OUTPUT, PHASE and MILESTONE are related to each other. In the absence of the BProjM domain model, the requirement study would have taken up a lot more time to identify the basic concepts of business project management, confirm the derived understanding with the stakeholders and get the consensus on the definitions before the exercise can proceed further.

Thus by applying the same techniques of extracting the domain-imposed requirements from the domain model for all the Use Cases, the first draft of the system requirements for the integrated P.M.I.S can be developed within a very short period of time. At the end of the requirement study, the original domain model was added with (a) 4 Use Case Diagrams which provide an overview of the overall system requirements; (b) 20 Sequence Diagrams which denote the series of actions that must be carried out by each of the Use Case / system functionality.

## 4    Summary, Conclusion and Implications

By using the case of business project management and P.M.I.S for business projects, we have demonstrated that domain-imposed requirements of an application system can be captured based on basic research findings which have been translated and expressed in the form of a UML model in a straight forward manner. Given the preview of the fundamental concepts in the BProjM domain model, the requirement analyst was able to validate the information they received during requirement study; add value by highlighting the deviations from standard practice for further discussions; check the completeness of the gathered requirements after the exercise etc. In addition, since the domain-imposed requirements are already expressed in semi-formal notations, the P.M.I.S system requirements were able to capture an accurate interpretation of the BProjM fundamentals which is backed by academic research, rather than extracting them from the descriptive and subjective statements of needs from the stakeholders. In summary, the devised method has effectively combined the goal-oriented requirements engineering (typically engaged by the problem world) and functional-oriented requirement engineering (typically engaged by the solution world) [19]. It is also conclusive that the resulting BProjM domain model has led to a more focused and productive process of eliciting requirements; as well as the delivery of the P.M.I.S system requirements which are well aligned with the business project management fundamentals.

This proposal however, is not without limitations. Theoretically, the method can be applied to other subject matter of interest, as long as its basic research findings can be expressed in the form of theoretical framework and the components can be expanded using prevailing standards. But the fact remains that it has only been tested in the case of business project management and P.M.I.S. There are still more work to be done for its generalization and fine-tuning.

Secondly, the use of the UML model as the source of domain-imposed requirements will work well only if the model is already a comprehensive representation of the domain knowledge. In other words, the basic research must be one that aims at producing a holistic understanding of the targeted domain, as in the case of the business project management research discussed in this paper. Otherwise, the resulting model is at best an isolated perspective of the domain knowledge which will not yield the desired results. Alternatively, the required comprehensiveness in the model could be achieved by cumulating findings of basic research of the same domain over time. The good news is that the method proposed by this paper has already built in the ability to reuse existing work which includes the resulting model itself. Thus the desired model can be constructed by applying the same series of steps to translate and incorporate new research findings into the UML specifications progressively. In which case, in addition to having an objective source to extract domain-imposed knowledge, the development of application system stands to benefit from the direct injections of on-going basic research findings that would help to close the gap between theories and practices [7]. The realization of this vision however, would require inter-disciplinary collaborations between the management researchers and the domain modellers; as well as between the academia and the software engineering industry. Furthermore, compliance to semantic specifications such as Business Vocabulary and Business Rules [38] introduced by OMG must be enforced in order to minimise the subsequent need to realign the terms. Standards organization such as OMG which already have a head start in this area would be in an ideal position to coordinate the required efforts.

Despite the limitations and the need for future work, the implication brought about by the proposal is potentially more profound; since it has partially answered March & Allen's [23] call for "ontology of the artificial" for the socially constructed world of business in order to facilitate conceptual modelling of system requirements. By applying the devised method to all other management areas, a multi-facet enterprise domain model can theoretically be developed in a similar manner, based on which domain-imposed requirements of enterprise systems can be captured. The attempt to create common domain model for the more mature management disciplines (such as human resources management, procurement management, financial management etc.) may appear challenging since most system developers or consulting houses already have some forms of reference models in place. It must be noted however, that most of these existing models are process based i.e. a product of design/solution world, constructed based on a subjective interpretations of the underlying problem domain. In contrast, the proposed UML model is an ontological view of the domain based on basic research to support the design of the solution. Thus the need for the proposed domain model for the more mature management disciplines should not be any different. The real challenge is how to make system developers adopt a domain model of a different nature and one which is other than their own. Incidentally, this is an issue similar to promoting the adoption of standards, i.e. the core mission of a standard organization, which we have proposed earlier to take on the lead in coordinating the domain modelling efforts.

When such an enterprise domain model successfully becomes a common reference for all system developers, it is envisaged that the software industry would become more efficient with shorter system development time. Given the shared ontological foundation, data interoperability issues across organizations should also become minimal. Last but not least with a more levelled playing field, the ability to meet standard domain-imposed requirements would become a pre-requisite for all enterprise systems. Consequently, systems selection exercises would rightfully place emphasis on evaluating the software's ability to meet user-defined requirements especially in terms of technical design and implementation flexibility.

# References

[1] Abels, S., Ahlemann, F., Hahn, A., Hausmann, K., Strickmann, J.: PROMONT – A Project Management Ontology as a Reference for Virtual Project Organizations. In: Meersman, R., Tari, Z., Herrero, P. (eds.) OTM 2006 Workshops. LNCS, vol. 4277, pp. 813–823. Springer, Heidelberg (2006)

[2] Berztiss, A.T.: Domain analysis for business software systems. Information Systems 24(7), 555–568 (1990)

[3] OMG, Business Motivation Model (BMM) Version 1.0. OMG Document Number: formal/2008-08-02 (2008), http://www.omg.org/spec/BMM/1.0/ (retrieved on May 10, 2009)

[4] OMG, Business Process Definition MetaModel II (BPDM): Process Definitions Version 1.0. OMG Document Number: formal/2008-11-04 (2008), http://www.omg.org/spec/BPDM/1.0 (retrieved on May 10, 2009)

[5] Brown, A.W., et al.: Introduction: Model, Modelling and Model Driven architecture (MDA). In: Model-Driven Software Development, pp. 1–16. Springer, Heidelberg (2005)

[6] Bryde, D.J.: Modelling project management performance. International of Quality & Reliability Management 20(2), 229–254 (2003)

[7] Bryde, D.J.: Project management concepts, methods and application. International Journal of Operations & Production Management 23(7), 775–793 (2003)

[8] Christenson, D., Walker, D.H.T.: Using vision as a critical success element in project management. International Journal of Managing Projects in Business 1(4), 611–622 (2008)

[9] Cicmil, S.J.K.: Critical factors of effective project management. The TQM Magazine 9, 390–396 (1997)

[10] CPPM, A Guide to the Project & Program Management Standard Certified Project Manager (CPM) and (CPP) (2007), http://www.iappm.org (retrieved on August 28, 2008)

[11] Cranefield, S.: Networked Knowledge Representation and Exchange using UML and RDF. Journal of Digital information 1(8) (2001)

[12] Dinsmore, P.: Why should project management matters to CEO (2006), http://www.chiefprojectofficer.com/column/175 (retrieved on April 12, 2007)

[13] Faisal, A.: Trapped in between: Realities of the project world. Trafford Publishing (2006)

[14] GAPPS, Global Alliance for Project Performance Standards - A framework for performance based competency standards for global level 1 and 2 project managers (2007), http://www.globalpmsstandards.org (retrieved on January 15, 2009)

[15] Geoghegan, L., Dulewicz, V.: "Do project managers' leadership competencies contribute to project success?". Project Management Journal 39(4), 58–67 (2008)

[16] Gómez-Pérez, A., et al.: Methodologies and methods for building ontologies. In: Ontology Engineering, pp. 107–197. Springer, Heidelberg (2006)

[17] Grangel, R., et al.: UML for enterprise modelling: basis for a Model-Driven Approach. In: Enterprise Interoperability - New Challenges and Approaches, pp. 91–101. Springer, London (2007)

[18] Gruber, T.: What is an ontology (2005), http://wwwksl.stanford.edu/kst/what-is-an-ontology.html (retrieved on April 12, 2008)

[19] Hull, E., et al.: Requirements engineering in the solution domain. In: Requirements Engineering, 2nd edn., Springer, Heidelberg (2005)

[20] ICB, IPMA Competency Baseline V3.0 (2006), http://www.ipma.ch (retrieved on August 28, 2008)

[21] OGC, Introduction to PRINCE2 (2008), http://www.ogc.gov.uk/prince/about_p2/about_intro.html (retrieved on May 2, 2008)

[22] Kotnour, T.: Organization learning practices in the project management environment. International Journal of Quality & Reliability Management 17, 393–406 (2000)

[23] March, S.T., Allen, G.N.: Challenges in Requirements Engineering: A Research Agenda for Conceptual Modeling. In: Lyytinen, K., Loucopoulos, P., Mylopoulos, J., Robinson, B. (eds.) Design Requirements Engineering. LNBIP, vol. 14, pp. 157–165. Springer, Heidelberg (2009)

[24] Maylor, H., et al.: From projectification to programmification. International Journal of Project Management 24, 663–674 (2006)

[25] Milosevic, D.Z., Srivannaboon, S.: A theoretical framework for aligning project management with business strategy. Project Management Journal 37(3), 98–100 (2006)

[26] Morris, P.W.G.: Research trends in the 1990s – the need now to focus on the business benefits of project management. In: The Frontiers of Project Management Research, Project Management Institute, ch. 2 (2003)

[27] Morris, P.W.G., Jamieson, A.: Moving from corporate strategy to project strategy. Project Management Journal 36(4), 5–18 (2005)

[28] Nagypal, G.: Ontology Development - Methodologies for ontology engineering. In: Semantic Web Services, pp. 107–134. Springer, Heidelberg (2007)

[29] OMG, Introduction to OMG's Unified Modelling Language (UML®) (2004), http://www.omg.org/gettingstarted/what_is_uml.html (retrieved on July 17, 2010)

[30] Omoronyia, I., Sindre, G., Stålhane, T., Biffl, S., Moser, T., Sunindyo, W.: A Domain Ontology Building Process for Guiding Requirements Elicitation. In: Wieringa, R., Persson, A. (eds.) REFSQ 2010. LNCS, vol. 6182, pp. 188–202. Springer, Heidelberg (2010)

[31] OSM, Organization Structure MetaModel 2nd Initial Submission Version 0.5. Submitted by 88Solutions, Adaptive, Borland Software, Data Access Technologies, EDS, Lombardi, Software., in response to: Organization Structure Metamodel RFP (OMG Document bei/2004-06-05) (2006)

[32] Pan, Y., Xie, G.T., Ma, L., Yang, Y., Qiu, Z., Lee, J.: Model-Driven Ontology Engineering. In: Spaccapietra, S. (ed.) Journal on Data Semantics VII. LNCS, vol. 4244, pp. 57–78. Springer, Heidelberg (2006)

[33] PIPC, PIPC Global Project Management Survey News Releases, London (December 2004)

[34] PMI, PMBOK® Guide : A Guide to the Project Management Body of Knowledge, 4th edn., Project Management Institute (2008)

[35] Reinhartz-Berger, I.: Towards automation Of domain modelling. Data & Knowledge Engineering 69, 491–515 (2010)

[36] Rolland, C.: From Conceptual Modeling to Requirements Engineering. In: Embley, D.W., Olivé, A., Ram, S. (eds.) ER 2006. LNCS, vol. 4215, pp. 5–11. Springer, Heidelberg (2006)

[37] Siau, K., Lee, L.: Are use case and class diagrams complementary in requirements analysis? An experimental study on use case and class diagrams in UML. Requirements Engineering 9, 229–237 (2004)

[38] SBVR, Business Vocabulary and Business Rules version 1.0. OMG Document Number: formal/2008-01-02 (2008), http://www.omg.org/spec/SBVR/1.0 (retrieved on May 11, 2009)

[39] Srivannaboon, S.: Linking project management with business strategy. Project Management Journal 37(5), 88–96 (2006)

[40] PMI, The Standard for Program Management. 2nd edn., Project Management Institute (2008)
[41] Vauquier, D.: Semantic Modelling Version 1.3. Praxeme Institute (2008)
[42] Winter, M., et al.: Focusing on business projects as an area for future research: an explanatory discussion of four perspectives. International Journal of Project Management 24, 699–709 (2006)
[43] Wong, S.M., et al.: Developing P.M.I.S for business projects based on social science research findings & ontology modelling. In: Proceedings of IADIS International Conference - Information Systems, Porto, Portugal (2010)
[44] Zwikael, O.: Top management involvement in project management: Exclusive support practices for different project scenarios. International Journal of Managing Projects in Business 1(3), 387–403 (2008)

# Biographies

**Siaw Ming Wong** is a certified management consultant who specializes in the management of large-scale regional and global business transformation programmes, with over 20 years of systems implementation and project management experience in both end-user and international consulting firms. A recipient of the first "Quality team of the year" conferred by the National Computer Board Singapore for her software engineering effort, she has recently completed her double doctorate under a joint programme between Open University Malaysia and Université de La Rochelle in France.

**Jean-Yves Lafaye** is a Professor in Computer Science at the University of La Rochelle and the head of the 'Data Semantics and Integration' team at the L3i laboratory\*, where he is responsible for a number of named scientific projects on Model Driven Engineering and Ontology Design. His present main scientific points of interest are system modeling, formal specification, proof and reasoning.
\*    http://www.univ-larochelle.fr/Laboratoire-Informatique-Image-et-Interaction-L3I.html

**Patrice Boursier** is a Professor in Computer Science at the University of La Rochelle and a Adjunct Professor at University of Malaya, Malaysia. He spends his time between France and Malaysia and currently supervises Joint PhD students from the two countries. His research interest is in the area of Database Systems and Information Systems, in particular, spatio-temporal database and information systems.

# BPR Methods Applied to a Manufacturer
# in the Domotics Sector

M. Bevilacqua, F.E. Ciarapica, and G. Giacchetta

Dipartimento di Energetica, Università di Ancona, via Brecce Bianche,
Ancona, Italy
{m.bevilacqua,f.ciarapica,g.giacchetta}@unian.it

**Abstract.** This paper describes a company reorganization plan developed using
business process reengineering (bpr) in a major enterprise operating in the
domotics sector, whose core business is the manufacture of cooker top extractor
hoods.

In the BPR process, our attention focused on two processes that, for reasons
specific to the company in question, had been pinpointed as high-priority, i.e. the
design and mass production of new products and production programming.

The work-up can be summarized in four main points, as follows:

- Delphi analysis in order to obtain individual opinions;
- complete analysis of the company's situation "As-Is" and preparation of
  IDEF0 diagrams describing the business processes;
- proposal of changes to the high-priority processes requiring
  reorganization and implementation of the new "To Be" diagrams;
- implementation of the new management and organizational solutions;

The outcome of the As-Is phase can be translated into the definition of:

a) functional relations describing the current situation, that correlate the
   activities forming the process in question and pinpoint the inputs, outputs,
   controls and resources characteristic of each activity;
b) random and/or temporal sequences of the stages comprising each activity;
c) performance and cost measurements;
d) lists of bottlenecks and superfluous activities.

After completing the analysis of the company's activities, the new network of
activities was designed using the "To-Be" models.

The aims of the redesign of the flow of activities were as follows:

1) to eliminate any activities that do not add value to the product/service in
   a manner perceptible to the customer;

M. Glykas (Ed.): Business Process Management, SCI 444, pp. 389–405.
springerlink.com

2)  to rationalize the essential activities, seeking to eliminate bottlenecks and reconsider the organization of actions that generate delays;

3)  to make the process more flexible and adaptable, concentrating where necessary on the handling of "exceptions to the rule";

4)  to prevent the risk of errors and returns, both by ensuring the adequacy of information and by involving the end-user of the service beforehand;

5)  to learn from examples of excellence, copying solutions that have already been tried and tested, and have proved effective.

An *Object-State Transition* approach was used to generate the new "To-Be" models, focusing our attention on the objects (inputs, outputs, controls and resources) and on how they change during the processes. The action taken on the company's organizational structure generated a series of important changes in its internal hierarchy, the reference roles and the responsibilities of the people involved.

**Keywords:** business process reengineering, domotic sector, supply chain management, idef methods, performance evaluation.

# 1 Introduction

Progressive changes in the markets, increasing globalization and the consequent new business prospects are currently imposing the need for a continuous reconsideration of business management methods. In this light, according to Mohanty and Deshmukh (2000) Business Process Reengineering (BPR) bas become a key factor, capable of facilitating the creation of an evolutionary structure that ensures effective organizational changes. The fundamental philosophy of the business process reengineering is an innovative approach to change management, resulting in best practices (Kam et al. 2003). An organization's business goals and strategies may have to be modified and targeted to raise the four performance standards, namely, conformance to standards, fitness for purpose, process cycle time, and process cost (US Department of Defense 1994) to match the changing customers' needs and market conditions (Lee and Chuah 2001).

The enterprise analyzed in this work operates in the domotics sector and its products are sold on markets virtually all over the world to industrial users (*B2B*) and end users (*Business to Consumers*). In the last decade, the company has experienced a period of economic growth, market expansion and business diversification. It has progressively succeeded in governing and managing a considerable increase in production volumes and, with time, it has become specialized in the development of other products. It has grown so rapidly, however, that it has been unable to achieve a parallel adaptation of the whole organizational and logistics system, and of the technical area, to the changing needs of the various company functions.

The design area has tended to depart from the needs of the commercial management, logistics have been finding it increasingly difficult to cope with procurement problems with old and new suppliers. Delivery backlogs and stock-outs have happened more and more often. To get the company's growth back on course, avoiding any imbalance between the company functions, and to make the *enterprising system* efficient, flexible and dynamic, the management opted to undertake a Business Process Reengineering scheme, i.e. to structure the organization and the company management placing the main focus on what generates value, i.e. the company processes.

The first step was to appoint a team to implement the BPR and to establish its top-priority objectives, what the scheme aimed to achieve within the first 4 months of the project:

- to bring technical production efficiency up to 70%;
- to reduce stock-outs to a sporadic phenomenon;
- to redesign the flow of activities involved in designing a new product and going into production;
- to reduce delivery backlogs to phenomena related only to contingencies beyond the company's control;
- to reorganize the technical area into a design office, a technical (ex industrialization) office and a laboratory;
- to succeed in budgeting with the real industrial costs, up to the second contribution margin, in mind;
- to define the fundamental economic indicators (contribution margins, value of sales and consumption, direct and indirect costs) in operational terms and then align them with the budgeted values (within a year).

## 2  Material and Methods

Since the early 1990s, BPR has become one of the most popular topics in organisational management, creating new ways of making business (Tumay, 1995). Davenport (1993) highlighted the roles of new organizational structures and human resource programs in developing process innovation. Since improving business performance was not achieved by automating existing business activities, many leading organisations have conducted BPR in order to gain a competitive advantage. The first wave of BPR was focused on the radical change of internal business processes. Furthermore, it was particularly suggested that TQM should be integrated with BPR (Al-Mashari and Zairi, 1999).

The second wave of BPR began in 1996 when the Internet and World Wide Web phenomenon took off and provided an IT infrastructure that enabled electronic business and new forms of Web-based business processes (El Sawy,

2001). To meet customer demand, companies depend on close cooperation with customers and suppliers. BPR driven by e-business should not be based only on the radical redesign of intra-organisational processes, but should also be extended to the entire business network (internal and external).

BPR tools support the *'re-thinking'* of business processes, and Workflow Management (WFM) systems are the software applications that make these re-engineered processes possible. Each of these tools requires an explicit representation of the business processes to hand. According to Davenport and Beer (1995) an ideal process modelling method for BPR would provide a simple but expressive modelling mechanism that reflects the customer orientation and cross-functional nature of BPR. Most of the business process modelling techniques used are vendor specific, i.e. they are supported by just one tool. Only a few tools use a generic technique such as Petri nets, Structured Analysis and Design Technique (SADT), Integration Definition for Function Modelling (IDEF0), or Event-driven Process Chains (EPCs).

The AI0WIN software developed on the standard IDEF0 (Integration DEFinition for function modeling 0) was used to model the business processes.

In IDEF0, the activities are graphically represented by boxes containing an indication of their name and number (fig. 1). The concepts graphically represent the interfunctional relationships between the activities comprising the model. They can be distinguished as follows:

⇒  Input: information or object needed to implement the activity;
⇒  Output: information or object achieved as a result of the activity;
⇒  Constraint: condition or circumstance governing the performance of the activity;
⇒  Mechanism: person or means conducting the activity.

Thanks to its modeling features, the IDEF0 enables us to proceed with a hierarchical breakdown of the activities and thereby "dissect" them in a sufficient degree of detail.

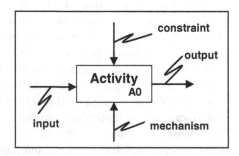

**Fig. 1** Graphic representation of activities

## 2.1 Delphi Method

The study of relations between the activities and the interviews with personnel has pointed out some problems during the modeling phase of the as-is process. The people involved in company are aware of their role, but they do not have a thorough understanding of the process as a whole, with a consequent weakness in communications between the parties, which makes it impossible to organize improvement plans. moreover, auditing practices are only undertaken at the tail end of the process and the entire procedure is carried through in a sub-optimal manner.

To overcome these obstacles a panel of experts was formed encouraging communicational exchanges and meetings where the operators could contribute their knowledge and information about the process. The panel was made of 8 participants, encompassing 2 academicians, whose research studies are mainly focused on operation management, 3 designers and 3 managerial operators involved in the processes analyzed. This number of participants, which at first sight may seem rather large, derives from the delphi technique (Linstone and Turoff 1975) adopted to operate with the panel. The delphi technique is a structured process which investigates a complex or ill-defined issue by means of a panel of experts. The methodology proves to be an appropriate research design for this type of research and permits to obtain individual opinions by a structure of a group during a communication process. The panel worked for a period of about two weeks, and the sessions were planned on a three round delphi process. In the first stage a series of statements concerning the requirements of processes were generated individually and anonymously by the experts. Then, all the statements were collected and delivered to the panel group participants, who were required to indicate their level of agreement; answers were finally feedback to the panel.

**Table 1** Design office

| STRENGTHS | WEAKNESSES |
|---|---|
| • Experience of office manager and staff<br>• Willingness of staff to revise working procedures | • Lack of visibility of shared objectives<br>• Poor use of IT support systems (project standardization)<br>• Activities not strictly related to the "natural mission" of the department |
| OPPORTUNITIES | RISKS |
| • Potential for improvement with an adequate focus and tidy organization | • Efforts that lack coordination and focus<br>• Activities unnecessary to the company because they are not focused |

The panel focused his attention on the processes carried out by Design and Product industrialization offices. These processes, i.e. the design and mass production of new products and production programming, have been pinpointed as high-priority by the company. The results of the Delphi procedure have been summarized in table 1 and 2.

**Table 2** Product industrialization office

| STRENGTHS | WEAKNESSES |
|---|---|
| • Professional expertise and in-depth knowledge of TW production processes<br>• Strong motivation of staff, keen to exploit individual and group capabilities | • Lack of visibility of shared objectives<br>• Personalized, individual use of IT support systems<br>• Excessively diversified and dissimilar activities; the office's "core business" is not clearly defined |
| **OPPORTUNITIES** | **RISKS** |
| • Active involvement in a new definition of the office's objectives and commitments, also in terms of a reliable cost accounting and effective management control | • Possible repercussions for production programming<br>• Imprecise cost accounting<br>• Difficult connection to the central Management Control system, if any |

## 3 Analysis of As-Is Procedures

The outcome of the As-Is phase can be translated into the definition of:

a)  functional relations describing the current situation, that correlate the activities forming the process in question and pinpoint the inputs, outputs, controls and resources characteristic of each activity;
b)  random and/or temporal sequences of the stages comprising each activity;
c)  performance and cost measurements;
d)  lists of bottlenecks and superfluous activities.

### 3.1  The Product Design and Industrialization Process

The design and mass production of new products were the first processes to be analyzed during the procedure BPR and are the main object of the present paper. The BPR team's entire operation on this process focused on the internal

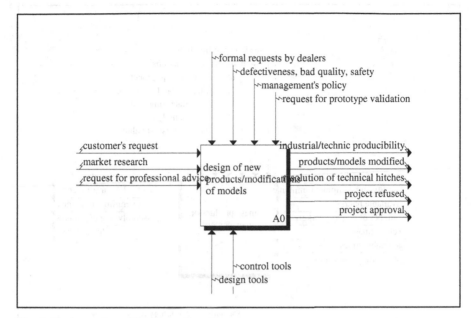

**Fig. 2** Design and modifications of products

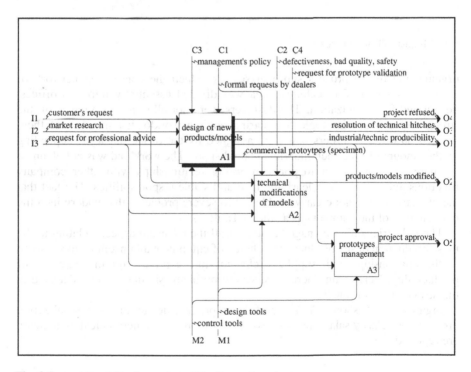

**Fig. 3** Second level_Design and modifications of products

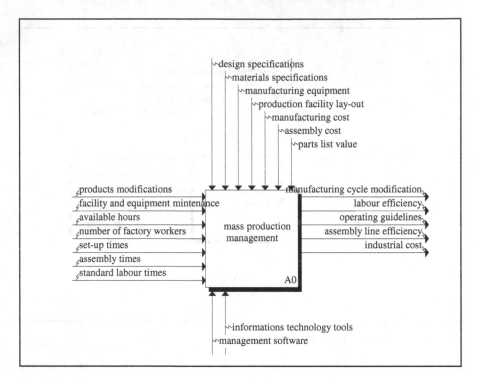

**Fig. 4** Industrialization process

organization of the process sub-matrixes, on which the consequent network of activities is based. The technical area was divided essentially into two offices, design and industrialization. The designers were not allocated to projects on the strength of any precise logic for sharing out the workload; almost all of them worked on different orders that were assigned by the office manager, depending on the amount of time each individual could spare. The workload was not planned, so there was constant confusion concerning relationships with other company functions due to a lack of clearly-defined and stable responsibilities. The fact that the objectives were not clearly visible led to severe problems that undermined the performance of the company as a whole (Table 1).

The industrialization managed a little of all the technical aspects relating to the mass production of the products, the choice of equipment and machine tools, some of the relationships with suppliers, changes to the project for the purposes of producibility. Here again, there were severe problems similar to those detected in the design office (Table 2).

Figures 2 to 5 show "As-Is" models of product design and industrialization process. For clarity sake only two levels of two processes hierarchical breakdown are reported.

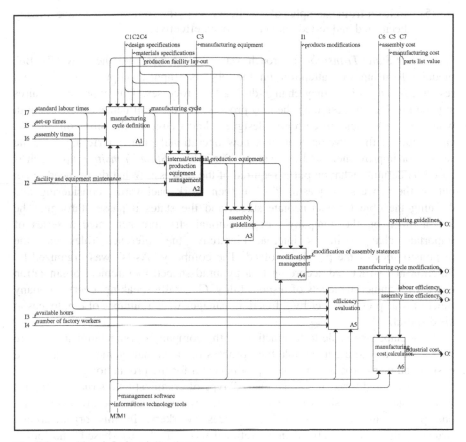

**Fig. 5** Second level_Industrialization process

# 4 To-Be Processes

After completing the analysis of the company's activities, the new network of
activities was designed using the "To-Be" models.
The aims of the redesign of the flow of activities were as follows:

1) to eliminate any activities that do not add value to the product/service in
a manner perceptible to the customer;
2) to rationalize the essential activities, seeking to eliminate bottlenecks and
reconsider the organization of actions that generate delays;
3) to make the process more flexible and adaptable, concentrating where
necessary on the handling of "exceptions to the rule";
4) to prevent the risk of errors and returns, both by ensuring the adequacy of
information and by involving the end-user of the service beforehand;

5) to learn from examples of excellence, copying solutions that have already been tried and tested, and have proved effective.

An *Object-State Transition* approach was used to generate the new "To-Be" models, focusing our attention on the objects (inputs, outputs, controls and resources) and on how they change during the processes. This approach requires anyone involved in designing the new processes to be uninfluenced by the way in which the processes are currently designed; knowledge of the situation "As-Is" can influence the development of a new operational plan, interfering with the search for improvement and innovation. The *Object-State Transition* approach is based on defining what output is required of the process: "What do we need to get out of the process in question?". We then work backwards, considering and defining the object's intermediate stages and the states it passes through. The action taken on the company's organizational structure generated a series of important changes in its internal hierarchy, the reference roles and the responsibilities of the people involved. The company "As-Is" was organized by function and structured according to a pyramid-shaped hierarchical organization chart, with numerous levels of responsibility. Generally speaking, every company function was presided over by a function manager with a number of employees at his disposal.

An attempt was made to take action on the company's organizational structure with a view to making it suitable for a process-oriented management. Both for the design and mass production of new products and for the production programming system, we opted for a "relational matrix" type of organizational structure (fig. 6). Work groups with a simplified hierarchy are created around the processes, with one process manager supported by process workers. In this organizational structure, that can be defined in operational terms as "matrix-based", no longer divided according to company functions, but according to business processes, the path covered by the Primary Processes (PP) and Secondary Processes (SP) meet at various points in the matrix, defining the *"nodes"* in the structure.

The full circles are the nodes in the matrix and represent the decision-making times, both in the matrix correlating the processes and in the *relational submatrixes* of the fundamental process workgroups (in design the submatrixes come under the process managers). The empty circles represent the decisional meeting points that are not necessarily always activated, but remain linked to the needs of the process objectives, which vary depending on the circumstances and may not always be active.

Developing this organization through the work groups, which become the "think tanks" of the processes, capable of fully exploiting the professional capabilities and expertise of the human resources involved, promoting a more strongly felt sense of shared responsibility for the performance of the process.

PRIMARY
PROCESSES

SECONDARY
PROCESSES

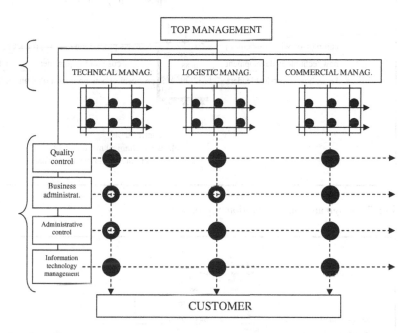

**Fig. 6** Relational matrix

## 4.1 Reengineering the Product Design and Industrialization Process

For the reorganization of the technical area (i.e. the design and industrialization processes), the approach adopted by the BPR team was to join them together synergically, constructing the load-bearing process that goes from design to mass production on each department's specific functions. The BPR team began by suitably redefining the roles of the staff, the relationships with the commercial area and the responsibilities for the process objectives. They concentrated their efforts on getting each employee to fully absorb the *trans-functional* nature of the process and sub-matrixes were defined in order to establish the new organizational architecture for internal process management. The introduction of the professional figures of _product owner_ and _product manager_ were placed at the hub of the entire sub-matrix organizational structure. The fundamental idea is that Design is what links the company reality with the outside world, with potential customers. By creating products, it acts as the *trait d'union* between the customers' requirements, market needs, technological innovation and the enterprise's profitability objectives.

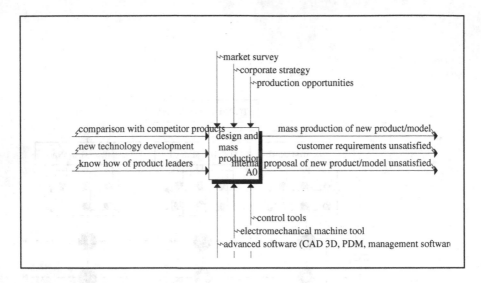

**Fig. 7.** Design and industrialization of products

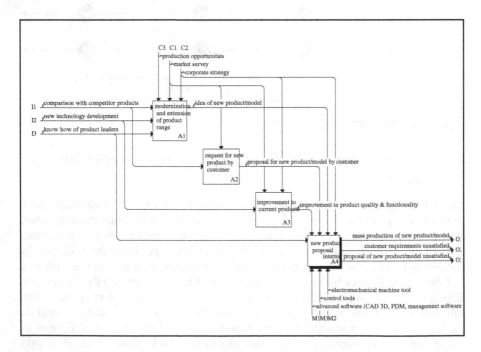

**Fig. 8** Second level_ Design and industrialization of products

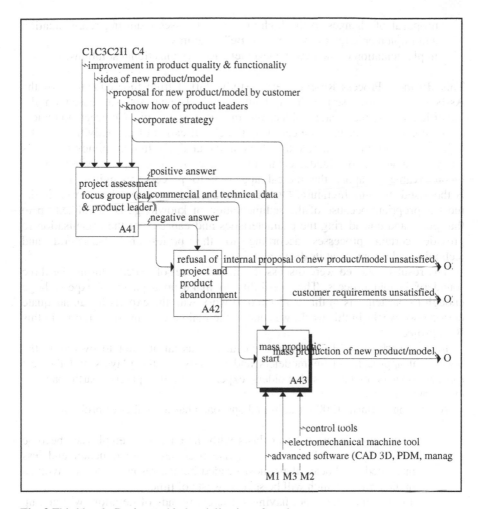

**Fig. 9** Third level_ Design and industrialization of products

Figures 7 to 9 show "To-Be" models of process that goes from design to mass production. For clarity sake only three levels of processes hierarchical breakdown are reported.

## 5  Discussion and Conclusions

The work-up can be summarized in four main points, as follows:

-       complete analysis of the company's situation "As-Is" and preparation of IDEF0 diagrams describing the business processes;

- proposal of changes to the high-priority processes requiring reorganization and implementation of the new "To Be" diagrams;
- implementation of the new management and organizational solutions;

This Business Process Re-engineering study started from the acquisition of the As-Is situation to describe the system. The As-Is analysis was based on the identification of the system objectives in order to set up a correct analytical approach. This has been made easier by the classification of the activities and the objectives according to the role they have in the analysis. In our opinion the As-Is analysis stage is of fundamental importance in a company where it is inconceivable to apply the radical approach to process re-engineering, as it is theorised in some literature. On the contrary, the incremental approach is the most appropriate because of the sedimentation of knowledge accumulated up to that point and considering the characteristics and capacity of the organisation to provide certain processes according to the professional, structural and technological resources at its disposal.

The results obtained were discussed with the panel of experts during the three round of Delphi process. The possibility to make up a panel of experts large enough (8 participants in this work) and to work with the experts for an adequate period (two weeks in this work) was one of the main problems to overcome in this BPR project.

In this work, using Delphi approach, there was an attempt to overcome the traditional approach, where modelers and analysts are the players and the rest (process owners, mangers, stakeholders, experts) are either passive participants or even absent from the scene.

According to Barjis (2009) traditional approach has a number of problems:

- as the interaction of the analysts with the enterprise employees become more and more often, the enterprise becomes more reluctant and less interested to allocate their most needed human resources to be involved in the project, which will be seen as waste of time.
- In turn, modellers, not having sufficient rounds of iteration, will end up with a model that is either incomplete, or there are many assumptions that are intuitively made by the modellers. As a result, the model may contain a lot of flaws. These flaws remain quite undetected as majority of enterprise process modelling is not based on formal semantics to check the models and simulate their dynamic behavior.

In literature, in order to address these challenges in enterprise modelling, innovative approaches have been discussed and introduced such as participative enterprise modelling (Persson, 2001). A central goal of enterprise modelling is to discover domain knowledge and document the enterprise existing business processes. The role of participative modelling is to represent this knowledge in a coherent and comprehensive model, create shared understanding, consolidate different stakeholder views, and in order to do so an extended participation of stakeholders is crucial (Stirna et al., 2007).

In order to expedite the modelling process and validity of the models, Barjis (2009) proposed an approach called collaborative, participative, and interactive modelling (CPI Modelling). The main objective of the CPI approach is to furnish an extended participation of actors that have valuable insight into the enterprise operations and business processes. Three aspects constitute the CPI Modelling approach, where each aspect is a dimension: the collaboration aspect represent the *Experts* (analysts) dimension; the participation aspect represents the *Users* (stakeholders) dimension; and the interaction aspect represents the *Technology* (tools) dimension.

In this work, the process of data collection, using Delphi method, was carried out from two academics that had "facilitators" role. The participants of panel of experts worked almost independently and together with the academics and thus reducing the amount of time to develop the models of the processes.

The use of Delphi methodology, in order to incorporate expert opinions in the re-engineering process, was one of the main points of this work. This technique allowed the academics to access to the positive attributes of interacting groups (knowledge from a variety of sources, creative synthesis, etc.), while pre-empting their negative aspects (attributable to social, personal and political conflicts, etc.). From a practical perspective, the method allowed to collect input from a larger number of participants than could feasibly be included in a group or committee meeting, and from members who are geographically dispersed. Important feature introduced by Delphi techniques was the anonymity of answers. Anonymity was achieved through the use of questionnaires. By allowing the individual group members the opportunity to express their opinions and judgments privately, undue social pressures – as from dominant or dogmatic individuals, or from a majority – were avoided. Furthermore, with the iteration of the questionnaire over a number of rounds, the individuals were given the opportunity to change their opinions and judgments without fear of losing face in the eyes of the (anonymous) others in the group. Between each questionnaire iteration, controlled feedback was provided through which the group members were informed of the opinions of their anonymous colleagues. In this phase additional information, regarding arguments from individuals whose judgments fall outside certain pre-specified limits, were provided. In this manner, feedback comprised the opinions and judgments of all group members and not just the most vocal.

The redesign of the flow was not the only field of action for the reengineering effort, however; it was also necessary to consider modifying the features of the product and dealing with the problem of training for the human resources involved in the process.

The leveling of the hierarchies induced by the matrix structure has made the individuals who previously tended to submit passively to high workloads more responsible; on the other hand, it has brought to the surface and highlighted professional capabilities that lay hidden amidst the restrictions of the previous pyramid structure, prompting a flourishing of the faith and enthusiasm among the employees. There was an evident improvement in the level of co-operation between company functions and company departments connected in the flow of activities of a given process. At the end of the BPR process, the management was able to coordinate the process managers effectively, providing them with flexible and responsible workgroups so that they can achieve the goals of the process.

The BPR team's last activity involved defining performance indicators. The need to measure and then compare results with objectives, be they of a global nature affecting the company as a whole or specific to a given process, prompted the BPR team and management to identify the focal nodes in the process-based business management, specifying which indicators to use to measure the results obtained on these nodes. This led to the creation of a sort of "*process control dashboard*", comprising certain measurements of the company's performance, from which to extrapolate the information needed for a straightforward, clear and rapid assessment of the company's situation.

The basic premise for the development of this economic/financial management control system was a regular comparison between budget and final balance figures relating to costs (direct and indirect), production efficiencies, average prices, equivalent production, quantities, contribution margins, and so on. This ongoing comparison between budget and final balance prompts a careful monitoring of the trend of the company processes.

# References

Al-Mashari, M., Zairi, M.: BPR implementation process: an analysis of key success and failure factors. Business Process Management Journal 5(1), 87–112 (1999)

Barjis, J.: Collaborative, Participative and Interactive Enterprise Modeling. In: Filipe, J., Cordeiro, J. (eds.) Enterprise Information Systems. LNBIP, vol. 24, pp. 651–662. Springer, Heidelberg (2009)

Davenport, T.H.: Process Innovation: Reengineering work through information technology. Harvard Business School Press, Boston (1993)

Davenport, T.H., Beer, M.: Managing information about process. Journal of management Information Systems 12(1), 57–80 (1995)

El Sawy, O.A.: Redesigning enterprise processes for e-Business. McGraw-Hill (2001)

Kam, W., Yung, C., Ting, D., Chan, H.: Application of value delivery system (VDS) and performance benchmarking in flexible business process reengineering. International Journal of Operations & Production Management 23(3), 300–315 (2003)

Lee, K.T., Chuah, K.B.: A Super methodology for business process improvement. An industrial case study in Hong Kong/China. International Journal of Operations & Production Management 21(5/6), 678–706 (2001)

Linstone, H.A., Turoff, M.: The Delphi Method Techniques and Application. Addison-Wesley, London (1975)

Mohanty, R.P., Deshmukh, S.G.: Reengineering of a supply chain management system: a case study. Production Planning & Control 11(1), 90–104 (2000)

Persson, A.: Enterprise Modelling in Practice: Situational Factors and their Influence on Adopting a Participative Approach. PhD dissertation, Stockholm University (2001)

Stirna, J., Persson, A., Sandkuhl, K.: Participative Enterprise Modeling: Experiences and Recommendations. In: Krogstie, J., Opdahl, A.L., Sindre, G. (eds.) CAiSE 2007 and WES 2007. LNCS, vol. 4495, pp. 546–560. Springer, Heidelberg (2007)

Tumay, K.: Business process simulation. In: Proceedings of the 1995 Winter Simulation Conference, Washington, DC, pp. 55–60 (1995)

US Department of Defense (DoD), Framework for managing process improvement. The Electronic College of Process Innovation, sec. 2, p. 6 (1994),
http://www.dtic.dla.mil/c3i/bprcd/3003s2.html

# Managing the SME Clustering Process Life-Cycle

Christina Ampantzi[1], Marianthi Psyllou[1], Evaggelia Diagkou[4], and Michael Glykas[1,2,3]

[1] Financial and Management Engineering Department, University of the Aegean, Greece
[2] Aegean Technopolis, The Technology Park of the Aegean Region, Greece
[3] Faculty of Business and Management, University of Wallangong, Dubai
[4] Department of Interior Design, Technological Education Institution, Athens, Greece

## 1 Introduction

Globalization has changed the nature of innovation requiring a multitude of skills for solving complex challenges; this is why the old closed innovation model, which dominated most of the 20th century, is hurt to death. To survive in the worldwide competitive economic system firms have to invest in Intellectual Capital, as economic activity has become inherently socially situated and embedded in networks of organisational and personal relationships and cooperation, thus forcing a change of paradigm from competition to cooperation or, increasingly, to coopetition.

For this purpose, partnerships and collaborative networks arise and symbiotic relationships are created between transnational companies, micro companies and public institutions. Nowadays firms intend to exchange knowledge in order to accomplish innovation. They can achieve that through clusters, a network tool giving access to information to every company that participates. The interaction through clusters helps firms to manage uncertainty by creating social norms, conventions and habits while the dynamic exchange of knowledge between them contributes in their growth and in certain cases even their survival. The recent economic crisis have made companies review clusters policy and programs summing up with 15 criteria that every world class cluster should observe and the major challenges challenges that it might face.

### 1.1 The Networked Economy and the New Dynamics of Competition

In the last decades there has been a systematic and fundamental change in the way companies carry out innovation activities (Zeng et al. 2010). The recognition of the increasing importance of external sources of knowledge and the demand for firms to adapt their strategies and business models accordingly has seriously threatened the closed model, which postulates the effectiveness of vertically

M. Glykas (Ed.): Business Process Management, SCI 444, pp. 407–456.
springerlink.com        © Springer-Verlag Berlin Heidelberg 2013

integrated R&D departments and the primacy of in-house developed technology2. It assumes a different dimension as firms scan global markets to complement and expand their knowledge base. However, as the knowledge base of an industry increases in complexity, being characterized by widely dispersed pools of expertise, the locus of innovation is more likely to be centred in networks of learning rather than in individual firms (Powell et al., 1996).Following Ahuja (2000) networking means that firms form linkages to obtain access to assets (Nohria and Garcia- Pont, 1991), learn new skills (Kogut, 1988); manage their dependence upon other firms (Pfeffer and Salancik, 1978), or maintain parity with competitors (Garcia-Pont and Nohria, 1998).

In turn, Chesbrough and Crowther (2006) define two types of open innovation strategies: inbound and outbound open innovation. While the former describes the situation by which in-house R&D is 'completed' with resources from outside the boundaries of the company, the latter emphasises the search for external organisations that are better suited to commercialise a given technology. Although none of these modes of open innovation is new, in recent years inbound open innovation (e.g. Laursen and Salter, 2006) has gained space. In a global and networked context (open) search strategies become both an enabler and driver of innovation as they influence the capability of the firm to identify and assess the value creation potential of certain external knowledge and relationships. In this purview, they play an important role in expanding –provided the firm has the necessary absorptive capacity – the firm's vision, knowledge base and scope of action or, conversely, they may prove costly (and eventually counterproductive) if it has no slack. Such variety provides opportunities for firms to choose among different technological paths, creating rich environments for the selection of different technological solutions (Metcalfe, 1994).

## 1.2   The Catalyst Role of Clustering in the Economic Development Process

Moreover, companies in clusters experience stronger growth and faster innovation than those outside clusters (Audretsch and Feldman, 1996; Swann et al., 1998; Baptista, 2000; Klepper, 2007). According to Porter (1998: 80), "Clusters affect competition in three broad ways: first, by increasing the productivity of companies based in the area; second, by driving the direction and pace of innovation, which underpins future productivity growth; and third, by stimulating the formation of new business, which expands and strengthens the cluster itself. A cluster allows each member to benefit as if it had greater scale or as if it had joined with others formally – without requiring it to sacrifice its flexibility". The alliances and networks established within the cluster and with parties external to it are an enabler and driver of the distributed (tacit) knowledge, stimulating positive feedback – also from spillovers – and expanding the absorptive potential and attractiveness of the companies involved.

## 1.3 SMEs in this Context: The (Network) Challenge of Muddling Through

Between 2002 and 2008, small businesses in the 27 Member States of the European Union grew strongly and turned out to be the job engine for much of the European economy with 9.4 million new jobs created during that period7. In particular, SMEs have proved to be important employment growth and innovation engines in high-tech sectors, both through existing firms and ''New Technology Based Firms'' (NTBFs) (Santarelli and Vivarelli, 2007).

Over the last two decades an extensive literature has emerged on the role of inter-firm networks and their impact upon innovation and firm performance (Tomlison, forthcoming; see a review in Hoang and Antoncic, 2003). The focus on SME networking has its origins in a mainstream of earlier studies that showed that SMEs were as innovative as larger firms despite employing less internal resources (Acs and Audrescht 1990). Nevertheless, not all SMEs embrace greater collaboration, as pointed out by Huggins (2001) who uncovered a deep rooted scepticism among many UK SMEs as to the benefits of greater cooperation with external partners for innovation. There are relatively few empirical studies that have captured the nature and intensity of network ties and their impact on innovation. Studies exploring the direct links between cooperative ties and innovation have tended to rely upon binary variables to indicate whether firms cooperate with external partners or not. The amount of social capital and the amount and density of the networks the catalyst SME is able to build at the very early stages of the cluster life cycle will greatly determine how fast the cluster will grow and, consequently, its chances of survival.

Thus far, (open) clusters and regional and global innovation networks have been viewed here as stimulating both growth and innovation from both an individual business perspective and a regional perspective. In particular, they appear to provide the adequate scope and space for SMEs through which to access external knowledge flows, learn and upgrade their IC base. On the other hand, SMEs, due to their dynamism, flexibility and knowledge of the markets, have proven in many cases the effective interface between research and the transformation of new ideas into successful, products, services and ultimately businesses, promoting social and economic development and prosperity. Entrepreneurs and small firms are indeed often able to spot where new technologies meet customer needs and can develop products that meet this demand. Below are mainly aimed at unveiling the elements at the firm level that affect SMEs' capacity to (co-)innovate and collaborate with the prospects of benefitting from and contributing to global networks (for instance, by fostering IC flows) as well as the supportive collaborative tools and methodologies and online services.

Before so doing, this study will address the dynamics and benefits emerging from clusters as well as their recent evolution with a special focus on the reasons for the transition towards bottom-up approaches This analysis will provide the necessary framework to understand further developments at the firm level. Finally, a few conclusions are raised.

## 2 Clusters: An Overview

### 2.1 The Concept of Cluster: Nature and Evolution

The social-network model argues that there is more order to inter-firm interactions and less order to intra-firm interactions than the economic models would imply (Granovetter, 1985). The territory concept could be described using the Camagni (2002: 2396) point of view as a system of localized technological externalities, social relations and local governance which unites the group.

The division between intentional (traded) and unintentional (untraded) is fundamental. While the unintentional is mainly covered by the untraded idea, the intentional can be referred to traded (pure or hybrid market-based transaction, as we will discuss below) flows of knowledge. Studies on clusters have also pointed out the social character of linkages mainly based on personal and informal interactions. In this view, clusters resources are obtained through informal inter-firm ties and personal interactions (e.g. Camagni, 1991; Capello and Faggian, 2005), as well as through contractual (formal) ties, whose network structure also affects the firm's performance as suggested by organizational theory (Gulati et al., 2000). There is little evidence of formal linkages on local contexts (Malmberg and Power, 2005). Traded interdependencies have received considerable attention in the field of strategic management alliances (e.g. Zaheer and Bell, 2005; Gulati et al., 2000; Gulati, 1999) but unfortunately much less in the cluster mainstream. Jenkins and Tallman (2010) argue that knowledge flows are more effectively conveyed between formal alliance partners – even within a cluster – since this type of channel is more productive than those based upon informal interactions (Almeida and Kogut, 1999; Gomes-Casseres et al., 2006). However, the informal relationships that exist around the formal ones moderate the transfer (Jenkins and Tallman, 2010: 613). Accordingly, research collaborations (i.e. formal networks) are also important mechanisms for the exchange of knowledge. Conversely, other scholars (Tsang, 2005) emphasise the minor importance of formal linkages relative to untraded exchanges (i.e. spillovers) such as the scanning of research publications and labour mobility. Conclusions and experiences cannot be easily extrapolated to other clusters, not only because the type of knowledge created (analytical vs. synthetic) is an important moderator (Moodyson, 2008), making it difficult to figure out the outcomes, but fundamentally because the knowledge exchanges that take place are strongly grounded on the individual's perceptions about the risk and opportunities these flow-exchanges might offer, trust being a key driver. In clusters which are mainly based on synthetic knowledge.

This brief introduction is an account of the complexities (and difficulties) that are involved in growing and assessing clusters, particularly as seen through the rich lens of social networks.

### 2.2 How Knowledge Flows in Clusters

Cluster resources and capabilities or higher-order capabilities (Foss, 1996) are the result of the combination and interaction of all the localized elements self-reproduced and self-reinforced in the spatial context, including the strategies (Porter, 1990) of co-located firms in generating competitive advantage capable of upgrading the territory. Nevertheless, the keystone consists in understanding how knowledge is formed and disseminated in order to contribute to the process of knowledge accumulation in clusters, and particularly, in clusters within a global value chain. According to Matusik and Hill (1998) and Henderson and Clark (1990) knowledge is classified into component knowledge and architectural knowledge. As Tallman et al. (2004) pointed out, when referring to clusters component knowledge is related to specific skills and associated technologies which occur in specific cluster sectors with no links to the whole spatial system or cluster. Component knowledge is potentially transferable among organizations and their members within clusters. Conversely, architectural knowledge refers to clusters as a whole, including a complex system of organizational routines which coordinate and integrate the knowledge components. If the component knowledge is transferred to other clusters through the role played, the possibility that it will be assimilated depends on the absorptive capacity and the resources and capabilities of the recipient cluster. Both of these moderate the combination and the knowledge accumulation process. In principle, the competitive advantage gained in the territory – would be kept within the cluster, although in the course of time it could end up being dispersed throughout the industry, becoming thus fully public for non-located firms (Lissoni, 2001).

One shared idea that all agents in clusters have social contacts (e.g. Huber, 2010), which is in part founded on the fact that social networks are exclusive and created by individuals (e.g. Lissoni, 2001; Dahl and Pedersen, 2004), has also been called into question. Informal networks are exclusive, not open to all employees in the local industry, and are even necessary in order to acquire knowledge from informal contacts with universities, being a pre-condition that the parties have participated in formal projects and been educated at the local university. It seems evident that most of the communities are built on personal ties of trust and reputation, rather than from inter-firm arrangements, and also, that many of them arise from commercial partnerships and deals. Lissoni's (2001) work on an Italian mechanical cluster argues that knowledge circulates just within few epistemic communities, rather than within clusters' boundaries. Thus, contrarily to the assumptions that homogeneity prevails in clusters, firms' resources matter a lot, i.e. heterogeneity does matter.

## 2.3 Main Benefits and Its Conditionings: Knowledge Spillovers, Technology Opportunities, Synergistic Effects and Absorptive Capacity

### 2.3.1 Externalities and Knowledge Spillovers

Externalities or KS have been considered in economic literature as dense tied networks which allow and promote tacit knowledge transmission and trust (Uzzi, 1996) and a paradoxical combination of cooperation and competition in the territory (Harrison, 1991). Grossman and Helpman (1992: 16) defined technological spillovers as firms who can access information created by others without paying in a market transaction and where the current owners have no effective recourse. The rationality of untraded or pure KS is that the geographical proximity provides unintentional contacts and interactions which foster knowledge creation and diffusion, thus enabling technological learning among the collocated firms to be achieved in a more satisfactory way. The agglomeration mainstream (e.g. Dumais et al., 2002) assumes that the localized social and institutional interactions produce KS and affect the productivity growth in a positive way. Put differently, the local informal networks and the knowledge theses interactions convey, based on frequent and repetitive interactions, support the idea of untraded flows.

Three main points appear as to deserve more attention by the academia in order to understand flows of knowledge in clusters. First, it seems that the externalities have to be analyzed more by the personal interactions rather than the inter-firm interactions. The latter has been the principal focus of the economic geography perspective (Grabher and Ibert, 2006). Second, formal and traded commercial partnerships and deals (with suppliers, universities, public labs, and so forth) are also assets or flows of knowledge available in clusters, apart from the common KS based on informal contacts. Therefore, externalities or KS are not the unique assets available in clusters. Thirdly, the granted assumption that the knowledge is in the air can arguably lead to naïve policy implications (Breschi and Lissoni, 2001) leading policymakers to focus on the cluster rather than on the networks or epistemic communities (Lissoni, 2001) in which externalities occur –as has been the case thus far.

### 2.3.2 Synergistic Effects of Combining Internal and External Resources

Relational capability allows access to information, technology and other assets which upgrade the (internal) technological capabilities (e.g. Teece, 1987; Lee et al., 2001). Nevertheless, this interaction effect seems to be a two-way street in the sense that there are positive relationships between internal and external variables in innovation (e.g. Cassiman and Veugelers, 2006) as well as negative ones (Laursen and Salter, 2006; Vega-Jurado, 2008). The authors specifically highlight synergistic effects in clusters in accord with the specificity of the environment in which firms maintain frequent and multiple relationships (e.g. Lawson, 1999; Maskell and Malmberg, 1999). In fact, the theory holds that geographic proximity increases not only the frequency of interactions between cluster firms, but also the effectiveness of knowledge exchanges through these interactions by facilitating face to face contact between the firm's members, and by contributing to the emergence of inter-firm trust and institutional norms of cooperation (Bathelt et al., 2004; Lawson and Lorenz, 1999; Maskell, 2001; Storper and Venables, 2004).

### 2.3.3 Technology Opportunities

On the one hand, the debate about relational assets and relational returns (e.g. Dyer and Singh, 1998) draws attention to the fact that external knowledge, such as a technological opportunity, improves innovation capacity and can be found in sources such as firm-university linkages, or relationships with suppliers or customers (e.g. Klevorick et al., 1995; Lee et al., 2001; von Hippel, 1987). Overall, the main sources of competitiveness from this perspective arise not from the firm, but from inter-firm sources of advantages (Dyer and Singh, 1998; Gomes-Cassares, 1984; Smith et al., 1995; Lavie, 2006). This idea is similar to the technological opportunity concept (Nelson and Winter, 1982). A firm incorporates external knowledge flows in learning and interaction processes, distinguishing between industry sources such as suppliers, and non-industry sources such as universities or technology centres (Klevorick et al.,1995). In addition, most of the innovation literature is in agreement with the relational view (e.g. Vega-Jurado, 2008; Huergo, 2006; Caloghirou et al., 2004; Caloghirou ct al., 2001).On the other hand, and in line with this chain of thought, Nelson (1959) and Arrow (1962), following Marshall's fundamentals, pioneered the concept of knowledge spillovers (KS) which occur in agglomerations.

### 2.3.4 Reinforcing Heterogeneity in Clusters

The fertile context found in the cluster provides a competitive advantage to the collocated firms which have access to and benefit from passive and collective efficiencies (Bell et al., 2009) in a restricted way (e.g. Porter, 1998; Saxenian, 1994). Put differently, the technological opportunities in clusters vis-à-vis scattered locations provide an advantage due to the existence of interactions (e.g. Keeble et al., 1999; Storper and Venables, 2004). This conversation assumes co-located firms as a homogeneous block (e.g. Molina, 2001, Cainelli, 2008), which is exactly the opposite of the heterogeneity concept claimed in the strategic management perspective (e.g. Nelson, 1991), also in clusters (Tallman et al., 2004; Jenkins and Tallman, 2010: 608). In essence, the firm-specific resources and capabilities or architectural knowledge moderate the access and exploitation of local knowledge making it necessary to turn to the strategic management perspective; i.e. innovation externalities from the cluster benefit members unequally. Therefore, understanding the real dynamics and complexities that take place in a cluster or networks context is a necessary step in order to identify what the key elements of the firm-cluster and inter-firm relationships are and how they intertwine.

### 2.3.5 Absorptive Capacity of Co-located Firms

Absorptive capacity is one of those fuzzy concepts that, despite its significance to business practice and theory, has not been subject to much rigorous (empirical) analysis. In addressing the firm-level, most papers cite the absorption capacity as a metaphor in a reified manner (e.g. Abbey et al., 2008) or without conducting or addressing empirical exercise (e.g. Tallman et al., 2004). The concept of absorptive capacity (Cohen and Levinthal, 1989; 1990: 128) is defined as "the ability of a firm to recognise the value of new external information, assimilate it and apply it to commercial ends". This concept stresses the idea that internal capabilities are central for a firm's technological capacity and enhance the firm's ability to assimilate and exploit external knowledge. Therefore, the firm's internal resources determine the possibility of using and exploiting external knowledge and, consequently, improving innovation in firms (e.g. Cohen and Levinthal, 1989, 1990; Klevorick et al., 1995; Arbussa and Coenders, 2007; Vega- Jurado et al., 2008; Escribano et al., 2009).

Literature has widely adopted the idea that absorptive capacity was something like a gift, whose obtainment was close to automatic. The idea prevailed that co-located firms, just for the sake of "being there" (Gertler, 1995), would take advantage of knowledge externalities (KS), assuming a direct causality between distance and access to local resources (e.g. Gertler, 1995; Grabher, 2002; Foss, 1996; Lawson, 1999; Maskell and Malmberg, 1999). This idea is in line with the cluster tradition of addressing co-located firms as a homogeneous block (e.g. Molina, 2001, Cainelli, 2008). This conception, however, has not been exempt from criticism. Indeed, the assumption reifying collocation and direct absorption of local resources has been contested with theoretical (Breschi and Lissoni, 2001; Malmberg and Maskell, 2002) and empirical evidence (Lissoni, 2001; Doring and Schnellenbach, 2006; Huber, 2010). a co-located firm, as any firm, will use differing strategic combinations of internal and external resources – available in the cluster and beyond it, at national or global scale – to build its own competitive (or collaborative) advantage (see Tallman and Jenkins, 2002; Tallman et al., 2004; McEvily and Zaher, 1999; Markussen, 1999). Put differently, firms exposed to the same potential amount of external knowledge flows, differ in their ability to recognize, acquire (learn) and exploit the available knowledge, which is ultimately moderated by its absorptive capacity (e.g. Escribano et al., 2009). On the other hand, traditional mainstream puts the emphasis just on the existence of social capital in firms as the enabler for tapping into local resources.Thus, while social capital is a must, other elements, such as absorptive capacity, are necessary to make use of external knowledge in way that would be fruitful to the firm.

## 2.4 Governance and Lock-Ins

Firms can lose competitive advantage because of emerging weaknesses in their environment (Porter, 1990). When the cluster in which a firm operates is not able to respond to new (technological) developments, the consequences for the cluster as well as for some of its actors can be severe. Cluster governance is about the intended, collective actions of cluster actors to upgrade a cluster in order to build and maintain a sustainable competitive advantage as a cluster. Cluster governance is specifically aimed at facilitating and improving processes of innovation (Bahlmann and Huysman, 2008). The decisions on the governance principles of the cluster are a fine-tuning strategic task of major importance. There is always the temptation to secure the knowledge created in the firm – and the cluster as well – by imposing tight control mechanisms and rules. Moreover, attentionshould be paid to the fact that "...different levels of governance relate not in a linear, power-imposing manner, but by evolving spheres of capability among which interactions occur by negotiation between parties of consequence to specific competence areas." (Cooke, 2005: 1136). Clusters and networks because of its very nature – involving multiple interrelationships and dimensions – determine that an effective governance system will be the result of a negotiated balancing set of principles and norms which shall (1) match the culture and values of the individual firms, (2) be tight enough to provide a safety environment for IC to flow and ideally loose enough to foster innovation and the expansion of the cluster.

## 2.5  Life Cycle and Performance Assessment

Often the life-cycle of any given cluster owes to new developments in technology and when a new technology or innovation process occurs, new or embryonic clusters emerge. Existence could be triggered by a variety of processes that lead to co-location (Maskell and Kebir, 2005). Certainly, all clusters of industrial activity do not necessarily follow this simple linear developmental path, especially those with global links or whose constituent companies are involved in various diverse networks. Empirical evidence argues that clustered companies outperform non-clustered companies at the beginning of the life cycle and have a worse performance at its end (Audretsch and Feldman, 1996a; Pouder and St. John, 1996). Basically, the cluster grows as long as the firms have some technological distance –i.e. are heterogeneous in the sense that they present different (related) technologies. This fact might increase their absorptive capacity to access external knowledge and new ideas, which is expected to rejuvenate the cluster's technological stock. Once the heterogeneity decreases and firms are more homogeneous, they imitate each other and this too specialization result in a negative lock-in which also affects the absorptive capacity of the cluster.

For clusters, particularly knowledge-based clusters, the sustainability challenge is to maintain the innovation cycle and the competitiveness of the resulting products and services even in the face of disruptive shifts in markets or knowledge. Renewal of knowledge cluster life cycles requires not only pursuing the benefits of the local cluster but also reaching out to global networks (Huggins, 2008). Broader network connections become important in part because of technological changes that blur the lines among industries and require knowledge transfers among actors that might not normally expect to gain from an exchange. The introduction of new ideas, new knowledge and new areas of technological specialisation become critical to rejuvenate the cluster and avoid exhaustion because firms have become too similar.

## 3  Top-Down *vs.* Bottom-Up Initiatives

The changes in the business context and the nature and scope of innovation have ensued a major redefinition of the cluster concept along the past twenty years, evolving from the classical conception of Porterian clusters as geographical aggregations of interconnected firms and related associations to rather loose groupings of firms or networks linked to global markets through global emergent pipelines. A similar evolutionary path can be traced through the various cluster policy approaches to economic growth. In this respect, a closer look to the context and the factors shaping this evolution could be of help to understand the reasons that might have prompted a top-down view – or what Solvell (2009) call the 'visible hand' – as well as its downturn and the claim for alternative bottom-up approaches as the response to whimsical economic growth and major societal challenges. The differentiating element in both these types is the initiator entity or organization, a governmental institution in the case of top-down clusters and, on the contrary, a firm (typically an MNC) or, more appropriately, a group of firms in the case of bottom-ups. Meier (2009) defines these categories as follows.

Bottom-up clusters: This approach focuses on fostering dynamic market functioning and removing market imperfections. They are typically characterised in that they emerged by a gathering of industrial and scientific partners to intensify mutual cooperation in order to gain competitive advantages for their daily business. The financing model might differ considerably between clusters. Fee based financing models urge the cluster organisations from the very beginning to provide demand oriented services and added values to cluster members.

Top-down externally initiated clusters: The installation of this type of cluster is typically supported by a clear mandate, and publicly funded by authorities on federal or federal state level. Political influence in these clusters is typically quite high, since policy makers consider these clusters as appropriate tools to successfully increase the innovation capability and competitiveness of a certain region.

Top-down internally initiated clusters: In this type of cluster, the main driving force is typically a specific organization, most likely a research institution or university but possibly also a company. This leading organization inherits the governance and management of the whole cluster, and also provides resources for cluster organisation. The initiator often follows objectives that are supposed to be pursued by means of cluster activities. In a later stage of cluster development the initiator is likely to dominate the activities and themes of the whole cluster.

## 3.1 Top-Down Initiatives: The Zero Sum of Policy Driven Initiatives

The acceleration and fierceness of globalization in the late 90s was certainly one key factor unleashing cluster-solution initiatives worldwide on the essence that they would drive inter-regional competitiveness. Increasingly lower costs of transport and communication, and the simultaneous liberalization of international trade revealed the weaknesses of regional economies and exposed them to global competition. At that time, cluster policy's underlying view of competition was positive sum in which productivity improvements and trade would expand the market and many locations could prosper provided they became more productive and competitive. At that time, cluster policy's underlying view of competition was positive sum in which productivity improvements and trade would expand the market and many locations could prosper provided they became more productive and competitive. Clusters were regarded as an effective environment in which it would be easier to realize the initiation of new products. The increasing awareness about its advantages and the publishing of the first case studies led the way to a myriad of initiatives aimed at lagging regions and sectors but also at stimulating activity in new high-tech and high-growth emerging sectors (e.g. biotechnology, nanotechnology, digital media, etc). However, what factors might have caused their decline? A comprehensive answer will encompass the analysis of too complex idiosyncratic elements and their interrelationships. In short, the main factors explaining this transition are:

✓  Too much (or sole) emphasis on input factors and the provision of competitive infrastructure.
✓  A tradition of a rather strong intervention putting forward market and, later on, systemic failures.
✓  The lack of appropriate measures and methodologies to assess cluster performance and adequately inform EU and country member policies.
✓  The deployment of whatever type of policy or model takes time before it starts to make its results and consequences evident –especially to its creator.

A particularly interesting finding of empirical studies was that localised clusters of similar economic activity are normally not 'locally defined industrial systems' (Ketels et al., 2008). In today's global economy, a large proportion of firms have few or no trading links with other local firms within the same cluster, even when there is a strong spatial clustering of a particular industrial sector. Clustering processes and dynamics are tremendously idiosyncratic and to tackle the real problematic the appropriate level of cluster deployment and core analysis would be the firm (or inter-firm), keeping the cluster as the analytical dimension. Bottom-up approaches seemed to better suit open integrated clusters.

## 3.2  Bottom-Up Initiatives: The Business-Led Approach to Cluster Creation

Clusters are the instrumental policy tool to deploy learning innovation systems and boost innovation and competitiveness, the concern is now with the appropriate mechanisms to leverage the impact of local and regional initiatives on a national scale. To the strong evidence – drawn on hard data – that clusters are true and effective engines of positive externalities, new research has confirmed that strong clusters foster innovation because of dense knowledge flows and spillovers that positively influence regional economic performance. So far, according to the recent Brookings Institution (BI) report, cluster initiatives have been both "too few" and "thin and uneven in levels of geographic and industry coverage, level and consistency of effort, and organizational capacity" (Mills et al., 2008: 3). The main focus is on targeted investments in data analysis, planning, and capital access to support regional innovation clusters (RICs) activities and incentivize regional collaboration across the diverse set of stakeholders. But, this initiative should not be interpreted as a shift towards a top-down approach to cluster creation or greater levels of intervention in the economy.

# 4  Building Robust Clusters from the Bottom-Up

## 4.1  SMEs' IC Strength and Dynamic Capabilities: Where To go from InCaS?

The recent focus on bottom-up clusters is ultimately the consequence of the relative failure of top-down configurations to create the conditions that would favour and foster value-adding collaborative networking within and across clusters and firms. In fact, the systematic perforation of clusters observed in the 2000s has been mainly the result of the partnerships and strategic alliances of clustered firms on a 'one-to-oneself' basis but not the consequence of the firm's (individual) strategy ideated from or supported by the cluster's shared vision or a collective mind. On the other hand, the complexities and specificities involved in inter-firm relationships plus the lack of (IC) preparedness of, especially, SMEs have certainly contributed to slow down the process of clustering and networking and the capacity of the SMEs to learn and extract value from them. A firm's networking capabilities are grounded on its current IC –human capital, structural capital, relational capital and its interactions – and its ability to systematically upgrade it through internal and external IC exchanges in rather heterogeneous and flexible socially distributed systems. It is the latter that fundamentally subscribes for the firm's IC potential and competitiveness. This fact assumes that equally important for a firm's innovation and networking capabilities, besides its current IC pool – is the flexibility and dynamism with which it is capable to reconfigure it. Therefore, dynamic capabilities, IC accumulation and performance enhancing are badly required. These capabilities are the fundamental drivers of the creation, evolution, and recombination of other resources to provide new sources of growth (Henderson and Cockburn 1994, Zander and Kogut 1995).

In particular, we will mention the following three dynamic capabilities because of its capital importance for SMEs:

- ✓ (Alliance) management capabilities.
- ✓ Organizational learning.
- ✓ Strategic flexibility.
- ✓ Dynamic relational capabilities (DRC).

Certainly, no cultural change is the result of one-stroke but we should be aware that the lack of systematisation of a reflective and learning culture especially in SMEs, is a serious impediment to capitalize on IC increasing returns and networking effects.

## 4.2  Managing Clusters along the Life Cycle

Successful clusters are those that are effective at building and managing a variety of channels for accessing relevant knowledge from around the globe (Bathelt et al., 2004). Assuming the relevance of networks (or alliances) in upgrading the clusters' technological and IC base, the question that follows is how to affect the latter on a continuous basis so as to create a foundation for sustainable growth and competitiveness. In parallel, the high failure rates exhibited by strategic alliances tell us about the management difficulties they involve, causing about half of the alliances to fail (Kale and Singh, 2009, 2007; Lunnan and Haugland, 2008; Hoffman and Schlosser, 2001). The striking aspect of the numbers apart, what we shall really raise awareness of are the economic, financial and social implications these percentages purport since they entail that forming such relationships has resulted in many cases in (significant) shareholder value destruction for the member companies (Kale et al., 2002), which unavoidably results in wealth destruction with negative impacts on GDP and employment. The failure of the alliance may act as a (cognitive) barrier, restraining the firm from participating in future collaborative arrangements, which, paradoxically, could have provided a way to overcome their weaknesses with regard to intellectual capital utilization and, ultimately, survival. Addressing these issues necessarily entails taking a (strategic) business management perspective on clusters. Firms should secure these dynamic alliance management capabilities throughout the cluster's life cycle in order to affect superior performance of both the firm itself and the cluster and networks it is part of. Before going onto the more specific issues involved in alliance management – e.g. what it means; what it entails; what its key success factors are, etc – the following section will briefly address a few essentialities of the social character of networks in the understanding that such an approach will provide the ground for developing and apprehending the fundamentals of 'managing networks along the life cycle'.

### 4.2.1 The Social Essence of Networks and Its Management Implications

Gulati (1999) pointed out that the accumulated network resources arising from participation in the network of accumulated prior alliances are influential in firms' decisions to enter into new alliances. The author highlights the importance of network resources that firms derive from their embedding in networks for explaining their strategic behaviour. Network resources result from the informational advantages they obtain from their participation in inter-firm networks that channel valuable information. A network of embedded ties accumulated over time can become the basis of a rich information exchange network that enables firms to learn about new alliance opportunities with reliable partners (Ahuja, 2000; Gulati, 1995b; Powell et al., 1996). Moreover, Gulati (1995a; 1998) stressed the key importance of structural embedding to refer to the social ties or networks which underpin alliances, channel information about the experiential learning of the network and foster new alliances (Gulati 1995a). Once firms begin to enter alliances, they can internalize and refine specific routines associated with forming such partnerships. Structural embedding, with its focus on the network as a whole and its relational and cognitive consequential effects, individually and collectively influence incremental and radical forms of entrepreneurial behaviour, thus affecting the development and performance of the network (Simsek et al., 2003).

Gulati (1998) pointed out that each network or alliance should be understood in a dynamic way due to the "social network structure" that economic relationships enable, meaning that repeated alliances and the processes resulting from prior interactions will influence the future of the alliances and also allow more interactions and the further development of alliances based on the social network structure. A social network can be defined as "a set of nodes (e.g., persons, organizations) linked by a set of social relationships (e.g., friendship, transfer of funds, overlapping membership) of a specified type" (Laumann et al., 1978: 458, in Gulati, 1998: 295). A company can be rewarded by the common network benefits – increasing in turn its private benefits – as long as it has other opportunities outside the scope of the alliance to apply what it has learnt. Similarly, it can be argued that, the greater the overlap between the alliance's scope and the firm's scope, the higher the potential for common benefits and thus the more difficult it will be to translate the learning and the IC amassed in the networking or alliances building and development processes to other network members, thus lowering the potentiality of each company to reap private benefits.

### 4.2.2 Alliance Management Capabilities or Helping the SMEs to Thrive through

The complexity and dynamics inherent to alliances has determined that the kernel of their success is the firm's ability to manage such configurations. But, unfortunately, only recently there has been an increasing number of academic works directed their research interest to the managerial aspects involved in taking alliances to fruition and the great bulk of these works have focused on large firms. Much of the literature on networks and alliances of the last fifteen years has been focused on either their social aspects or on some partial elements contributing to their success, especially at the initiation phase

In general, to participate in a productive network of interactions, a firm has to take under consideration its own criteria, such as its social capital assets, its relational assets or indirect capabilities, its reciprocity or strategic and social interdependence, its absorptive capacity, the distance management, its similar cognitive schemes, the balance of weak and strong ties, the national and/or global networks to offset inertia, the traded and untraded interactions, the personal and inter-firm interactions, the accumulated network resources, the importance of history and experience, its learning skills, the reputation building and the management of the dilemma between private and common benefits.

Following Gulati (1998), the success of an alliance depends on some key factors whose relevance varies as the alliance evolves. These may be the formation phase, the design phase and the post-formation phase. From all the above, it is evident that the concentration of critical success factors in the early stages of the alliance life cycle suggests that systematic preparation and careful planning from both partners are key for its success. However, the full value of an alliance can only be developed as it evolves. In this view, evidence suggests that the strategic rationale of the collaboration, the fit of the partners and the chosen configuration of the alliance together form the foundation for its development (Kale et al., 2009). As it seems the overall success of an alliance depends greatly on the firms' management abilities to run it throughout its life cycle which can be built through recursive learning experiences in similar endeavours.

In this respect, the work of Hoffman and Schlosser (2001) on 164 Austrian SMEs showed that both content and process factors are critical for alliance success. Among the content factors they identified "Precise definition of rights and duties", "Establishing required resource" and "Contributing specific strengths" to be key. On the other hand, process-oriented factors like "Deriving alliance objectives from business strategy" and "Speedy implementation and fast results," proved decisive for alliance success. Especially relevant here is Kale and Singh's (2009) model of key success factors for single alliances. The main interest of their framing lies in the fact that the identification of the key factors is spread along the three main stages of the alliance life cycle, the alliance formation and partner selection, the alliance governance and design and the post-formation alliance management. Schreiner et al. (2009) developed further the capabilities common to the post-formation phase. The authors conceptualise these capabilities as a multidimensional construct comprising three distinct but related aspects or skills in managing a given individual alliance after its setup and once it is running: coordination. Communication – credibly convey relevant information and knowledge to the partner, including "the formal as well as informal sharing of meaningful and timely information between firms".

Kale and Singh (2009) highlight three main building blocks as underlying the development of alliance capability in firms: prior alliance experience, creation of a dedicated alliance function and implementation of firm-level processes to accumulate and leverage alliance management know-how and skills. Hoang and Rothaermel (2005) suggest, conversely, that structural mechanisms like a dedicated alliance function are more effective in building alliance capability in large firms than in small firms. More recently, following Nonaka's knowledge cycle, Kale and Singh (2007: 54) contend that firms can also develop alliance capability by implementing "alliance learning processes".

Evidence suggests that the strategic rationale of the collaboration, the fit of the partners and the chosen configuration of the alliance together form the foundation for its development (Kale et al., 2009; Schreiner et al., 2009; Hoffman, 2007; Ahuja, 2000). A well conceived alliance structure and agreement provide a potential frame for generating benefits in the post-formation stage. But they are of limited value in actually realizing that benefit if they are not complemented by appropriate post-formation management practices that address the almost inevitable and unforeseeable contingencies that arise in the dynamic process of managing that alliance on a ongoing basis. Much of the success

Finally, Hoffman (2007) and Kale and Singh (2007) have introduced the concept of the alliance portfolio capability which refers to the organizational skills and ability to form new alliances that do not compete with others, to carefully select partners, to set up an appropriate firm-level evaluation mechanism to monitor the portfolio and to coordinate activities and knowledge flows across individual alliances in the portfolio and others.

## 4.3  Collaboration and Coopetition

The recent emphasis on clusters raised voices in the Academia which assume that collaboration is the source of competitive advantage. Business practices, however, revealed that companies were pushed forward building up collaborative networks even with competitors. Therefore, a new value creation system appeared, grounded on coopetition and based on the idea that all management activities should aim for the establishment of a mutually beneficial partnership relationship with other actors in the system, including competitors (Zineldin, 1998). Coopetition means cooperation and competition that merge together to form a new kind of strategic interdependence between firms, giving rise to a coopetitive system of value creation. By forming alliances with strategically chosen competitors, companies can share financial risks, improve their organizational learning and increase their access to markets.

The case of Microsoft-Intel-Apple describes a common path or trajectory of many other alliances or partnerships and shows that for firms to get involved in and develop effective coopetitive relationships they should dedicate time and energy to in-depth understanding and management of the motivation, interaction, vision and the learning drivers that underlie these relationships. Microsoft wanted computer hardware to be inexpensive whereas Intel wanted software to be the cheaper component. By building on their common base of technological innovation, they were able to cooperate constructively in the design of both microprocessors and software, so that Microsoft's increasingly complex software could be easily handled by Intel's advancing processors, allowing customers the benefits of both hardware and software advancement. Instead of seeking advantages over other firms as postulated in the competitive paradigm, coopetitive firms can seek and exploit new cooperative opportunities by synthesizing and transforming their alliance competencies and capabilities, changing a cooperative relationship in a coopetitive one. Moreover, an open and equal opportunity environment and just and fair systems should be provided to all partners that should share mutual dependency even though they do not share mutual interests but must always believe in mutual benefits.

Coopetition between firms is necessary nowadays. The more traditional the company, the bigger the need for collaboration is. Failure to see and apprehend the benefits of collaboration, which is very much conditioned by the SME's absorptive capacity and learning and alliance management capabilities, could mean the failure of the cluster initiative. Cooperative strategies also require an environment of reciprocity. To lead to a successful business outcome, alliances should support and leverage each participant's strategic strengths, including competencies, knowledge and resources. Cultural similarities and strategic complementarities are not very important in the success of alliances.

According to Moss Kanter (1994), integration at the following five levels will be necessary to achieve productive relationships: strategic integration, tactical integration, operational integration, interpersonal integration and cultural integration.

As threats have been identified that private interests might overtake the alliances, Moss Kanter's levels of integration can provide a useful framework to cope with potential negative externalities arising from collaboration, the most evident one being the (fear of) loss of know-how.

## 5 The Way Ahead

In the edge of 2000 it was considered that the existence of clusters gave an advantage to companies who enabled them compared to non cluster firms. But this explanation fell short to account for the uneven rates of innovation elicited by clusters and cluster firms worldwide. The answer is far from straightforward, involving a myriad of factors as well as various analytical dimensions and its interactions. To provide an adequate framework upon which to judge the potential of building strong, dynamic and innovative clusters, this section starts by just briefing the major challenges clusters may face in the pathway to catch-up World-Class Clusters (WCC).The emphasis must be on those challenges and/or principles with management implications for the cluster and/or the firms within it.

### 5.1  The Challenges on the Way to WCC

To achieve world-class excellence clusters should overcome 3 challenges and their guiding principles which are the projection on to the world stage and moving up the value chain needs the ability to develop a global strategic vision, needs securing a position on the global market or generating new integrating markets on a global scale and also needs to know how to control one's own value chain – 'know yourself in order to act'. The second challenge is an inward projection to become an internal player which means to combine attractiveness and competitiveness following the art of combining all three branches of the triple helix, the art of combining governance with the dynamics of cluster members and the art of combining attractiveness with competitiveness. The final challenge is to harness the potential, which is the ability to implement strategies for cooperation and mutual reinforcement with reinforcing one another through global cooperation, developing a structure on the basis of world-class value chains, and coming of age to become the actors of a new worldwide industrial policy. There is a need of adaptation to the changes in the nature of innovation and to the entering of new and fierce players in the global markets and the shift of the centre of gravity from the West to the East. Prior to thinking up strategies it is more important for the firm to be fully aware of its own competence map as a clusters first challenge will therefore be to control its own value chain of productive innovation in order to optimize its performance, from the available resources to the market.

Once cleared the external challenges, the internal strategies should be identified. In the context of an emergent collaborative knowledge economy this entails developing the arts of combining all three branches of the triple helix, the governance of the cluster with the dynamics of its cluster members, and the attractiveness of a vigorous region that hosts cutting-edge cluster initiatives with an "ecosystem cluster" for achieving global competitiveness. So such management needs strategic planning towards emerging on the world stage and a capacity for tactical monitoring internally, virtuous dynamics between all three pillars of the knowledge economy. Finally it is needed more cooperation for European clusters in order to make a great cluster instead of many constituent.

## 5.2  Cluster Management Framework

The approach to clusters has undergone significant changes in the past two years, touching openly for the first time on aspects of the strategic management of clusters formed bottom-up. In this respect there are other concerns besides global competitiveness of clustered firms. In order to be competitive worldwide and in respect on the idea of technological spillovers there must be a concept of absorptive capacity efficiency. In addition, the pressure to innovate in shorter cycles has pushed firms to forge strategic alliances along their value chains, fragmented their business models and imposed new –fundamentally management – demands on their need for adaptability. Networking offers opportunities for new relationships, links or markets and allows access to new or complementary competencies and technologies but particularly for cluster SMEs, this assumes having to develop essential alliance management capabilities and alliance learning processes, and the acceptance that breeding WCC (e.g. power clusters) requires also having to cope with an overarching business model appropriate to the reality of the cluster's knowledge flows and business environment.

The Cluster Management Framework is intended to aid the cluster and the cluster SMEs thrive through towards WCC. To achieve this, organizations should start common activities around an idea, individual goals should be translated into common goals and there should be rules of interaction emerge. So to initiate a cluster they should define the goals and rules of interaction and translate the common goals in individual and company benefits. To support requirements they ought to find potential founder members for good IC flow, support cluster building process, start IC-Flow between members and provide Web 2.0 tools.

There are two dimensions that any management system intended to support the creation and the development of an SME cluster has to take account of, a structural dimension –made up of the 'actants' in the network and a temporal dimension –which accounts for the evolutionary trajectory of the cluster along its life cycle. First and foremost, the success of the cluster will result from the collective and pertinent actions of its SME members. It is also crucial that due attention is paid to individual level. The extent to which the cluster – and the SMEs in it– shows capable of yielding results will strongly depend on the capacity of the SME to identify who the key people are and involve them from the onset.

The temporal dimension portrays the evolution of the cluster from its formation to its reconfiguration or decline and death. In essence, this dimension accounts for the life cycle and the dynamics of the cluster. Addressing the clustering challenge suggests that each SME joining the cluster will have to relearn the ways it apprehends its ecosystem and does business. While there are costs and benefits associated with collaboration (Das and Teng, 2000), the competitive landscape demands that firms learn to do it well (Doz and Hamel, 1998).

The Initiation Cluster Guideline and the subsequent Cluster Initiation Meeting are intended to unveil the environment and scope for the potential cluster actionable areas, the experience of the catalyst with collaborative networks, its willingness and the degree of commitment with the initiative, and a very rough idea of the competencies and IC to be pooled and the commonalities and more striking differences or fears to be smoothed out or contested, all four with the final objective of finding a common topic or, very roughly, some common goals for the potential cluster that would set the route for co-innovation and superior performance.

A process evaluation system and ongoing monitoring has been foreseen by the CMF in the progress of the cluster and the SMEs within it. The clustering process should be enabled and facilitated by means of tailored tools and methodologies. Also, there must be a process evaluation to ensure trust building and the enhancement of collective learning and efficient collaboration. But, for recursive and (alliance) learning processes to occur communication shall flow naturally through both formal and informal channels involving also the individual. These factors can achieve strategic effectiveness and collaborative efficiency. This entails that the SME will be able to transcend its own cognitive strategic mapping and affect its routines and processes, effectively and efficiently, with the purpose to generate win-win situations that will reinforce trust and the commitment of each of the involved parties with the sustainable superior performance of the network/cluster. The locus of the support initiative resides at the SME level (bottom-up), which, like the Cluster Management Framework, co-evolves in continuous exchange with the environment to shape and reshape the cluster ecosystem.

# 6  Web 2.0 Platform and Online Services - Enabling Change

Web 2.0 was coined by Tim O' Reilly (2005) with the aim of understanding how new integrated web technologies could enable business opportunities. Thus firms can enable services, not packaged software, with cost-effective scalability, control over unique, hard-to-recreate data sources that get richer as more people use them, trusting users as co-developers, harnessing collective intelligence, leveraging the long tail through customer self-service, software above the level of a single device and lightweight user interfaces, development models, and business models. The Web Platform supports collaborative and creative authoring of innovative outcomes, enabling change, while protecting IP and forging productive networks both online and offline.

## 6.1  Web Platform Overview

The platform provides networking channels and pathways dynamically linked for cluster enhancement and growth focuses. It allows highly flexible interactions between cluster members where the provided tools, electronic workspaces with whiteboards and discussion panels, will define a common ground for all the participating SMEs. Using a shared tool can reduce the possibility of misunderstandings and, as a consequence, the likelihood of conflicts. This will help avoid any problems with the uptake and use of such facilities.

## 6.2 Networking Hub Implemented through LinkedIn

Each cluster will be able to set up a LinkedIn group comprising its members, from where they can also access their personal networks on LinkedIn. Members will have personal and collective access, free of charge, to their Cluster Workspace on Huddle together with their own personal workspace.

## 6.3 Cluster Workspaces Implemented through Huddle

Each Cluster workspace will be accessible both stand-alone on Huddle and from within LinkedIn, with targeted functionality. It provides white boards and discussion panels, activities in CLEs in the 'real world' multimedia documents and a large variety of applications. Furthermore, the data held in a given workspace is 'cluster-specific'.

## 6.4 Justification for the Choice of Huddle for the Implementation on Cluster Workspaces

Huddle.net was able to respond to criteria that other firms such as Microsoft or Google could not satisfy. These criteria are a component-based system allowing external scalability, a Web Platform, a gateway system with variable privileges, a possible graded access, a possibility of inter-connectivity between different organizations at no cost, a determinate rationale upon needs of given company or cluster of companies, a knowledge gathering and disseminating system, a decision making supported by global best practice, an off-line integration through workshops and training and a connectivity with other networks of value.

## 6.5 Benchmarking and Its Contributing Potential to Collaborative Networks and Clusters

It was first introduced by XEROX in 1979. It is a technique of comparing products and processes and it can be used in many areas as well as the area of strategic orientation of companies. One common definition describes Benchmarking as "the search for best practices that lead to superior performance". It may be used as a tool to detect convoluted corporate structures or as an instrument to identify potential areas for improvement, in business processes as well as in areas concerning the strategic orientation of companies.

There are 3 types of Benchmarking:

1.  The Benchmarking of companies
2.  The Benchmarking of sectors
3.  The Benchmarking of the environment

The Benchmarking of companies has been spread most widely. The Benchmarking of Sectors compares the performances of individual sectors. The Benchmarking of the Environment gains increasing importance. In the future, countries can compare the political, social or economic environment.

# 7 Progress Beyond the State of the Art

'Cluster' is neither a statistical term nor a static concept. A proof of this is the array of different concepts and typologies10 that have flooded the academic journals in the last ten years. Recently, a new Progress beyond the state-of-the-art 'Cluster' is neither a statistical term nor a static concept. A proof of this is the array of different concepts and typologies10 that have flooded the academic journals in the last ten years. Recently, a new wave of concepts such as knowledge clusters, virtual clusters and virtual business environments (VBE) as well as opposing dyads such as open vs. closed clusters, soft vs. hard clusters, second vs. first generation clusters, or social networks vs. pure agglomeration and industrial complex have emerged to highlight the importance of connectivity, relations and 'knowledge' spikes to competitiveness and business success. Adding more confusion to this panorama is the relatively recent policy and business strategists' interest in networks or inter-firm cooperation prompted by the success of new industrial districts (NID) and the quick pace of IT-driven transformations and increased globalisation. To sum up, there is no overriding cluster theory per se and clusters are generally viewed as an economic development process rather than a definite development theory.

One effect of the present development towards a global economy is that many previously localized capabilities and production factors have become commodities. What is not commoditized, however, is the non-tradable/noncodified knowledge creation. Thus, the possibility to tap into tacit knowledge is a key determinant of the geography of innovation. Accordingly, the flexible specialization school12 supports the idea that the fundamental reason behind the promotion of clusters worldwide is 'untraded interdependencies' –i.e. formal and informal collaborative and informational networks, interactions and shared customs and rules for enabling communication and interpreting knowledge. The major argument is that an attempt to characterize industry clusters using traded relationships or tangible flows fails to fully capture inter-firm relations that explain the agglomeration phenomena. From this perspective, a very thick network of knowledge sharing, which is supported by close social interactions and institutions, building trust, and informal relations among actors, is considered to be central to enhancing clusteredfirms' competitiveness and economic development13.

On the other hand, the new paradigm on innovation supports the idea that the enhanced competitiveness and superior performance of a SME depends greatly on its ability and capacity to enter into collaborative and dynamic networks in open and increasingly virtual business environments14. The speed and quality of product and process innovations required to compete in an increasingly knowledge-based economy has had the inevitable consequence of pushing business organisations to enter into both local and global collaborative networks simply to keep up. Flexible forms of production, associated with the existence of agglomerations of small, innovative, dynamic local producers, represent a necessary element in the 'supply architecture' for learning and innovation. It is now suggested that the benefits of collaboration can overcome the negative externalities of fear competition and diseconomies of scale. In support of this statement, the 'communities of practice' view15 asserted that tacit knowledge may also flow across regional and national boundaries as long as organizational or 'virtual community' proximity is strong enough. In other words, learning – and its counterpart, the sharing of tacit knowledge – need not be spatially constrained provided relational proximity is present. This view highlights the importance of relationships and the strength of underlying similarities rather than geographical proximity per se in determining the effectiveness of knowledge sharing between economic actors.

In this context, to become part of far-ranging collaborative networks and to develop their organisational capacity to benefit from networking is a key strategic issue and a matter of survival for SMEs, especially those located in open and small economies, which must seek out different alternatives of differentiation. As change accelerates, the firm's stocks of knowledge assets (and physical assets as well) depreciate at a more rapid rate. The other side of this view is that flows of new knowledge become critical to competitive success and these flows occur only in the context of relationships. Thereby, successful strategies will depend on the ability of the firm to sense its business ecosystem for opportunities and new knowledge and bw wave of concepts such as knowledge clusters, irtual clusters and virtual business environments (VBE) as well as opposing dyads such as open vs. closed clusters, soft vs. hard clusters, second vs. first generation clusters, or social networks vs. pure agglomeration and industrial complex have emerged to highlight the importance of connectivity, relations and 'knowledge' spikes to competitiveness and business success. Adding more confusion to this panorama is the relatively recent policy and business strategists' interest in networks or inter-firm cooperation prompted by the success of new industrial districts (NID) and the quick pace of IT-driven transformations and increased globalisation. To sum up, there is no overriding cluster theory per se and clusters are generally viewed as an economic development process rather than a definite development theory11.

One effect of the present development towards a global economy is that many previously localized capabilities and production factors have become commodities. What is not commoditized, however, is the non-tradable/noncodified knowledge creation. Thus, the possibility to tap into tacit knowledge is a key determinant of the geography of innovation. Accordingly, the flexible specialization school12 supports the idea that the fundamental reason behind the promotion of clusters worldwide is 'untraded interdependencies' –i.e. formal and informal collaborative and informational networks, interactions and shared customs and rules for enabling communication and interpreting knowledge. The major argument is that an attempt to characterize industry clusters using traded relationships or tangible flows fails to fully capture inter-firm relations that explain the agglomeration phenomena. From this perspective, a very thick network of knowledge sharing, which is supported by close social interactions and institutions, building trust, and informal relations among actors, is considered to be central to enhancing clusteredfirms' competitiveness and economic development13.

On the other hand, the new paradigm on innovation supports the idea that the enhanced competitiveness and superior performance of a SME depends greatly on its ability and capacity to enter into collaborative and dynamic networks in open and increasingly virtual business environments14. The speed and quality of product and process innovations required to compete in an increasingly knowledge-based economy has had the inevitable consequence of pushing business organisations to enter into both local and global collaborative networks simply to keep up. Flexible forms of production, associated with the existence of agglomerations of small, innovative, dynamic local producers, represent a necessary element in the 'supply architecture' for learning and innovation. It is now suggested that the benefits of collaboration can overcome the negative externalities of fear competition and diseconomies of scale. In support of this statement, the 'communities of practice' view15 asserted that tacit knowledge may also flow across regional and national boundaries as long as organizational or 'virtual community' proximity is strong enough. In other words, learning – and its counterpart, the sharing of tacit knowledge – need not be spatially constrained provided relational proximity is present. This view highlights the importance of relationships and the strength of underlying similarities rather than geographical proximity per se in determining the effectiveness of knowledge sharing between economic actors. In this context, to become part of far-ranging collaborative networks and to develop their organisational capacity to benefit from networking is a key strategic issue and a matter of survival for SMEs, especially those located in open and small economies, which must seek out different alternatives of differentiation. As change accelerates, the firm's stocks of knowledge assets (and physical assets as well) depreciate at a more rapid rate. The other side of this view is that flows of new knowledge become critical to competitive success and these flows occur only in the context of relationships. Thereby, successful strategies will depend on the ability of the firm to sense its business ecosystem for opportunities and new knowledge and be capable of extending its core capabilities in a continuous basis within a rich network of relationships, both long term and ad hoc. This calls for the essentiality of 'dynamic specialization'16 – making hard choices about what business activities to focus on while at the same time making strong commitments to build capability rapidly around these areas of specialization, focused on providing creative ways to connect highly specialized participants from around the world in global process networks. The pressure to innovate systematically and in short cycles, together with the increased risk of global markets, is definitely pushing firms towards higher levels of flexibility in the mobilisation of resources. In most countries, however, SMEs have not yet been able to fully capture the benefits17 of increased external linkages and knowledge sharing. Many SMEs are still unaware of the opportunities offered by cooperation with different actors – firms, universities, research institutes and professional communities – in multidisciplinary and strategic global nets.

On the side of regional or developmental economics, most studies have limited their focus to describe and analyse the elements involved in the design, development and implementation of industry clusters. Also common are those studies that attempt to disentangle the fundamental reasons behind geographical concentration of economic activities (clusters, NID, regional innovation systems) or to describe the enablers and the barriers encountered in initiating and growing such configurations. Meanwhile, when the issue at stake is the performance of the cluster and the economic benefits it yields, analysis has remained focused fundamentally on input-output econometric models and descriptive case studies. As Wolfe18 rightly argues, there is an imbalance between theory and empirical studies and a lack of a unified theoretical and methodological framework for cluster analysis.

Likewise, when the perspective is firm-centred research studies have evolved around the operationalisation of the new organisational configurations and strategies (e.g. open innovation, virtual clusters, "Internet Work Enterprises", business ecology), the motivations for networking, and, only recently, on the impact that such strategies may have on the firm19 and cluster's business model and overall performance20. Most of this research follows the knowledge-based view of the firm and assume a behavioural approach to focus on the reasons, enablers and barriers for networking at the firm and inter-firm level; the effectiveness of networks and the needs and practices of SMEs with respect to networking. So far, none of it has focused on IC-flows that, if properly understood and managed, could convert such networks into truly value-adding ones.

## 7.1 Roles and Activities

In order to match the SME perspective (demand side) with the Cluster Facilitation Services (offer side) on a caseto-case basis, there are two generic roles that lead the whole cluster evolution process aiming atproviding maximum benefit for the cluster and its members by utilising efficient support services:

### 7.1.1 Roles

- "Cluster Relation Manager (CRM)": This role is defined at the individual SME level: Every SME participating in cluster activities has to assign the role "Cluster Relation Manager" to one or more persons representing the SME. The CRM focuses on the interface between the company level and the cluster level. Also taking into account personal interests and motivation (individual level), the CRM acts upon the SME's interests engaging in the process of translating individual and company goals into cluster goals and matching interests with potential and new partners (company-to-cluster level). On the other hand, the CRM is responsible to incorporate learnings and results from the cluster level into the company by filtering and translating results from collaboration activities of the cluster according to the needs, language and culture of the company . In order to maximize benefits for the single company the CRM has to ensure the absorptive capacity of the SME to absorb and utilise external knowledge for own purposes. As a representative of his company, a CRM can initiate any kind of cluster activity.

- Role: "Cluster Facilitator (CF)": This firstly describes a generic role within the system as a whole and as a result from the project being implemented and applied by the SME-AGs in the project as well as by further users afterwards. Within the project, the cluster facilitator in each core country is supported by a representative of the RTD partners in order to ensure the success of the implementation in each country. This team will be called "Cluster Facilitation team". It comprises a County Coach, from the RTD partner responsible for implementing offline and online services in the cluster, and a moderator (Cluster Facilitator) from the SME-AG responsible for the cluster. The Country Coach and the moderator work closely together in implementing the CF role. Provision is made for a moderator to be able to gain accreditation as a Cluster Facilitator through the Training Facility, involving both field experience in a cluster Facilitation team and short intensive training course. The CF has the key responsibility to ensure that cluster events, activities and processes triggered by the SME members are well designed and executed, thus promoting cooperation and trust in networking among the SMEs in the cluster according to their needs and opportunities. Thus, the CF focuses on each cluster for which he or she is responsible to offer support for any kind of cluster activity. Utilising support services according to the specific demand, the CF coordinates the bundling and implementation of methods, tools and services to meet this specific cluster's demand, determined bottom-up. The CF then offers a concrete and customized bundle of cluster support services to facilitate the specific cluster activity and negotiates this offer with the cluster. Within the project this process of bottom-up triggered support will be tested in order to develop a sound business model after the funding-phase

## 7.1.2 Activities

Assigning this important role to a "bottom-up" member of the cluster ensures that the necessary rules always match the real needs of the cluster members and their activity. Following the bottom-up and activity-based approach of cluster facilitation, are categorized into three generic cluster core activities:

- Initiation Activities are those activities that a catalyst SME wants to do in order to initiate something new on a cluster level. This can range from initiating a totally new cluster with new topics, initiating sub-groups to initiating any other activity, such as an IC Benchmarking cooperation workshop or even a project within the cluster.

- Orientation Activities comprise all activities concerned with gathering information about one or more cluster from the perspective of one company, e.g. an SME wants orientation within one cluster or within the whole network (between clusters) concerning profiles of the members or topics being discussed, or an IC Benchmarking is triggered by a catalyst SME helping the cluster to increase transparency within the cluster's resource base and strengths and weaknesses of the cluster members to have an orientation about opportunities of collaboration or about which competencies are missing so far and should be integrated by finding new members for the cluster who could fill the gap.

- Collaboration Activities is the generic term for all kinds of actual collaboration within a cluster from the perspective of the participating SMEs, e.g. companies taking part in a cluster event to define common-goals, or companies taking part in IC Benchmarking events to learn from each other and to solve actual problems in their intellectual resource base, or companies work together on a specific project in a sub-group of the cluster producing an added value for clients or developing new technologies or prototypes which provide a benefit for each participating company.

## 7.2  SME Network

According to  bottom-up approach, the core activities (Orientation, Initiation, Collaboration) are to be understood as generic categories of demand, which are a bundle of specific cluster activities in practice driven by SMEs within the cluster, as shown in figure 1:

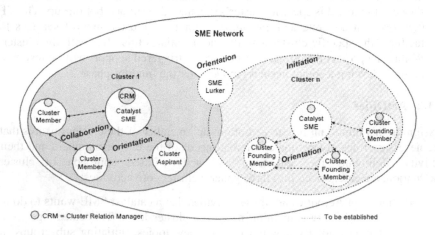

**Fig. 1** Cluster facilitation and networking

Therefore, Cluster Facilitation services will be offered to SMEs according to their demand, i.e. the cluster activity they want to achieve in a specific situation. Such activities are driven bottom-up by SMEs within the cluster. The crucial role is that of each SME's "Cluster Relation Manager (CRM)" who may request support for a specific cluster activity. This request is directed towards the role "Cluster Facilitator" (CF). The Cluster Facilitator is then offering suitable Cluster Support Services, coordinating all necessary sub-services, i.e. providing special tools and methods or sub-contracting a expert in special fields . This can also be enlarged to other specific consulting services outside the actual service portfolio, then being a mediator between SME clusters' demand for services and suitable consulting services (allowing for further revenue streams).

# 8 Case Study

Second Life is a complex system with many inter-operating components, from simulators and databases to the viewer you run, and the Internet connections over which data flows. Although it is intended to be as <u>loosely coupled</u> as possible and resilient against problems, events can still occur which lead to service disruptions.

In some cases the failures are beyond the control of Linden Lab. However, in nearly all cases there is active work being done to mitigate the disruptions - either prevent them from happening or significantly reduce the impact.

When a disruption occurs, the following sequence usually occurs:

- a system stops responding
- automated notifications go off, alerting our operations team
- residents often notice immediately, and alert in-world support, who confirm the problem to our operations team
- the operations team identifies the system that is the root cause of the problem
- the communications team is notified, and asked to provide information about the disruption to the blog
- if the disruption lasts more a few minutes, updates are made to the blog regularly
- once the problem is solved, an "all clear" is reported to the blog.

Many companies are looking at the benefits of virtual worlds and create their own. However the most successful virtual environments listed in the following Table1 and described their main features.

Note that in many cases a disruption may be solved before any information can make it to the blog explaining the details of the problem. One purpose of this document is to provide a "clearing house" for types of service disruptions, so that in the event of a system failure the blog post can reference this page. Moreover, these are the systems which have been known to cause service disruptions.

**Table 1** Comparison of Virtual Environments

| Comparison of Virtual Environments | | | | | | | | | |
|---|---|---|---|---|---|---|---|---|---|
| Virtual World | OS | Cost per month | Target user & style | Edit avatars? | Build or design content? | Script content? | Own land or sell items? | Education ready? | Comm. Events? |
| Active Worlds | PC & Linux | Free / $6.95 | General; Exploration | ✓ | ✓ | ✓ | ✓ No selling | Can code | ✓ |
| Barbie Girls | PC | Free | Young girls; Fashion, social | ✓ | ✓ | No | Neither | No | ✓ |
| Club Penguin | PC & Mac | Free / $5.95 | Kids; Games and Activities | ✓ | No | No | Neither | No | ✓ |
| Forterra Systems | PC | Contract | Training, E-Learning, Serious Games | ✓ | Can code | ✓ | ✓ | Can code | Can code |
| Gaia Online | PC & Mac | Free | Social; Top-down overview, sprites | ✓ | No | No | Sorta | No | ✓ |
| Habbo Hotel | PC & Mac | Free | Teens; Social | ✓ | ✓ | No | Neither | No | ✓ |
| Kaneva | PC & Mac | Free | Teens; Social | ✓ | ✓ | No | Sorta | No | ✓ |
| Neopets | PC & Mac | Free / $7.95 | Kids and teens; Mini-games, social | ✓ | No | No | ✓ ✓ | No | ✓ |
| Teen Second Life | PC & Mac | Free / $9.95 | Teens only; 3D, Creation, social | ✓ | ✓ | ✓ | ✓ | Can code | ✓ |
| Second Life | PC & Mac | Free / $9.95 | 18+ only; 3D, Creation | ✓ | ✓ | ✓ | ✓ | Can code | ✓ |
| The Sims Online | PC | $9.95/mo. | General; Social; 3D | ✓ | ✓ | No | ✓ No selling | No | ✓ |
| There | PC | Free / $9.95 | General; Social | ✓ | ✓ | Limited | ✓ | No | ✓ |
| Webkins | PC | Free | Social | ✓ | No | No | Neither | No | ✓ |
| Whyville | PC & Mac | Free | Kids and teens; 2D sprites; Educational | ✓ | ✓ | No | Neither | ✓ | ✓ |
| Zwinktopia | PC & Mac | Free | Teens; Social | ✓ | No | No | Neither | No | ✓ |

## Asset Storage Cluster

A cluster of machines that form a whopping huge WebDAV (think "web-based disk drive") storage mechanism with terabytes of space for storing assets, including uploaded textures, snapshots, scripts, objects taken into inventory, script states, saved region states (simstates), etc that make up Second Life. The technology (software and hardware) is licensed from a third party.

The system should be resilient against single node failures. In the case of multiple disk failures, software upgrades, removing problem nodes or adding new nodes, some or all of cluster can fall offline. If this happens, asset uploads and downloads fail - this causes texture uploads and simstate saves to fail. Since transient data during region crossings (attachment states, etc) are written as assets, region crossings will also often fail.

When detected, we often disable logins and message in-world (if possible) to help avoid data loss. Failed nodes can be taken out of rotation. A restart of other nodes may be necessary. When upgrading the software on the nodes, the grid is usually closed to prevent data loss during any inadvert outages.

## Central Database Cluster

A cluster of databases that store the core persistent information about Second Life - including Resident profiles, groups, regions, parcels, L$ transactions and classifieds.

The database can become loaded enough during normal operations that some fraction of transactions fail and either must be manually retried or are automatically retried. Hardware failure and software bugs in the database code can also cause the database to crash or stop responding. Logins will fail, transactions in-world and on the web site will fail, and so forth.

If the primary database fails, we swap to one of the secondaries. If the database load is high but hasn't failed we can turn off services to try and reduce the load.

## Agent ("Inventory") Database Cluster

Storage for most agent-specific data such as the inventory tree is partitioned across a series of databases. Each agent is associated with a particular inventory partition (a primary database and its secondary backups). At the time of this writing, we have approximately 15 agent database partitions. The initial use of these agent-partitioned databases was for inventory, so they are often referred to as "inventory databases" by Lindens, but this is no longer the extent of what agent-specific data is stored within them.

Hardware or software failure can affect the primary database within the partition, so that it either stops responding to queries or becomes excessively slow.

When an agent database fails we can swap to the backup within that partition, which takes a few minutes. If this will not happen immediately or if problems are encountered, that particular agent partition is "blacklisted" temporarily; this causes logins of agents who are associated with that partitions to be blocked and any agents logged in are "kicked", while the fix is in progress. This will affect some fraction of the grid, but not everybody.

## Other Database Clusters

There are a handful other database clusters in use. One is used for logging data. Hardware or software failures can take a database cluster offline. There should be no in-world effect from one of these other database clusters failing, but occasionally a software design flaw does introduce a dependency that is not caught. For example, logins used to require a successful connection to the login database to record the login and viewer statistics, but this dependency has been removed.

All databases act in clusters with a primary machine and several secondaries. In case of failure, a secondary can be swapped into place as the new primary.

## Login Server Cluster

A cluster of servers represents the first service that the viewer connects to when attempting to log in. This validates the resident's credentials, checks the viewer version for possible updates, ensures the latest Terms of Service have been updated. Assuming those check out, it sends the viewer an initial overview of the resident's inventory folders and a few other chunks of data. Finally, it negotiates with the simulator for the requested start location and lets the viewer know which simulator to talk to.

If one drops offline, some percentage of logins will fail. Additionally, since the login sequence is database-intensive, if the central database or inventory database cluster are having problems then logins will also fail. Finally, after a major disruption that leads to many Residents being kicked or unable to connect, there may be more Residents trying to connect than our Second Life can handle (roughly 1000 logins/minute); this can appear to Residents trying to log in as though the login service is failing, even though it is fully functional and just at maximum capacity.

If a login server itself fails, we take it out of rotation. If the problem is in another system or service, we fix it there.

## Web Site

A cluster of machines that serve the web pages and web services exposed to the public - including secondlife.com, lindenlab.com, slurl.com, etc.

Hardware failures can slow down or shut down a machine in the web cluster. In that case, a load balancer should automatically redirect web traffic away from machines that are performing poorly, but the load balancer itself may have bugs (e.g. it may not detect such failures properly, or itself become blocked up). Web site bugs can be introduced by code updates to the web site, which are made daily. In addition, the web site relies on the central database cluster for many service actions, so failures there will affect web site actions such as the LindeX and transaction history, land store, friends online, and so forth.

Problematic hardware can be taken out of rotation to restore the responsiveness of the web site. Problems in other systems such as the central database cluster need to be addressed there.

## Linden Network

Linden is the connections between our co-location facilities ("colos"), e.g. SF and Dallas, but also the plumbing within colos. This includes "VPNs", switches, routers, and other esoteric stuff. Some of this is Linden equipment, some of this is leased equipment (e.g. we pay a third party to have dedicated use of their "tubes" between our colos), and public Internet pipes are also used.

A component can go bad, for example, a router can start dropping packets (stop sending data). This often appears as one of the other problems (asset storage, database, simulators, logins) since the systems can no longer talk to each other.

Isolate the affected component and take it out of service or replace it as quickly as possible. If this is a leased component we need to talk to our provider.

## Internet

Internet is <u>series of tubes</u> that bring Second Life to your computer, from the large trans-oceanic and trans-continental pipes that link the world down to high-speed connection to your home from your Internet Service Provider (ISP).

Failures occur on several levels. If this happens at a high level - for example, a major Internet trunk to Europe drops offline - thousands of Residents can be disconnected from Second Life.

This is usually beyond our control. If we can isolate the problem we can report it to network contacts, but otherwise we just need to wait for the issue to get fixed, like the residents.

## *8.1 3D Version of the Technology Park*

We created a 3D version of the technology park that offers direct networking, information, communication and evaluation of human factors and advertising company, with low cost. Apart from the virtual building of the Aegean Technopolis, we created the right software that meets the requirements of a modern company based in the immediate dissemination of information etc. in strong cooperation with other similar companies. This virtual building was designed and programmed to generate online meetings that will reduce costs, provide direct information, assess the human factor and the knowledge and finally to offer reduced-cost advertising and an increasingly growing number users second life. All these were carried out with the amount of 120 euro without taking into account the man hours and fixed costs assistance on the Internet.

When concluded that it may find application in modern business and meet their objectives targeted from the beginning. This application can be used in conferences and seminars as well as advertising for the Technopolis Aegean and the island of Chios where located. It is also the first sample business venture that aims to create a virtual building in the Greek community in second life.

The following figure 2 and figure 3 shows the final form of the virtual building decorated and furnished.

**Fig. 2** Southeast outside virtual building of the Aegean Technopolis

**Fig. 3** Northwest outside virtual building of the Aegean Technopolis

In the above figures, it is visible the Chian architecture and around are the three most common species of plants of the island the mastic trees, the citrus trees and olives trees. We also observed that in the main entrance we have the label of the Aegean Technopolis from the side entrance we have the label of the Aegean University.

Moreover, we will refer to one of the internal spaces of the virtual building. In this internal space (figure 4) was placed a virtual projector, a table and 9 chairs. This space can be used for meetings with many participants for that reason there are not so many furniture (free spaces for avatars).

**Fig. 4** Inside view of upstairs

Generally, the virtual building covers all the aims as:

- the building resembles the original building because its design was based on the original building,
- the visitors can find information and can enter to the website,
- the building has an educational role (Conference rooms),
- the building has private rooms,
- there are labels for advertising,
- available secretarial support for instant information.

**Fig. 5** The presentation of the virtual environment

Also the virtual building presented at the European Congress of European Federation of small and medium-sized enterprises on 1 October 2010. The conference was attended by individuals who were in the rooms of the conference and representatives virtual educational institutes of second life. Here are the pictures of the virtual environment (figure 5).

# 9  Conclusion

## 9.1  Concluding Remarks on Clusters and Networks and SMEs

Globalization has dramatically transformed the trading and business environments worldwide. Competitiveness demands flexible forms of production, associated with the existence of agglomerations of small, innovative and dynamic local producers linked globally. The advantage shifts to smaller producer working in flexible, market-responsive, knowledge, rich collaborative relationships. Coopetition has become the key of success.

The concept of cluster has changed and evolved distributing knowledge through rather loose and flexible arrangements. There has also been a distinction between focusing on competitiveness, power clusters, but without neglecting the regional drive, area clusters.

The possibility for clusters to achieve higher levels of innovation and global standard competitiveness hangs necessarily on their becoming thoughtfully – i.e. the result of a consistent strategy – and increasingly integrated to global networks of production. So clusters and the organizations making it up will have to expand their networks and learn to manage and muddle through a new complex and flexible (sometimes volatile) set of heterarchical relationships.

The role of linkages external to the region should be emphasized. Firms will have to rely on an active set of relationships to keep abreast with cutting-edge innovation, and ensure individual and collective efficiency and competitiveness. Companies should enable shared visions and coordinated action using synergistic dynamics in collaborative networks.

Clustering today is both about proximity and global interconnectivity; bottom-up processes and world-class dynamic relational and alliance or clustering management capabilities. Every 'actant' shall be called into the game and learn to play its particular role in an environment that is reconfigured in short cycles by the dynamic interplay of (collaborative) networks.

## 9.2  On Dynamic Capabilities and Coopetition

In this world, the primary value of assets lies in their ability to help us build and sustain relationships. Unbundling the firm enables even more rapid growth.

There is a co-evolution of local and global linkages and networks, therefore, new interactive modes of knowledge creation are needed. The value of networking for innovation is the fast composition of a complex knowledge base and diffusion system of innovations through streamlining IC flows. Networking offers upgrading options for SMEs, opportunities for new relationships, links or markets and allows access to new or complementary competencies and technologies.

SMEs are forming symbiotic collaborative relationships as a way to keep up with or to access unique or cutting-edge resources and to achieve efficiencies and access markets. Firms looking to be involved in collaborative networks need to develop new and flexible IC management processes.

Pipelines, refer to channels of communication used in distant interaction, between firms in clusters and knowledge-producing centers located at a distance. Important knowledge flows are generated through network pipelines. The effectiveness of these pipelines depends on the strength of pre-established social relationships and the quality of trust that exists between the firms in the different nodes involved. The advantages of global pipelines derive from the integration of firms located in multiple selection environments, each of which is open to different technical potentialities.

The new collaborative environment demands that the SME counts with the support of the appropriate methodologies and tools, and roles, to guide it in the experience of building robust, flexible and competitive clusters and networks from the bottom-up.

The CMF provides methods and tools for cluster enhancement and growth, all of which can be used by participants on a distributed basis. These supporting methodologies and tools all complement each other in the sense that they might provide the key input for developing the selection partners' criteria or identifying the IC synergies, interdependencies and complementarities which lay at the foundation of coopetition. In this same purview, the collaborative and safe environment will provide all partners with an open and equal opportunity environment and with just and fair systems which are the ground for effective coopetitive relationships.

A construction process operating from the bottom-up presupposes that each SME has made clear its strengths and core competencies and on the value adding or core competencies each of them will contribute to the cluster. This exercise of transparency and putting in common strengths and weaknesses, and fears, will set the ground for trust building and opportunities. The full value of the partnership has to be developed as it evolves.

## 9.3  On Research and Business and Policy Implications

It is important for companies to have knowledge and experience in leading with collaborative (and eventually coopetitive) environments and assuming the responsibility of a cluster construction process.

On the policy side, the implications are many, or better, could be many depending on the ability and capacity to affect real change at the business. This renders the capacity to influence or inform policy to lie in its ability to generate set of methodologies and tools that is a rich, easy to use and replicate in fresh contexts, and to develop from the inside-out.

The opportunity and timing of the initiative is important. The ground has already been seeded as the last reports on clusters recognize that the cluster is the policy tool from which to boost the economic growth and employment.

Finally, the main challenge is to be ready to ask the appropriate questions and open enough to unveil potential and alternative real pathways to value as a result.

# References

Abbey, J., Gareth Davies, G., Mainwaring, L.: Vorsprung Durch Technium: Towards a system of innovation in South-West Wales. Regional Studies 42(2), 281–293 (2008)

Acs, Z.J., Audretsch, D.B.: Innovation and Small Firms. MIT Press, Cambridge (1990)

Ahuja, G.: The duality of collaboration. Strategic Management Journal 21, 317–343 (2000)

Ahuja, G., Polidoro, F., Mitchell, W.: Structural homophily or social asymmetry? The formation of alliances by poorly embedded firms. Strategic Management Journal 30, 941–958 (2009)

Albaladejo, M., Romijn, H.: Determinants of innovation capabilities in small UK Firms: An empirical analysis, Working paper 00.13. Eindhoven Centre for Innovation Studies, Eindhoven (2000)

Allen, J., James, A.D., Gamlen, P.: Formal versus informal knowledge networks in R&D: A case study using social network analysis. R&D Management 37(3), 179–196 (2007)

Almeida, P., Dokko, G., Rosenkopf, L.: Startup size and the mechanisms of external learning: Increasing opportunity and decreasing ability? Research Policy 32, 301–315 (2003)

Almeida, P., Kogut, B.: Localization of knowledge and the mobility of engineers in regional networks. Management Science 45, 905–917 (1999)

Antonelli, C., Patrucco, P.P., Quatraro, F.: Productivity growth and pecuniary knowledge externalities: An empirical analysis of agglomeration economies in European regions. Economic Geography (forthcoming)

Arbussa, A., Coenders, G.: Innovation activities, use of appropriation instruments and absorptive capacity: Evidence from Spanish firms. Research Policy 36(10), 1545–1558 (2007)

Arikan, A.T.: The interfirm knowledge exchanges and the knowledge creation capability of clusters. Academy of Management Review 34(3), 658–676 (2009)

Arora, A., Fosfuri, A., Gambardella, A.: Markets for technology and their implications for corporate strategy. Industrial and Corporate Change 10(2), 417–449 (2001)

Arrow, K.J.: Economic welfare and the allocation of resources for innovation. In: Nelson (ed.) 1962 The Rate and Direction of Inventive Activity (1962)

Asheim, B., Coenen, L.: Knowledge bases and regional innovation systems: comparing nordic clusters. Research Policy 34, 1173–1190 (2005)

Asheim, B.T., Coenen, L., Svensson-Henning, M.: Nordic SMEs and Regional Innovation Systems, Final Project Report, Nordic Industrial Fund, Oslo, Norway (2003)

Audretsch, D.B.: Agglomeration and the location of innovative activity. Oxford Review of Economic Policy 14(2), 18–29 (1998)

Audretsch, D.B.: Research issues relating to structure, competition and performance of small technology-based firms. Small Business Economics 16, 37–51 (2001)

Audretsch, D.B.: The dynamic role of small firms: Evidence from the US. Small Business Economics 18, 1–13 (2002)

Audretsch, D.B., Feldman, M.P.: Innovative clusters and the industry life cycle. Review of Industrial Organization 11, 253–273 (1996a)

Audretsch, D.B., Feldman, M.P.: R&D spillovers and the geography of innovation and production. The American Economic Review 86, 630–640 (1996b)

Bagshaw, M., Bagshaw, C.: Co-opetition applied to training. A case study. Industrial and Commercial Training 33(4/5), 175–177 (2001)

Bahlmann, M.D., Huysman, M.H.: The emergence of a knowledge-based view of clusters and its implications for cluster governance. The Information Society 24, 304–318 (2008)

Bamford, J., Gomes-Casseres, B., Robinson, M.: Envisioning Collaboration: Mastering Alliance Strategies. Jossey-Bass, San Francisco (2004)

Baptista, R.: Clusters, innovation and growth. In: Swann, P., Prevezer, M., Stout, D. (eds.) The Dynamics of Industrial Clustering: International Comparisons in Computing and Biotechnology. Oxford University Press, Oxford (1998)

Baptista, R.: Do innovations diffuse faster within geographical clusters? International Journal of Industrial Organization 18, 515–535 (2000)

Bathelt, H.: Cluster relations in the media industry: Exploring the 'distanced neighbour' paradox in Leipzig. Regional Studies 39, 105–127 (2005)

Bathelt, H., Malmberg, A., Maskell, P.: Clusters and knowledge: Local buzz, global pipelines and the process of knowledge creation. Progress in Human Geography 28(1), 31–56 (2004)

Baum, J.A.C., Dutton, J.E. (eds.): The Embeddedness of Strategy (Introduction). Advances in Strategic Management, vol. 13. JAI Press, Greenwich CT (1996)

Becattini, G.: The Marshallian industrial district as a socioeconomic notion. In: Pyke, F., Beccatini, G., Sengenberger, W. (eds.) Industrial Districts and Inter-Firm Co-Operation, pp. 37–51. International Institute for Labour Studies, Geneva (1990)

Becattini, G.: Dal 'settore' industriale al 'distretto' industriale: Alcune considerazioni sull'unità d'indagine dell'economia industrial. en Rivista di Economia e Politica Industriale 5(1), 7–21 (1979)

Bell, S.J., Tracey, P., Heide, J.B.: The organization of regional clusters. Academy of Management Review 34(4), 623–642 (2009)

Bellandi, M.: Industrial clusters and districts in the New Economy: Some perspectives and cases. In: Sugden, R., Cheung, R.H., Meadows, G.R. (eds.) Urban and Regional Prosperity in a Globalised New Economy, pp. 196–219. Edward Elgar, Cheltenham (2003)

Belussi, F., Gottardi, G.: Evolutionary Patterns of Local Industrial Systems: Towards a Cognitive Approach to the Industrial District. Ashgate, Aldershot (2000)

Belussi, F., Sedita, S.R.: Life cycle vs. multiple path dependency in industrial districts. European Planning Studies 17(4), 505–528 (2009)

Belussi, F., Sammarra, A., Sedita, S.: Learning at the boundaries in an "open regional innovation system": A focus on firms' innovation strategies in the Emilia Romagna life science industry. Research Policy 39, 710–721 (2010)

Bergman, E.M.: Cluster life-cycles: An emerging synthesis, Institut für Regional- und Umweltwirtschaft, SRE 2007/04 (2007) (access online October 13), http://epub.wu-wien.ac.at/

Best, M.H., Forrant, R.: Creating industrial capacity: Pentagon-led versus production-led industrial policies. In: Michie, J., Grieve Smith, J. (eds.) Creating Industrial Capacity: Towards Full Employment, pp. 225–254. Oxford University Press, Oxford (1996)

Bhargava, H.K., Power, D.J., Sun, D.: Progress in Web-based decision support technologies. Decision Support Systems 43(4), 1083–1095 (2007)

Bonel, E.: Complementarities in action: Modelling complementarity thresholds in enacting a coopetition strategy. In: Proceedings of the 8th Global Conference on Business & Economics, Florence, October 18-19 (2008)

Bönte, W.: R&D and productivity: Internal vs. external R&D - Evidence from West German manufacturing industries. Economics on Innovation and New Technology 12, 343–360 (2003)

Boschma, R.: Proximity and innovation: A critical assessment. Regional Studies 39, 61–74 (2005)

Boschma, R.A., Lambooy, J.G.: Knowledge, market structure and economic coordination: dynamics of industrial districts. Growth and Change 33(3), 291–311 (2002)

Boshuizen, J., Geurts, P., Van Der Veen, A.: Regional social networks as conduits for knowledge spillovers: Explaining performance of high-tech firms. Tijdschrift Voor Economische en Sociale Geografie 100, 183–197 (2009)

Bougrain, F., Haudeville, B.: Innovation, collaboration and SMEs internal research capacities. Research Policy 31, 735–747 (2002)

Brandenburger, A.M., Nalebuff, B.J.: Co-opetition, Harper Collins Business. Hammersmith, London (1996)

Breschi, S., Malerba, F.: The geography of innovation and economic clustering: Some introductory notes. Industrial and Corporate Change 10(4), 817–833 (2001)

Bresnahan, T., Gambardella, A., Saxenian, A.: Old economy inputs for new economy outcomes: Cluster formation in the New Silicon Valleys. Industrial and Corporate Change 10(4), 835–860 (2001)

Brusco, S.: The Emilian model: Productive decentralisation and social integration. Cambridge Journal Economics 6, 167–184 (1982)

Buchko, A.A.: Barriers to strategic transformation: Interorganizational networks and institutional forces. Advances in Strategic Management 10B, 81–106 (1994)

Burt, R.: Structural Holes. Harvard University Press, Cambridge (1992)

Cainelli, G.: Spatial agglomeration, technological innovations, and firm productivity: Evidence from Italian business districts. Growth and Change 39, 414–435 (2008)

Caloghirou, Y., Tsakanikas, A., Vonortas, N.: University-Industry cooperation in the context of the European framework programmes. Journal of Technology Transfer 26, 153–161 (2001)

Caloghirou, Y., Kastelli, K., Tsakanikas, A.: Internal capabilities and external knowledge sources: Complements or substitutes for innovative performance? Technovation 24, 29–39 (2004); Camagni, R.: Innovation Networks: Spatial Perspectives. In: Carnagni, R. (ed.) Belhaven, London (1991)

Camagni, R.: On the concept of territorial competitiveness: Sound or misleading? Urban Studies 39(13), 2395–2411 (2002)

Camagni, R.: Local milieu, uncertainty and innovation networks: Towards a new dynamic theory of economic space. In: Camagni, R. (ed.) Innovation Networks: Spatial Perspectives, pp. 121–142. Belhaven Press, London (1991)

Camp, R.C.: Benchmarking: The Search for Industry Best Practises that Lead to Superior Performance. ASQP Quality Press, Milwaukee (1989)

Capaldo, A.: Network structure and innovation: The leveraging of a dual network as a distinctive relational capability. Strategic Management Journal 28, 585–608 (2007)

Capello, R.: Spatial transfer of knowledge in high technology milieux: Learning versus collective learning processes. Regional Studies 33(4), 353–365 (1999)

Capello, R., Faggian, A.: Collective learning and relational capital in local innovation processes. Regional Studies 39(1), 75–87 (2005)

Carbonara, N.: Information and communication technology and geographical clusters: Opportunities and spread. Technovation 25, 213–222 (2005)

Carlsson, B., Jacobsson, S., Holmen, M., Richne, A.: Innovation Systems: Analytical and methodological issues. Research Policy 31, 233–245 (2002)

Cassiman, B., Veugelers, R.: In search of complementary in innovation strategy: Internal R&D and external knowledge acquisition. Management Science 52, 62–68 (2006)

Cassiman, B., Di Guardo, M.C., Valentini, G.: Organising R&D projects to profit from innovation: Insights from coopetition", References and further reading may be available for this article. To view references and further reading you must purchase this article. Long Range Planning 42(2), 216–233 (2009)

Chaminade, C., Vang, J.: Innovation policy for Asian SMEs: Exploring cluster differences. Science, Technology and Society 13(1), 61–94 (2008)

Chang, H.-J.: Kicking away the ladder: Infant industry promotion in historical perspective. Oxford Development Studies 31(1), 21–32 (2003)

Chesbrough, H.: Open innovation: the new imperative for creating and profiting from technology. Harvard Business School Press, Boston (2003)

Chesbrough, H., Crowther, A.K.: Beyond high tech: Early adopters of open innovation in other industries. R&D Management 36(3), 229–236 (2006)

Child, J., Faulkner, D.: Strategies of Co-operation, Managing Alliances, Networks, and Joint Ventures. Oxford University Press, UK (1998)

Chin, K.-S., Chan, B.L., Lam, P.-K.: Identifying and prioritizing critical success factors for coopetition strategy. Industrial Management & Data Systems 108(4), 437–454 (2008)

Coenen, L., Moodysson, J., Asheim, B.T.: Nodes, networks and proximities: On the knowledge dynamics of the Medicon Valley biotech cluster. European Planning Studies 12(7), 1003–1018 (2004)

Cohen, W.M., Levinthal, D.A.: Innovation and learning: The two faces of R&D. The Economic Journal 99, 569–596 (1989)

Cohen, W.M., Levinthal, D.A.: Absorptive capacity: A new perspective on learning and innovation. Administrative Science Quarterly 35(1), 128–152 (1990)

Coleman, J.S.: Social capital in the creation of human capital. American Journal of Sociology 94(suppl.) (1988)

Cooke, P.: Introduction: Origins of the concept. In: Braczyk, H., Cooke, P., Heidenreich, M. (eds.) Regional Innovation Systems: The Role of Governances in a Globalized World. UCL Press, London (1998)

Cooke, P.: Regionally asymmetric knowledge capabilities and open innovation. Exploring 'Globalisation 2', a new model of industry organisation. Research Policy 34, 1128–1149 (2005)

Cooke, P., Morgan, K.: The Associational Economy. Oxford University Press, Oxford (1998)

Cooke, P., Clifton, N., Olega, M.: Social capital, firm embeddedness and regional development. Regional Studies 39(8), 1065–1077 (2005)

Crevoisier, O.: The innovative milieus approach: Toward a territorialized understanding of the economy? Economic Geography 80, 367–380 (2004)

Cross, R., Nohria, N., Parker, N.: Six myths about networks and how to overcome them. Sloan Management Review 43(3), 67–76 (2002)

Csíkszentmihályi, M.: Good Business: Leadership, Flow, and the Making of Meaning. Penguin Books, New York (2003)

Dagnino, G.B., Padula, G.: Coopetition strategy: A new kind of interfirm dynamics for value creation. Paper presented at EURAM–The European Academy of Management Second Annual Conference, Innovative Research in Management, Stockholm, May 9-11 (2002)

Dahl, M., Pedersen, C.: Knowledge flows through informal contacts in industrial clusters: Myth or reality? Research Policy 33(10), 1673–1686 (2004)

Das, T.K., Teng, B.: Instabilities of strategic alliances: An internal tensions perspective. Organization Science 11(1), 77–101 (2000)

De Propis, L.: Types of innovation and inter-firm co-operation. Entrepreneurship and Regional Development 14, 337–353 (2002)

Dibben, M.R., Harris, S.: Social relationships as a precursor to international business exchange. In: Proceedings of the 17th IMP Conference, Oslo (September 2001)

Diez, J.D.: Innovative networks in manufacturing: some empirical evidence from the metropolitan area of Barcelona. Technovation 20, 139–150 (2000)

Doring, T., Schnellenbach, J.: What do we know about geographical knowledge spillovers and regional growth? A survey of the literature. Regional Studies 40(3), 375–395 (2006)

Dosi, G.: Technical Change and Industrial Transformation-The Theory and an Application to the Semiconductor Industry. Macmillan, London (1984)

Doz, Y.: The evolution of co-operation in strategic alliances: Initial conditions or learning processes. Strategic Management Journal 17, 55–83 (1996)

Doz, Y., Hamel, G.: Alliance advantage: The Art of Creating Value through Partnering. Harvard Business School Press, Boston (1998)

DTI, Competing in the global economy: The innovation challenge, DTI Economics Paper No. 7. DTI, London (2003),
  http://webarchive.nationalarchives.gov.uk/tna,
  http://www.dti.gov.uk/files/file12093.pdf

Dumais, G., Ellison, G., Glaeser, E.L.: Geographic concentration as a dynamic process. Review of Economics and Statistics LXXXIV(2), 193–204 (2002)

Dussauge, P., Garrette, B., Mitchell, W.: Learning from competing partners: Outcomes and durations of scale and link alliances in Europe, North America and Asia. Strategic Management Journal 21(2), 99–126 (2000)

Dyer, J.H., Singh, H.: The relational view: Cooperative strategy and sources of interorganizational competitive advantage. The Academy of Management Review 23(4), 660–680 (1998)

EC, Integrated Guidelines for Growth and Jobs (2005-2008), COM (2005) 141 Final, 141, Commission of the European Communities, Brussels (2005)

EIS, European Innovation Scoreboard (2009), http://www.Proinno-Europe.Eu/Page/European-Innovation-Scoreboard-2009

Eisenhardt, K., Martin, J.: Dynamic capabilities: What are they? Strategic Management Journal, Special Issue 21(10-11), 1105–1121 (2000)

Ellison, G., Glaeser, E.L.: The geographic concentration of industry: Does natural advantage explain agglomeration? American Economic Review 89(2), 311–316 (1999)

Enkel, E., Gassmann, O., Chesbrough, H.: Open R&D and open innovation: Exploring the phenomenon. R&D Management 39(4), 311–316 (2009)

Escribano, A., Andrea Fosfuri, A., Tribó, J.: Managing external knowledge flows: The moderating role of absorptive capacity. Research Policy 38, 96–115 (2009)

EU Cluster Observatory, The EU Cluster Memorandum: Promoting European innovation through clusters (November 2008) (access online, October 12, 2010),
  http://www.clusterobservatory.eu/upload/European_Cluster_Memorandum.pdf

European Cluster Alliance, http://www.proinno-europe.eu/eca

European Cluster Excellence Initiative, http://www.cluster-excellence.eu/

European Cluster Observatory,
  http://www.clusterobservatory.eu/index.html

European Community, Business networks and the knowledge-driven economy: An empirical study carried out in Europe and Canada. Enterprise Directorate General, Geneva (2000) (access online November 3, 2010),
  http://www.insme.org/documents/Industrial%20aspects%20of%20the%20information%20society.pdf

European Innovation Platform for Clusters (Cluster-IP),
  http://www.proinnoeurope.eu/action/european-innovation-platform-clusters-cluster-ip

Foss, N.J.: Higher-order industrial capabilities and competitive advantage. Journal of Industry Studies 3, 1–20 (1996)

Freel, M.S., Harrison, R.T.: Innovation and cooperation in the small firm sector: Evidence from Northern Britain. Regional Studies 40, 289–305 (2006)

Freeman, C.: Networks of innovators: A synthesis of research issues. Research Policy 20, 499–514 (1991)

Fromhold-Eisebith, M., Eisebith, G.: How to institutionalize innovative clusters? Comparing explicit top-down and implicit bottom-up approaches. Research Policy 34, 1250–1268 (2005)

Furre, H.: Cluster policy in Europe: A brief summary of cluster policies in 31 European countries. Oxford Research AS, Norway (2008) (unpublished paper)

Galvagno, M., Di Guardo, M.C.: The dynamic capabilities view of coopetition: The case of Intel, Apple and Microsoft. In: EIASM–2nd Workshop on Coopetition, September 14-15. Bocconi University, Milan (I) (2006)

Garcia-Lorenzo, L.: Innovation and knowledge processes in inter-organizational collaboration. In: Adam, F., et al. (eds.) Creativity and Innovation in Decision Making and Decision Support. Ludic Publishing (2006)

Garcia-Pont, C., Nohria, N.: Local versus Global: The dynamics of alliance formation in the automobile industry. Strategic Management Journal 23(4), 307–321 (1998)

Gertler, M.: Being there: Proximity, organisation, and culture in the development and adoption of manufacturing technologies. Economic Geography 71, 1–26 (1995)

Ghemawat, P.: Games Business Play: Cases and Models. MIT Press, Cambridge (1997)

Giuliani, E.: The selective nature of knowledge networks in clusters: evidence from the wine industry. Journal of Economic Geography 7(2), 139–168 (2007)

Giuliani, E., Bell, M.: The micro-determinants of meso-level learning and innovation: Evidence from a Chilean wine cluster. Research Policy 34(1), 47–68 (2005)

Gnyawali, D., Madhavan, R.: Cooperative networks and competitive dynamics: A structural embeddedness perspective. Academy of Management Review 26(3), 431–445 (2001)

Gomes-Casseres, B.: Group versus group: How alliance networks compete. HBR 62(4), 4–11 (1984)

Gomes-Casseres, B., Hagedoorn, J., Jaffe, A.: Do alliances promote knowledge flows? Journal of Financial Economics 80(1), 5–33 (2006)

Gordon, I.R., McCann, P.: Industrial clusters: Complexes, agglomeration and/or social networks? Urban Studies 37(3), 513–532 (2000)

Grabher, A.: Cool projects, boring institutions: Temporary collaboration in social context. Regional Studies 36(3), 205–214 (2002)

Grabher, G., Ibert, O.: Bad Company? The ambiguity of personal knowledge networks. Journal of Economic Geography 6, 251–271 (2006)

Granovetter, M.: Economic action and social structure: the problem of embeddedness. American Journal of Sociology 91(3), 481–510 (1985)

Griliches, Z.: Issues in assessing the contribution of research and development to productivity growth. Bell Journal of Economics, The RAND Corporation 10(1), 92–116 (1979)

Grimpe, Sofka: Search patterns and absorptive capacity: Low- and high-technology sectors in European countries. Research Policy 38(3), 495–506 (2009)

Grossman, G.M., Helpman, E.: Innovation and Growth in the Global Economy. MIT Press, Cambridge (1992)

Gruca, T., Nath, D.: Regulatory change, constraints on adaptation and organizational failure: An empirical analysis of acute care hospitals. Strategic Management Journal 15, 345–363 (1994)

Gulati, R.: Social structure and alliance formation patterns: A longitudinal analysis. Administrative Science Quarterly 40(4), 619–652 (1995a)

Gulati, R.: Does familiarity breed trust? The implications of repeated ties for contractual choice in alliances. Academy of Management Journal 38(1), 85–112 (1995b)

Gulati, R.: Alliances and networks. Strategic Management Journal 19(4), 293–317 (1998)

Gulati, R.: Network location and learning: The influence of network resources and firm capabilities on alliance formation. Strategic Management Journal 20(5), 397–420 (1999)

Gulati, R., Nohria, N., Zaheer, A.: Strategic networks. Strategic Management Journal 21, 203–215 (2000)

Hagedoorn, J., Schakenraad, J.: The effect of strategic technology alliances on company performance. Strategic Management Journal 15(4), 291–309 (1994)

Hamel, G.: Competition for competence and inter-partner learning within international strategic alliances. Strategic Management Journal 12, (special issue), 83–104 (1991)

Hansen, M.T., Nohria, N.: How to build collaborative advantage. MIT Sloan Management Review 46(1), 22–30 (2004)

Harrison, B.: Industrial districts: Old wine in new bottles? Regional Studies 26, 469–483 (1991)

Heimeriks, K.H., Duysters, G.M.: Alliance capability as mediator between experience and alliance performance: An empirical investigation into the alliance capability development process. White paper, Eindhoven University of Technology (2003)

Helfat, C.: Relational capabilities: Drivers and implications. In: Helfat, C.E., Finkelstein, S., Mitchell, W., Peteraf, M., Singh, H., Teece, D.J., Winter, S.G. (eds.) Dynamic Capabilities: Strategic Change in Organizations, pp. 65–80. Blackwell, Oxford (2007)

Helfat, C., Raubitschek, R.: Product sequencing: Co-evolution of knowledge, capabilities and products. Strategic Management Journal 21, 961–979 (2000)

Henderson, R., Cockburn, I.: Measuring competence? Exploring firm effects in pharmaceutical research. Strategic Management Journal 15, 63–84 (1994)

Henderson, R.M., Clark, K.B.: Architectural innovation: The reconfiguration of existing product technologies and the failure of established firms. Administrative Science Quarterly 35, 9–30 (1990)

Henricks, M.: Joining forces: Work with your competitors – not against them – and soon you'll be succeeding with the enemy. Entrepreneur, 76–79 (1996)

Henry, N., Pinch, S.: Neo-Marshallian nodes, institutional thickness and Britain's 'Motor Sport Valley': Thick or thin? Environment and Planning 33(7), 1169–1183 (2001)

Hervas-Oliver, J.L., Albors-Garrigos, J.: The role of the firm's internal and relational capabilities in clusters: When distance and embeddedness are not enough. Journal of Economic Geography 9(2), 263–283 (2009)

Hoang, H., Antoncic, B.: Network-based research in entrepreneurship: A critical review. Journal of Business Venturing 18, 165–187 (2003)

Hoang, H., Rothaermel, F.: The effect of general and partner-specific alliance experience on joint R&D project performance. Academy of Management Journal 48(2), 332–345 (2005)

Hoffmann, W.H.: Strategies for managing a portfolio of alliances. Strategic Management Journal 28(8), 827–856 (2007)

Hoffmann, W.H., Schlosser, S.: Success factors of strategic alliances in small and mediumsized enterprises: An empirical survey. Long Range Planning 34, 357–381 (2001)

Howells, J.: Research and technology outsourcing. Technology Analysis & Strategic Management 11(1), 17–29 (1999)

Huber, F.: Do clusters really matter for innovation practices in information technology? Questioning the significance of technological knowledge spillovers. Druid Congress (2010), http://www.Druid.Dk

Huergo, E.: The role of technological management as a source of innovation: Evidence from Spanish manufacturing firms. Research Policy 35(9), 1377–1388 (2006)

Huggins, R.: Innovative Action Report. IA Centre, Glasgow (2005)

Huggins, R.: Inter-firm network policies and firm performance: Evaluating the impact of initiatives in the United Kingdom. Research Policy 30, 443–458 (2001)

Huggins, R.: The evolution of knowledge clusters: Progress and policy. Economic Development Quarterly 22(4), 277–289 (2008)

Humphreys, P.C., Jones, G.A.: The evolution of group support systems to enable collaborative authoring of outcomes. World Futures 62, 1–30 (2006)

Ibrahim, S.E., Fallah, M.H., Reilly, R.R.: Localized sources of knowledge and the effect of knowledge spillovers: An empirical study of inventors in the telecommunications industry. Journal of Economic Geography 9, 405–431 (2009)

Imai, M.: Kaizen: The Key to Japan's Competitive Success. McGraw-Hill, Irwin (1986)

Jaffe, A.: Technological opportunity and spillovers of R&D: Evidence from firms' patents, profits, and market value. The American Economic Review 76(5), 984–1001 (1986)

Jaffe, A.B., Trajtenberg, M., Henderson, R.: Geographic localization of knowledge spillovers as evidenced by patent citations. Quarterly Journal of Economics 108, 577–598 (1993)

Jenkins, M., Tallman, S.: The shifting geography of competitive advantage: Clusters, networks and firms. Journal of Economic Geography 10(4), 599–618 (2010)

Johannisson, B., Ramírez-Pasillas, M., Karlsson, G.: The institutional embeddedness of local inter-firm networks: A leverage for business creation. Entrepreneurship and Regional Development 14, 297–315 (2002)

Jones, G.A., Humphreys, P.C.: Spaces, processes and event design for facilitating innovation, creativity and communication in complex organizational contexts. London Multimedia Lab. LSE and The Ludic Group LLP (2007)

Juan-Li, J., Poppo, L., Zheng, K.: Relational mechanisms, formal contracts, and local knowledge acquisition by international subsidiaries. Strategic Management Journal 31, 349–370 (2010)

Kale, P., Singh, H.: Building firm capabilities through learning: The role of the alliance learning process in alliance capability and success. Strategic Management Journal 28(10), 981–1000 (2007)

Kale, P., Singh, H.: Managing strategic alliances: What do we know now, and where do we go from here? Academy of Management Perspectives, 45–62 (August 2009)

Kale, P., Singh, H., Pelmutter, H.: Learning and protection of proprietary assets in strategic alliances: Building relational capital. Strategic Management Journal 21(3), 217–237 (2000)

Keeble, D., Wilkinson, F.: Collective learning and knowledge development in the evolution of regional clusters of high-technology SMEs in Europe. Regional Studies 3, 295–303 (1999)

Kesidou, E., Marjolein, C.J., Romijn, H.A.: Local knowledge spillovers and development: An exploration of the software cluster in Uruguay. Industry & Innovation 16(2), 247–272 (2009)

Ketels, C., Lindqvist, G., Sölvell, Ö.: Clusters and clusters initiatives. Center for Strategy and Competitiveness. Stockholm School of Economics (June 2008)

Khanna, T., Gulati, R., Nohria, N.: The dynamics of learning alliances: Competition, cooperation and scope. Strategic Management Journal 19(3), 193–210 (1998)

Kim, J., Mahoney, J.T.: How property rights economics furthers the resource-based view: Resources, transaction costs and entrepreneurial discovery. International Journal of Strategic Change Management 1(1), 40–52 (2006)

Klepper, S.: The evolution of geographic structures in new industries. In: Frenken, K. (ed.) Applied Evolutionary Economics and Economic Geography, pp. 69–92. Edward Elgar, Cheltenham (2007)

Klevorick, A., Levin, R., Nelson, R., Winter, S.: On the sources of significance of interindustry differences in technological opportunities. Research Policy 24, 185–205 (1995)

Kogut, B.: Joint ventures: Theoretical and empirical perspectives. Strategic Management Journal 9(4), 319–332 (1988)

Kohl, H.: Integriertes Benchmarking für kleine und mittlere Unternehmen: Eine Methode zur Integration von Best-Practice-Informationen in das interne Unternehmenscontrolling. Dissertation, Technische Universität Berlin (2007)

Kor, Y.Y.: Experience-based top management team competence and sustained growth. Organization Science 14(6), 707–719 (2003)

Kreuz, W., Herter, M.: Benchmarking – mit diesem wirkungsvollen Management-Instrument werden Spitzenleistungen erzielt. In: Kreuz, W. (ed.) Mit Benchmarking zur Weltspitze Aufsteigen. Landsberg, Lech (1995)

Lado, A.A., Boyd, N.G., Wright, P., Kroll, M.: Paradox and theorizing within the Resource-Based View. Academy of Management Review 31, 115–131 (2006)

Lagendijk, A.: Learning from conceptual flow in regional studies: framing present debates, unbracketing past debates. Regional Studies 40(4), 385–399 (2006)

Lane, P., Koka, B., Pathak, S.: The reification of absorptive capacity: A critical review and rejuvenation of the construct. Academy of Management Review 31(4), 833–863 (2006)

Lasker, R.D., Weiss, E.S., Miller, R.: Partnership synergy: A practical framework for studying and strengthening the collaborative advantage. The Milbank Quarterly 79(2), 179–205 (2001)

Laso-Ballesteros, I.: Future and emerging technologies and paradigms for collaborative working environments. In: 5th Collaboration@Work Expert Group Report, European Commission (2006)

Latour, B.: Reassembling the Social- An Introduction to Actor-Network-Theory. Oxford University Press (2005)

Laursen, K., Salter, A.: Open for innovation: the role of openness in explaining innovation performance among UK manufacturing firms. Strategic Management Journal 27(2), 131–150 (2006)

Lavie, D.: The Competitive advantage of interconnected firms: An extension of the resourcebased view. Academy of Management Review 31(3), 643–647 (2006)

Lawson, C.: Towards a competence theory of the region. Cambridge Journal Economics 23, 151–166 (1999)

Lawson, C., Lorenz, E.: Collective learning, tacit knowledge and regional innovative capacity. Regional Studies 33, 305–317 (1999)

Lee, C., Lee, K., Pennings, J.: Internal capabilities, external networks, and performance: A study on technology-based ventures. Strategic Management Journal 22, 615–640 (2001)

Leonard-Barton, D.: Core capabilities and core rigidities: A paradox in managing new product development. Strategic Management Journal 13, 111–125 (1992)

Lissoni, F.: Knowledge Codification and the geography of innovation: The case of Brescia Mechanical Cluster. Research Policy 30, 1479–1500 (2001)

Lundvall, B.A. (ed.): National System of Innovation: Towards a Theory of Innovation and Interactive Learning. Pinter, London (1992)

Lundvall, B.A.: Innovation as an interactive process: from user-producer interaction to the national system of innovation. In: Dosi, G., et al. (eds.) pp. 349–369 (1988)

Lundvall, B.-Å., Borras, S.: The globalising learning economy: Implications for innovation policy, Brussels, DG XII (1999)

Lundvall, B.-Å., Nielsen, P.: Competition and transformation in the learning economy–the Danish case. Revue d' Economie Industrielle 88, 67–90 (1999)

Lundvall, B.-A., Intakumnerd, P., Vang, J. (eds.): Asian Innovation Systems in Transition. Edward Elgar, Cheltenham (2006)

Lunnan, R., Haugland, S.: Predicting and measuring alliance performance: A multidimensional analysis. Strategic Management Journal 29(5), 545–556 (2008)

Maillat, D.: SMES, innovation and territorial development. European Summer Institute of The Regional Science Association, Arco (1989)

Malmberg, A., Maskell, P.: The elusive concept of localization economies: Towards a knowledge-based theory of spatial clustering. Environment and Planning 34(3), 429–449 (2002)

Malmberg, A., Power, D.: On the role of global demand in local innovation processes? In: Shapira, P., Fuchs, G. (eds.) Rethinking Regional Innovation and Change: Path Dependency or Regional Breakthrough? Springer, New York (2005)

Malmberg, A., Power, D.: (How) do (firms in) clusters create knowledge? Industry and Innovation 12(4), 409–431 (2005)

Markusen, A.: Sticky places in slippery space. In: Barnes, T.J., Gertler, M.S. (eds.) The New Industrial Geography, pp. 98–126. Routledge, London (1999)

Martin, S., Maier, H.: Sustainability, clusters, and competitiveness. Economic Development Quarterly 22(4), 272–276 (2008)

Martin, S., Sunley, P.: Deconstructing clusters: Chaotic concept or policy panacea? Journal of Economic Geography 3(1), 5–35 (2003)

Martins, B., Alwert, K., Humphreys, P.: Is there a future for SME clusters in highly dynamic business landscapes? In: Proceedings of the IFKAD, 5th edn., Matera, June 26-28 (2010)

Martins, B., Jorcano, J.: SMEs' shortcomings in managing intellectual capital: The InCaSSpain experience. In: Proceedings of the IFKAD, 3rd edn., Matera, June 24-25 (2008)

Maskell, P.: Knowledge creation and diffusion in geographic clusters. International Journal of Innovation Management 5(2), 213–237 (2001)

Maskell, P., Kebir, L.: What qualifies as a cluster theory? DRUID Working Papers 05-09, DRUID. Copenhagen Business School (2005)

Maskell, P., Malmberg, A.: Localised learning and industrial competitiveness. Cambridge Journal of Economics 23, 167–185 (1999)

Maskell, P., Malmberg, A.: Myopia, knowledge development and cluster evolution. Journal of Economic Geography 7(5), 603–618 (2007)

Matusik, S.F., Hill, C.W.L.: The utilization of contingent work, knowledge creation, and competitive advantage. Academy of Management Review 23, 680–697 (1998)

McEvily, S.K., Chakravarthy, B.: The persistence of knowledge based advantage: an empirical test for product performance and technological knowledge. Strategic Management Journal 23, 285–306 (2002)

Meier, G.: Clusters in Germany: An empirical based insight view on emergence, financing, management and competitiveness of the most innovative clusters in Germany, 2nd edn. Institute for Innovation and Technology (2009)

Menzel, M.-P., Fornahl, D.: Cluster life cycles: Dimensions and rationales of cluster evolution. Industrial and Corporate Change 19(1), 205–238 (2009)

Mertins, K., Kohl, H.: Benchmarking: Leitfaden für den Vergleich mit den Besten, vol. 1. Symposion Publishing, Düsseldorf (2004)

Mertins, K., Rabe, M., Sauer, O.: Produktionsstrukturierung auf Basis des strategischen Eigenfertigungsanteils. ZwF 89(1-2), 27–29 (1993)

Metcalfe, J.: The economics of evolution and the economics of technology. Policy Economic Journal 104, 931–944 (1994)

Mills, K., Reynolds, E., Reamer, A., et al.: Clusters and Competitiveness: A new federal role for stimulating regional economies. Brookings Institution, Washington (2008)

Milberg, J., Schuh, G.: Erfolg in Netzwerken. Springer, Heidelberg (2002)

Molina, F.X.: European industrial districts: Influence of geographic concentration on performance of the firm. Journal of International Management 7, 277–294 (2001)

Molina-Morales, F.X., Martinez-Fernandez, T.M.: How much difference is there between industrial district firms? A net value creation approach. Research Policy 33, 473–486 (2004)

Molina-Morales, F.X., Martinez-Fernandez, T.M.: Social networks: Effects of social capital on firm innovation. Journal of Small Business Management 48(2), 258–279 (2010)

Moodysson, J., Coenen, L., Asheim, B.: Explaining spatial patterns of innovation: Analytical and synthetic modes of knowledge creation in Medicon Valley Life Science Cluster. Environmental and Planning 40, 1040–1056 (2008)

Moodysson, J.: Principles and practices of knowledge creation: On the organization of 'buzz' and 'pipelines'. Life Science Communities Economic Geography 84(4), 449–469 (2008)

Moodysson, J., Jonsson, O.: Knowledge collaboration and proximity: The spatial organization of biotech innovation projects. European Urban and Regional Studies 14(2), 115–131 (2007)

Morris, M.H., Kocak, A., Özer, A.: Coopetition as a small business strategy: Implications for performance. Journal of Small Business Strategy 18(1), 35–55 (2007)

Moss Kanter, R.: Collaborative advantage: The art of alliances. HBR, 96–108 (July-August 1994)

Moulaert, F., Sekia, F.: Territorial innovation models: A critical survey. Regional Studies 37(3), 289–302 (2003)

Mowery, D.C.: The relationship between intrafirm and contractual forms of industrial research in American manufacturing, 1900–1940. Exploration in Economics History 20, 351–374 (1983)

Muro, M., Katz, B.: The new 'cluster moment': Regional innovation clusters can foster the next economy. Brookings (September 2010),
http://www.brookings.edu/papers/2010/
0921_clusters_muro_katz.aspx

Muthusamy, S.K., White, M.A.: Learning and knowledge transfer in strategic alliances: A social exchange view. Organization Studies 26(3), 415–441 (2005)

Nachum, L., Keeble, D.: MNE linkages and localised clusters: Foreign and indigenous firms in the media cluster of central London. Journal of International Management 9, 171–192 (2003a)

Nachum, L., Keeble, D.: Neo-Marshallian clusters and global networks. Long Range Planning 36, 459–480 (2003b)

Nadvi, K., Halder, G.: Local clusters in global value chains: exploring dynamic linkages between Germany and Pakistan. Entrepreneurship and Regional Development 17(5), 339–363 (2005)

Nahapiet, J., Ghoshal, S.: Social capital, intellectual capital and organizational advantage. The Academy of Management Review 23(2), 242–266 (1998)

Nelson, R.: The economics of invention: A survey of the literature. Journal of Business 32, 101–127 (1959)

Nelson, R.: Diffusion of development: Post-World War II convergence among advanced industrial nations. American Economic Review, American Economic Association 81(2), 271–275 (1991)

Nelson, R., Winter, S.: An Evolutionary Theory of Economic Change. Harvard University Press, Cambridge (1982)

Nieto, M., Quevedo, P.: Absorptive capacity, technological opportunity, knowledge spillovers, and innovative effort. Technovation 25, 1141–1157 (2005)

Nohria, N., Garcia-Pont, C.: Global strategic linkages and industry structure. Strategic Management Journal, Summer Special Issue 12, 105–124 (1991)

Nonaka, I., Takeuchi, H.: The Knowledge-creating Company. Oxford University Press (1995)

O'Reilly, T.: What is Web 2.0: Design patterns and business models for the next generation of software (2005), http://www.oreilly.com/pub/a/oreilly/tim/news/2005/09/30/what-is-web-20.html (consulted December 2010)

OECD, Boosting Innovation: the Cluster Approach. Organisation for Economic Cooperation and Development, Paris (1999)

OECD, Competitive Regional Clusters: National Policy Approaches. OECD, Paris (2008a)

OECD, Open Innovation in Global Networks (2008b) ISBN 978-92-64-04769-3

OECD, New Nature of Innovation (2009), http://www.newnatureofinnovation.org/full_report.pdf (access)

Ostergaard, C.: Knowledge flows through social networks in a cluster. Structural Change and Economic Dynamics 20, 196–210 (2009)

Owen-Smith, J., Powell, W.: Knowledge networks as channels and conduits: Spillover in the Boston biotechnology community. Organization Science 15, 5–21 (2004)

Padula, G., Dagnino, G.P.: On the nature and drivers of coopetition, Working paper, SSRN-id791667 (2007), http://www.ssrn.com

Pfeffer, J., Salancik, G.R.: The External Control of Organizations: A Resource Dependence Perspective. Harper and Row, NY (1978)

Pisano, G.: Knowledge, integration, and the locus of learning: An empirical analysis of process development. Strategic Management Journal 15 (Special Issue), 85–100 (1994)

Porter, M.: The Competitive Advantage of Nations. The Free Press, NY (1990)

Porter, M.: Clusters and the new economics of competition. Harvard Business Review, 77–90 (November-December 1998)

Porter, M.E., Stern, S.: Innovation: Location matters. Sloan Management Review 4(4), 28–36 (2001)

Pouder, R., John, C.: Hot spots and blind spots: Geographic clusters of firms and innovation. Academy of Management Review 21(4), 1192–1225 (1996)

Powell, W.W., Koput, K.W., Smith-Doerr, L.: Interorganizational collaboration and the locus of innovation in biotechnology. Administrative Science Quarterly 41(1), 116–145 (1996)

Putnam, R.D.: The prosperous community: Social capital and public life. The American Prospect 13, 35–42 (1993)

Rao, B.P., Swaminathan, V.: Uneasy alliances: Cultural incompatibility or culture shock? In: Proceedings of the Association of Management, 13th Annual International Conference, Vancouver, Canada, pp. 1–13 (August 1995)

Rijamampianina, R., Carmichael, T.: A framework for effective cross-cultural coopetition between organisations. Problems and Perspectives in Management 4, 92–103 (2005)

Ring, P.S., Van de Ven, A.H.: Developmental process of cooperative interorganizational relationships. Academy of Management Review 19, 90–118 (1994)

Rodriguez-Pose, A., Refolo, M.C.: The Link between local production systems and public and university research in Italy. Environment and Planning 35(8), 1477–1492 (2003)

Romijn, H., Albaladejo, M.: Determinants of innovation capability in small electronics and software firms in Southeast England. Research Policy 31, 1053–1067 (2002)

Ronde, P., Hussler, C.: Innovation in regions: What does really matter? Research Policy 34, 1150–1172 (2005)

Rothaermel, F., Boeker, W.: Old technology meets new technology: Complementarities, similarities and alliance formation. Strategic Management Journal 29(1), 47–77 (2008)

Rumelt, R.P.: Towards a strategic theory of the firm. In: Lamb, R.B. (ed.) Competitive Strategic Management, pp. 556–570. Prentice-Hall (1998)

Sallet, J., Paisley, E., Masterman, J.: The geography of innovation: The Federal Government and the growth of regional innovation clusters (2009), http://www.SCIENCEprogrESS.org

Sanchez, R.: Strategic flexibility in product competition. Strategic Management Journal 16, 135–159 (1995)

Santarelli, E., Vivarelli, M.: Entrepreneurship and the process of firms' entry, survival and growth. Industrial and Corporate Change 16(3), 455–488 (2007)

Saxenian, A.: Regional Advantage: Culture and Competition in Silicon Valley and Route 128. Harvard University Press, Cambridge (1994)

Schmitz, H.: Increasing returns and collective efficiency. Cambridge Journal of Economics 23, 465–483 (1999)

Schreiner, M., Kale, P., Corsten, D.: What really is alliance management capability and how does it impact alliance outcomes and success? Strategic Management Journal 30, 1395–1419 (2009)

Scitovsky, T.: Two concepts of external economies. Journal of Political Economy 62, 143–151 (1954)

Sherer, S.: Critical success factors for manufacturing networks as perceived by network coordinators. Journal of Small Business Management 41(4), 325–345 (2003)

Simsek, Z., Lubatkin, M.H., Floyd, S.: Inter-firm networks and entrepreneurial behaviour: A structural embeddedness perspective. Journal of Management 29(3), 427–442 (2003)

Smith, K.G., Carroll, S.J., Ashford, S.J.: Intra- and interorganizational cooperation: Toward a research agenda. Academy of Management Journal 38, 7–23 (1995)

Sölvell, Ö.: Clusters: Balancing Evolutionary Constructive Forces. Ivory Tower, Stockholm (2009)

Sölvell, Ö., Lindqvist, G., Ketels, C.: The Cluster Initiative Greenbook. Ivory Tower, Stockholm (2003)

Spithoven, A., Clarysse, B., Knockaert, M.: Building absorptive capacity to organise inbound open innovation in traditional industries. Technovation 30, 130–141 (2010)

Storper, M., Venables, A.: Buzz: face to face contact and the urban economy. Strategic Management Journal 22, 287–306 (2004)

Storper, M.: Regional technology coalitions an essential dimension of national technology policy. Research Policy 24(6), 895–911 (1995)

Sundali, J.A., Seale, D.A.: Is reciprocity necessary for coopetition? Some experimental evidence. Journal of Behavioral and Applied Management 4(1), 68–87 (2002)

Swann, G.M.P., Prevezer, M., Stout, D.: The Dynamics of Industrial Clustering– International In Comparisons. Computing and Biotechnology. Oxford University Press, New York (1998)

Swann, G.M.P.: Towards a model of clustering in high-technology industries. In: Swann, G.M.P., Prevezer, M., Stout, D. (eds.) The Dynamics of Industrial Clustering. Oxford University Press, Oxford (2002)

Tallman, S., Jenkins, M., Henry, N., Pinch, S.: Knowledge, clusters, and competitive (2004)

Tallman, S., Jenkins, M.: Alliances, knowledge flows, and performance in regional clusters. In: Contractor, F., Lorange, P. (eds.) Cooperative Strategies and Alliances, pp. 163–188. Elsevier Science, Oxford (2002)

Teece, D.: Profiting from technological innovation: Implications for integration, collaborating, licensing, and public policy. In: Teece, D. (ed.) The Competitive Challenge, pp. 185–219. Harper Collins, New York (1987)

Teece, D.J., Pisano, G., Shuen, A.: Dynamic capabilities and strategic management. Strategic Management Journal 18(7), 509–533 (1997)

Tijssen, R.J.W.: Quantitative assessment of large heterogeneous R&D networks: The case of process engineering in the Netherlands. Research Policy 26, 791–809 (1998)

Tödtling, F., Kaufmann, A.: The role of the region for innovation activities of SMEs. European Urban and Regional Studies 8(3), 203–215 (2001); Tomlinson, P.: Strong ties, substantive embeddedness and innovation: Exploring differences in the innovative performance of small and medium sized firms in UK manufacturing. Economics of Innovation and New Technology (forthcoming)

Trippl, M., Tödtling, F., Lengauer, L.: Knowledge sourcing beyond buzz and pipelines: Evidence from the Vienna software sector. Economic Geography 85(4), 443–462 (2009)

Tsang, D.: Growth of indigenous entrepreneurial software firms in cities. Technovation 25, 1331–1336 (2005)

Uzzi, B.: The sources and consequences of embeddedness for the economic performance of organizations: The network effect. American Sociological Review 61, 674–698 (1996)

Uzzi, B.: Social structure and competition in interfirm networks: The paradox of embeddedness. Administrative Science Quarterly 42, 35–67 (1997)

Van Geenhuizen, M., Reyer-Gonzales, L.: Does a clustered location matter for high. Technological Forecasting & Social Change 74, 1681–1696 (2007)

Vega-Jurado, J., Gutierrez-Gracia, A., Fernández-De-Lucio, I., Manjarrés-Henríquez, L.: The effect of external and internal factors on firms' product innovation. Research Policy 37, 616–632 (2008)

Veugelers, R., Cassiman, B.: Make and buy in innovation strategies: Evidence from Belgian manufacturing firms. Research Policy 28(2), 63–79 (1999)

Von Hippel, E.: Lead users: A source of novel product Concepts. Management Science 32(7), 91–805 (1986)

Von Hippel, E.: Cooperation between rivals: informal know-how trading. Research Policy 16(6), 291–302 (1987)

ter Wal, L.J., Boschma, R.A.: Applying social network analysis in economic geography: Framing some key analytic issues. The Annals of Regional Science 43(3), 739–756 (2009)

Waters, R., Lawton Smith, H.: Social networks in high-technology local economies: The cases of Oxfordshire and Cambridgeshire. European Urban and Regional Studies 15, 21–37 (2008)

Weissenberger-Eibl, M., Schwenk, J.: Lifeblood knowledge: Dynamic relational capabilities (DRC) and knowledge for firm innovativeness and competitive advantage. Measuring Business Excellence 13(2), 7–16 (2009)

Wenger, E., Mcdermott, R., Snyder, W.M.: Cultivating Communities of Practice, 1st edn. Harvard Business Press (2002) ISBN 978-1578513307

Wever, E., Stam, E.: Clusters of high technology SMEs: The Dutch case. Regional Studies 33(4), 391–400 (1998)

Whitley, R.: Dominant forms of economic organisation in market economies. Organisation Studies 15, 153–182 (1994)

Wolfe, D.A., Gertler, M.S.: Clusters from the inside and out: Local dynamics and global linkages. Urban Studies 41(5-6), 1071–1093 (2004)

Yu, A., Humphreys, P.: Intellectual Capital and support for collaborative decision making in small and medium enterprises. Journal of Decision Systems 17, 41–61 (2008)

Zaheer, A., Bell, G.: Benefiting from network position: Firm capabilities, structural holes and performance. Strategic Management Journal 26(9), 809–825 (2005)

Zahra, S.A., George, G.: Absorptive capacity: A review, reconceptualization and extension. Academy of Management Review 27(2), 185–203 (2002)

Zander, U., Kogut, B.: Knowledge and the speed of transfer and imitation of organizational capabilities: An empirical test. Organization Science 6(1), 76–82 (1995)

Zeng, S.X., Xie, X.M., Tam, C.M.: Relationship between cooperation networks and innovation performance of SMEs. Technovation 30(3), 181–194 (2010)

Zhang, G., Xu, Q., Liu, X.: Knowledge diffusion within the Datang Sock-Manufacturing Cluster in China. Regional Studies, 1–20 (2010)

Zineldin, M.A.: Towards an ecological collaborative relationship management. A 'coopetive' perspective. European Journal of Marketing 32(11/12), 1138–1160 (1998)

Zollo, M., Winter, S.: Deliberate learning and the evolution of dynamic capabilities. Organization Science 13, 339–351 (2002)

Zollo, M., Reuer, J.J., Singh, S.: Inter-organizational routines and performance in strategic alliances. Organization Science 13(6), 701–713 (2002)

Zucker, L.G., Darby, M.R., Armstrong, J.: Geographically localized knowledge: Spillovers or markets? Economic Inquiry 36, 65–86 (1998)

# Author Index